Benjamin M. Legum, Amber R. Stiles, Jennifer L. Vondran
Engineering Innovation

D1284979

Also of interest

Technoscientific Research
R. Morawski, 2019
ISBN 978-3-11-058390-8, e-ISBN 978-3-11-058406-6,
e-ISBN (EPUB) 978-3-11-058412-7

Human Forces in Engineering
A. Atrens, A. D. Atrens, 2018
ISBN 978-3-11-053472-6, e-ISBN 978-3-11-053512-9,
e-ISBN (EPUB) 978-3-11-053526-6

The Science of Innovation
K. Löhr, 2016
ISBN 978-3-11-034379-3, e-ISBN 978-3-11-034380-9,
e-ISBN (EPUB) 978-3-11-039658-4

Innovation Technology
L. Schramm, 2017
ISBN 978-3-11-043824-6, e-ISBN 978-3-11-042917-6,
e-ISBN (EPUB) 978-3-11-042925-1

Benjamin M. Legum, Amber R. Stiles,
Jennifer L. Vondran

Engineering Innovation

From Idea to Market Through Concepts and Case Studies

DE GRUYTER

Authors
Benjamin M. Legum
Keystone Scientific, Inc.
El@keystonescientific.net

Amber R. Stiles
Jennifer L. Vondran

ISBN 978-3-11-052101-6
e-ISBN (PDF) 978-3-11-052190-0
e-ISBN (EPUB) 978-3-11-052115-3

Library of Congress Control Number: 2019941362

Bibliographic information published by the Deutsche Nationalbibliothek
The Deutsche Nationalbibliothek lists this publication in the Deutsche Nationalbibliografie; detailed bibliographic data are available on the Internet at http://dnb.dnb.de.

© 2019 Walter de Gruyter GmbH, Berlin/Boston
Typesetting: Integra Software Services Pvt. Ltd.
Printing and binding: CPI books GmbH, Leck
Cover image: DigitalStorm/iStock/thinkstock

www.degruyter.com

Preface

Engineering Innovation is intended to bridge the gap between engineering and entrepreneurship. Our goal was to take everything we learned from our work experiences and everything we wish we had learned in school and put it into an easy to understand book that can be used by engineers, doctors, business professionals, and the aspiring entrepreneur (whether you are still in school, a graduate with less than ten years of work experience, a seasoned professional, or just wanting to learn something new). The takeaways from this book include:

1. Everyone has to start someplace.
2. The most important part of developing a new technology and creating a business is perseverance.
3. There is no such thing as a perfect way to do something.
4. It is an uplifting endeavor to create new products and services that are useful for the world.

Engineering Innovation is a comprehensive guide for individuals to gain an understanding of how to bring a product or idea from the laboratory, garage, computer, college dorm room, or basement to market by explaining concepts and providing illustrative case studies. This book is intended to be a structured toolset that shows how and when business development and product development align, overlap, and integrate. One of the best pieces of advice that we can offer is to find the right people to help you achieve the success that you are after and empower those people to help you achieve that success. Nothing is more result producing and effective for achieving your goals than having a team of the right people assigned to the right roles in your enterprise.

As you read this book, you will discover that we have provided pertinent examples of tools and resources that you can use to help you along your innovation journey. You will find many references to medical devices, life sciences, and healthcare examples throughout this book. This is because much of our experience has been in these industries; however, the guidance and information contained herein have been prepared in such a way that it is practical and applicable for nearly any industry. We believe that from this book you will be able to extract the relevant information that is applicable to your product or industry.

https://doi.org/10.1515/9783110521900-201

Contents

Preface — V

List of Figures — XIV

About the Authors — XVIII

Introduction — 1

Part A: The Business Side of Innovation

1 Starting Out with Due Diligence and Early-Stage Market Research — 15

1.1 The Life Cycle of a Company — **15**
1.1.1 Life Cycle Deviations — **18**
1.2 What is Due Diligence? — **18**
1.2.1 Why Conduct Due Diligence? — **20**
1.3 Types of Due Diligence — **23**
1.4 Performing Market Research — **24**
1.4.1 Utilizing Market Reports and Other Resources — **25**
1.4.2 Bringing It All Together with an Example — **27**
1.4.3 Identifying Target Consumers — **29**
1.4.4 Closely Scrutinize Competitors — **31**
1.5 At the Conclusion of Due Diligence and Market Research, Prepare A Value Proposition — **32**

2 Validating Your Business Model — 34

2.1 Business Validation — **34**
2.1.1 Will the Market Support the Business Model? — **36**
2.1.2 SWOT Analysis — **39**
2.1.3 Validate Business Models with Stakeholders — **40**
2.2 Concept Validation — **40**
2.2.1 Defining User Requirements — **41**
2.2.2 Early-Stage Market Research Overlaps with Concept Validation — **43**
2.2.3 Validate Throughout Development — **43**
2.2.4 Final Thoughts on Concept Validation — **45**
2.3 Stakeholder Analysis — **46**
2.3.1 Stakeholder Feedback on Human Factors — **48**

2.4 Carefully Analyze the Voice of Your Prospective Customer – Survey and Focus Group Design — **49**
2.4.1 Getting Started — **50**
2.4.2 Psychology of VOC — **51**
2.4.3 Designing the Survey/Focus Group — **51**
2.4.4 Drawing Conclusions About Human Factors (Regarding the Concept/Product) from VOC Results — **53**
2.4.5 Conduct Customer Surveys — **54**
2.4.6 Extracting Useful Information from VOC Data Through Statistical Analysis — **56**
2.4.7 Two Practical Examples — **58**

3 Forming a Business Around Your Technology — 65
3.1 What Is a Business? — **65**
3.1.1 Sole Proprietorship — **68**
3.1.2 Partnership — **68**
3.1.3 Corporation — **70**
3.1.4 Limited Liability Company (LLC) — **71**
3.2 Support Structures for Startups and Small Companies — **72**
3.2.1 Small Business Administration (SBA) — **73**
3.2.2 State Programs — **74**
3.2.3 Small Business Development Centers (SBDCs) — **74**
3.2.4 Economic Development Corporations (EDCs) — **75**
3.2.5 Incubators — **75**
3.2.6 Accelerators — **76**
3.3 De-risking Innovation and De-risking Investment — **76**
3.3.1 De-risking Innovation — **76**
3.3.2 De-risking Investment — **77**
3.4 Establishing a Team — **78**
3.4.1 Leadership of the Team — **80**
3.4.2 Mentors, Advisors, and Board Members — **81**

4 Laying the Groundwork for a Business Plan and Preparing a Business Plan — 83
4.1 The Process of Developing a Business Plan – Preparation for a Business Plan — **83**
4.2 Developing a Marketing Strategy and a Sales Strategy — **84**
4.2.1 Marketing Strategy: Using the Four Ps of Marketing to Achieve Marketing and Promotion Objectives — **85**
4.2.2 Marketing Strategy: Conduct Competitive Analysis — **86**
4.2.3 Marketing Strategy: Technology-Based Considerations — **88**

4.2.4 Documenting Your Marketing Strategy and Sales Strategy in a Marketing Plan and a Sales Plan — 89
4.3 Expansion Strategy: Planning for Entry into Future Markets — 97
4.4 IP Strategy and Plan — 98
4.5 Technology Plan — 98
4.5.1 Technology Life Cycle — 99
4.5.2 When Cutting Edge Innovation Satisfies Unmet Consumer Need, Commercial Success Isn't Far Behind — 103
4.5.3 The Importance of Technology Forecasting — 105
4.5.4 Developing a Technology Portfolio — 106
4.5.5 Developing a Realistic Timeline — 107
4.5.6 Writing a Technology Plan — 109
4.5.7 Transitioning from Research to Development and Beyond — 114
4.5.8 Planning for Special Considerations for Your Technology — 117
4.6 Preparing a Business Plan — 117
4.6.1 Executive Summary — 120
4.6.2 Company Description — 120
4.6.3 Marketing Plan — 121
4.6.4 Sales Plan — 121
4.6.5 Technology Plan — 122
4.6.6 Organizational Chart and Management Team — 122
4.6.7 Operating Procedures — 124
4.6.8 Financial Models — 125
4.6.9 Attachments — 128

5 Understanding Intellectual Property as It Relates to Your Business — 129
5.1 Patents — 131
5.1.1 Provisional and Non-Provisional Patent Applications — 133
5.1.2 Example Questions to Ask Your Patent Attorney — 135
5.2 Trademarks and Service Marks — 135
5.3 Copyrights — 137
5.4 Trade Secrets — 137
5.5 Developing an Intellectual Property Strategy and IP Portfolio — 139
5.5.1 Start by Deciding on Which IP Rights You Need — 140
5.5.2 Developing a Timeline — 140
5.5.3 Developing an International IP Strategy — 140
5.5.4 Perform an IP Audit to Determine if You are Missing Any Key IP Rights — 143
5.5.5 Monitor Your Intellectual Property Rights, and Your Competitors' — 143

5.6 The Importance of Freedom to Operate — 144
5.6.1 FTO Legal Opinions — 144
5.7 Licensing Agreements — 145

6 Funding Methods — 147
6.1 Start with the End in Mind: Know Your Exit Strategy — 147
6.1.1 Stages of Funding — 150
6.2 Funding — 152
6.2.1 Risk and De-Risking — 152
6.2.2 Pre-seed Funding/Bootstrapping — 153
6.2.3 Types of Funding: Equity, Debt, and Non-dilutive Funding — 154
6.3 Lessons Learned from Pitching to Prospective Investors — 164
6.4 Understanding Investor Expectations — 172
6.5 Common Early Startup Challenges — 173

7 Launching Your Innovation to Market – Strategy then Implementation — 175
7.1 Introduction — 175
7.1.1 Strategy versus Implementation — 177
7.1.2 Strategic Implementation Plan (SIP) — 178
7.2 Hiring Strategy Planning and Implementation — 178
7.2.1 Identifying Types of Resources — 178
7.2.2 Hiring Strategy Planning — 181
7.2.3 Hiring Strategy Implementation — 183
7.3 Vendor Management — 187
7.3.1 Assessing Vendors — 188
7.3.2 Vendor Risk Management — 192
7.3.3 Assessing Service Providers and Contract Manufacturers — 194
7.3.4 Scope Creep — 195
7.3.5 The Importance of Vendor Management — 196
7.4 Marketing and Sales Strategy and Implementation — 197
7.4.1 Marketing and Sales Strategy — 199
7.4.2 Marketing and Sales Implementation — 202
7.5 Postlaunch: Evaluating Effectiveness and Planning for Further Expansion — 203
7.5.1 Postlaunch Surveillance and Analytics Strategy — 205
7.5.2 Postlaunch Surveillance and Analytics Implementation — 207

8 Technology Project Management — 209
8.1 Introduction — 209
8.2 Tools for “First-Time” Project Managers — 212
8.2.1 Scope Management — 213

8.2.2 Risk Management and Scenario Planning — 216
8.2.3 Project Communication — 221
8.3 Managing Technical and Strategic Risk and Creating Realistic Project Timelines — 228
8.3.1 Managing Technical Risk — 228
8.3.2 Managing Strategic Risk — 231
8.3.3 Creating Realistic Project Timelines — 232
8.4 Effective Technology Management — 234
8.4.1 Establish Central Governance — 234
8.4.2 Utilize a Project Management Office — 234
8.4.3 Design an Employee Performance Incentive Program — 235
8.4.4 Execute Best Practices in Project Management — 236
8.5 Change Management — 236
8.5.1 Step 1: Assess and Understand the Current State of the Project — 237
8.5.2 Step 2: Research, Analyze, and Agree to the Problem — 237
8.5.3 Step 3: Communicate a Vision — 238
8.5.4 Step 4: Design and Implement a Project to Deliver the Change — 238
8.5.5 Step 5: Assess How the Change is Impacting the Organization — 240
8.5.6 Table Top Workshops – A Live Example of Executing Change Management — 240
8.5.7 Final Thoughts on Change Management — 241
8.6 Problem-Solving Techniques — 241
8.6.1 What is the Problem? — 241
8.6.2 Finding Solutions — 242
8.7 Brainstorming (Ideation) Techniques — 243
8.7.1 Do your Homework — 243
8.7.2 Facilitation Techniques — 244
8.7.3 Do Not Settle with a List, Take it Further! — 245
8.7.4 Follow Through, Be Proactive! — 246
8.8 Conclusion — 247

Part B: **Engineering the Innovation**

9 Needs Finding, Concept Generation, and Prototyping — 252
9.1 Needs Finding Process — 257
9.1.1 Observations, Unmet Needs, and Needs Statements — 257
9.1.2 Need Statement Selection — 260
9.2 Concept Generation Process — 263

9.3 Preparing for Needs Finding Process and Concept Generation Process — **268**
9.3.1 Preparedness — **268**
9.3.2 Alternative Decision-Making Models — **272**
9.4 Decision-Making Walkthrough — **273**
9.5 Needs Finding and Concept Generation for Startups — **276**
9.6 How a Concept Becomes a Product — **278**
9.6.1 Prototype Development — **279**
9.6.2 Design for Manufacturing (DFM) and Design for Assembly (DFA) — **280**
9.6.3 Sourcing — **283**
9.7 Finding Resources — **284**
9.7.1 Hardware and Physical Components — **284**
9.7.2 Computer-Aided Design (CAD) and Computer-Aided Manufacturing (CAM) — **285**
9.7.3 3D Printing — **285**
9.8 Moving on to Product Development — **287**

10 Product Development and Manufacturing — 288
10.1 Product Development through Manufacturing — **288**
10.1.1 Minimum Viable Product (MVP) — **289**
10.1.2 Planning How to Develop a Product/Technology with a Technology Development Plan — **290**
10.1.3 Product Development — **296**
10.1.4 Transfer to Manufacturing — **306**
10.2 Documentation — **307**
10.2.1 Design Controls — **309**
10.2.2 Revision Histories — **316**
10.2.3 Bill of Materials — **316**

11 An Introduction to Quality — 319
11.1 A World of Quality — **319**
11.1.1 Making Consistent Quality Products: An Overview of Quality Management — **320**
11.2 Quality Management Systems (and Examples) — **322**
11.2.1 Types of Quality Management Systems — **322**
11.2.2 Design of Quality Management Systems — **324**
11.2.3 Checks and Balances within Quality Management Systems — **327**
11.3 Total Quality Management (TQM) Tools — **330**
11.4 Six Sigma and Lean Six Sigma — **333**
11.4.1 Lean — **333**

11.4.2 Six Sigma — 337
11.4.3 Lean Six Sigma — 339

12 Medical Devices — 341
12.1 Introduction — 341
12.1.1 Impact of Medical Devices — 341
12.1.2 Overview of Medical Technologies — 344
12.2 Unique Management Aspects of Bringing Medical Devices to Market — 346
12.2.1 Medical Device Value Propositions — 346
12.2.2 Medical Device Business Models — 349
12.2.3 Medical Device Fundraising — 352
12.2.4 Medical Device Leadership — 354
12.2.5 Medical Device Startup Operations — 356
12.3 Unique Technical Aspects of Bringing Medical Devices to Market — 357
12.3.1 Medical Device Research and Development: Transitioning from Bench to Bedside — 357
12.3.2 Medical Device Validation Testing Requirements — 359
12.4 Other Unique Aspects of Bringing Medical Devices to Market — 364
12.4.1 Medical Device Regulatory Strategy — 364
12.4.2 Medical Device Labeling — 370
12.4.3 Medical Device Clinical Trials — 374
12.4.4 Medical Device Reporting — 377
12.4.5 Medical Device Marketing — 379
12.4.6 Medical Device Reimbursement in the United States — 380
12.5 Closing — 382

List of Acronyms and Initialisms — 384

Further Reading — 387

Subject Index — 389

List of Figures

Figure A Graphic representation of the various processes involved in simultaneously developing a business and bringing a product/technology to market —— 3
Figure B Chapter 1 focuses on Due Diligence and Early-Market Research —— 4
Figure C The first portion of Chapter 2 focuses on Business Validation —— 5
Figure D Chapter 3 addresses business incorporation and establishing a team —— 5
Figure E Chapter 4 focuses on the various business strategies and plans involved in preparing a Business Plan —— 6
Figure F Chapter 5 discusses the types of IP rights that are available and IP strategy during the Business Development Process —— 7
Figure G Chapter 6 discusses the types of funding that are available and provides tips on how to pitch to prospective investors during the Business Development Process —— 7
Figure H Chapter 7 discusses strategic implementation during the Business Development Process —— 8
Figure I Chapter 9 focuses on the ideation, concept generation, and early-stage prototyping stages of the Product/Technology Development Process —— 9
Figure J Chapter 10 focuses on the Product Development and Transfer to Manufacturing stages of the Product/Technology Development Process —— 10
Figure 1.1 Typical life cycle stages of a business —— 16
Figure 1.2 Engineering innovation life cycle: funding, business, and technology development —— 17
Figure 1.3 Business Development timeline —— 19
Figure 1.4 Samantha's technical due diligence process —— 29
Figure 1.5 Value proposition for PetSpec —— 32
Figure 2.1 Business Development timeline —— 35
Figure 2.2 SWOT model —— 39
Figure 2.3 Word Cloud – What problems exist with the current solution on the market? —— 42
Figure 2.4 Product Development timeline —— 44
Figure 2.5 Example of upstream and downstream gauze bandage stakeholders —— 47
Figure 3.1 Incorporation and establishing a team steps in the Business Development Process —— 66
Figure 3.2 Support structures for startups —— 72
Figure 3.3 Technology ecosystem and investment —— 78

https://doi.org/10.1515/9783110521900-202

Figure 4.1 A number of Business strategies and Plans must be prepared and then captured in a Business Plan — **84**
Figure 4.2 SWOT analysis template/example — **91**
Figure 4.3 Example market expansion plan — **97**
Figure 4.4 Technology life cycle stages — **100**
Figure 4.5 High-level technology development timeline — **108**
Figure 4.6 Sample Gantt chart — **115**
Figure 4.7 Sample budget — **116**
Figure 4.8 Examples of different company Organizational Charts — **123**
Figure 4.9 Example Roles and Responsibilities Matrix — **124**
Figure 5.1 Securing IP rights and developing an IP portfolio is an important part of the Business Development Process — **131**
Figure 5.2 Typical US utility patent life cycle and associated fees — **134**
Figure 5.3 Example diagram of related intellectual property rights in an IP portfolio — **141**
Figure 6.1 Seeking funding and pitching to prospective investors are key steps of the Business Development Process — **148**
Figure 6.2 Business life cycle stages and corresponding funding stages — **151**
Figure 7.1 Strategic Implementation planning activities throughout Business Development — **176**
Figure 7.2 Relation of Business Plan to Strategic Implementation Plan — **177**
Figure 7.3 Example Gantt chart — **179**
Figure 7.4 Using Project Roadmaps to create resource hiring plans — **180**
Figure 7.5 Vendor Management Process — **189**
Figure 7.6 Marketing and Sales communication channels — **200**
Figure 7.7 Postlaunch Surveillance and data analytics process flow — **204**
Figure 8.1 Important components of successfully managed projects — **211**
Figure 8.2 Tools for “First-Time” project managers — **212**
Figure 8.3 Cause-and-effect sequence of risk and mitigation events — **216**
Figure 8.4 RADIO Terminology — **218**
Figure 8.5 Manufacturing Process Analysis Storyboard — **219**
Figure 8.6 Lessons learned template — **221**
Figure 8.7 The three parts of every communication – sender, message, and receiver — **222**
Figure 8.8 Project meeting notes template — **223**
Figure 8.9 Communication plan template — **224**
Figure 8.10 SCRUM board template — **225**
Figure 8.11 Example scorecard for reporting status of project schedule critical path — **226**
Figure 8.12 Example project dashboard template — **227**
Figure 8.13 Example project dashboard template — **228**

Figure 8.14 Example startup company roadmap for an FDA regulated class II medical device — **229**
Figure 8.15 Example PMO organizational chart — **235**
Figure 8.16 Diagram showing an example of five WHY's — **237**
Figure 8.17 Diagram Showing an example of Fishbone — **238**
Figure 8.18 Example of a RACI matrix — **239**
Figure 8.19 Brainstorming poster set example — **245**
Figure 8.20 Important components of successfully managed projects — **247**
Figure 9.1 The Needs Finding and Concept Generation pathway — **254**
Figure 9.2 Needs Finding observation pathway — **258**
Figure 9.3 Map of the 4Ps after an observation — **259**
Figure 9.4 Needs Refinement decision-making process for selecting an unmet need — **262**
Figure 9.5 Concept Generation pathway — **263**
Figure 9.6 Basic example of functions, subfunctions, and base functions within a Functional Decomposition diagram — **264**
Figure 9.7 Example of the Concept Generation Process — **267**
Figure 9.8 Process for developing a Pugh Matrix — **269**
Figure 9.9 Functional Decomposition of partially automated car wash — **275**
Figure 9.10 Comparison of Functional Decomposition of (A) a partially automated car wash to (B) a fully automated car wash — **276**
Figure 9.11 Process for extrapolating and validating an unmet need from a preconceived product used by startups — **277**
Figure 9.12 Progression of the Development Processes for donuts depicting the steps of (A) Concept Generation, (B) Proof-of-Concept, (C) Prototype/Product Development, (D) Minimum Viable Product, and (E) a Product Portfolio — **278**
Figure 10.1 Product Development and Manufacturing stages of the Product/ Technology Development timeline — **290**
Figure 10.2 Product Development through Transfer to Manufacturing timeline — **291**
Figure 10.3 Progression of the Product Development of a donut depicting (A) Early-Stage (α)-prototypes, (B) Beta (β)-prototypes, (C) the Transfer to Manufacturing Process for donuts, (D) a Minimum Viable Product, and (E) Product Portfolio — **296**
Figure 10.4 Example Validation Report table of contents — **305**
Figure 11.1 Four types of quality management systems — **323**
Figure 11.2 QMS governance framework — **324**
Figure 11.3 Interdepartmental execution of QMS processes — **327**
Figure 11.4 Pareto chart – reasons for customer-returned merchandise — **335**
Figure 11.5 Pareto chart – merchandise nonwoven process defects — **336**
Figure 11.6 Example Bell curves — **338**

Figure 12.1 Summary of chapter content — **342**
Figure 12.2 Primary medical device startup valuation milestones — **342**
Figure 12.3 Key manufacturing, IDE, and IRB milestones in clinical trial preparation — **377**

About the Authors

Our professional and educational backgrounds are diverse but complementary. Benjamin Legum[1] is a biomedical and materials science engineer, who has experience working as a manufacturing and process engineer, and has a proven track record in developing prototypes and streamlining complex manufacturing processes. He is a cofounder and partner at Keystone Scientific, which is a technology and management consulting firm. Jennifer Vondran[2] is also a cofounder and partner at Keystone Scientific. Jennifer has a background in biomedical engineering and material science engineering, quality, and regulatory, and is a certified project management professional (PMP). She has over a decade of experience as a professional technology and management consultant, specializing in medical devices, life sciences, and healthcare. Amber Stiles[3] not only has a background in biomedical engineering, but also has earned an MBA and a law degree. Amber is a registered patent attorney, works as a patent examiner at the United States Patent and Trademark Office (USPTO), and also provides business consulting and technical writing services to startup companies.

We have known each other for more than 15 years, and we have helped one another on many various projects. This book is a culmination of our combined experiences, with emphasis on techniques and methods that worked well for us in the past. We hope that stylizing *Engineering Innovation* as a how-to guide is useful to our readers.[4]

---

1 Benjamin Legum would like to acknowledge Dr. Stanley Legum, Fred Marroni, Dr. Kevin Roth, and Cindy Nellis for their expertise, insight, and patience.

2 Jennifer Vondran would like to acknowledge (1) God for His many blessings and gifts, (2) my husband, Richard, for being my best friend and soul mate and for always supporting my dreams; (3) my mother, Louise, for her love, courage, and strength; (4) my father, Joseph, for his unique sense of humor and wisdom; and (5) my younger sisters, Erika and Andrea, for giving me a sense of purpose and making life worth living.

3 Amber Stiles would like to acknowledge Graham Stiles, Cynthia N. Stiles, and David Carroll for their love, support, and insight.

4 *Engineering Innovation* is meant to convey general information, guidance, and useful tips based on our collective experiences concerning the topics discussed. The book in no way constitutes legal advice, and the opinions contained in this book are our own.

https://doi.org/10.1515/9783110521900-203

Introduction

What Is Engineering Innovation?

Let's start by looking at the terms "engineering" and "innovation." An engineer is someone who approaches situations and problems in a methodical and systematic way. Due to this style of problem solving, engineers are often considered rigid and inflexible.

On the other hand, "innovation" brings to mind words such as creativity, brilliance, novelty, and inspiration. "Innovation" is a vague term that refers to anything that has not been done before. By way of example, "innovation" could mean developing a software program that merges information from multiple sensors in a new way, developing a platform service that revolutionizes the way people perform daily tasks, or developing a noninvasive technique to deliver gene therapy to specific cells in the body. But innovation does not need to be as complicated or technical as these examples. Innovation could simply be developing a cell phone case that also holds credit cards or a no-drip, homemade popsicle stick.

An innovation does not have to be one significant or disruptive technological development that will make billions of dollars. Rather, innovations can be small or incremental, but more importantly, innovations change how the process was done before. Innovation is often a response to the grueling redundancy of performing the same duties day in and day out in the workplace. As discussed in this book, innovation can also be formulated to increase the likelihood of commercial success.

Engineering Innovation outlines what it takes to transition a creative and inspiring new concept into a successful product/process/service using a methodical, structured plan, and approach.

What Does It Mean to Be an Innovator?

Innovation is the creation of new insight and new ideas that can be used to solve problems. Without a problem, there is no reason to innovate. Problems do not have to be obvious. It is remarkable when an innovator can "read between the lines" into what customers or users say they *want* and give them what they actually *need*. Innovators have the potential to create a market demand where there was not one before.

Innovators think about problems from different angles to understand root cause(s), and can break down impossibly complex problems into manageable questions that can be answered. Innovators do not create solutions that are more complicated than they need to be. Instead, clarity of thought and simplicity of approach and design are much more elegant – yet challenging – tactics for realizing a solution.

https://doi.org/10.1515/9783110521900-001

Stunning examples of elegant and simple innovations include the humble paperclip and the ubiquitous arch. A paperclip is a simple design and is inexpensive to manufacture, but the utility of a paperclip is immense. Similarly, in architecture the arch is a simple shape to behold, and yet when implemented effectively, invisible forces directed in multiple axes hold the arch in place and support vast amounts of weight situated above the arch.

When innovating around a technology, the fundamental principles of science and engineering should be the "backbone" of any product. Additionally, quantitative data can go a long way to help prove that:

- a problem exists with the current way things are done,
- there is a need for that problem to be solved (to which primary and secondary customers can testify),
- your idea can solve what is causing the problem,
- there is a market that finds value in your idea, and
- there is a consumer/economic demand for your idea.

Typically, the hard work that goes into bringing innovation to the public is not rewarded instantly. It is not uncommon for there to be many iterations of an idea or method before a solution is reached and innovation is successfully achieved. Throughout this iterative process, much is learned and new insight is gained. Be persistent and enthusiastic, and explore every idea with rigor. **Innovation is always possible.**

Overview of Sections and Chapters

Engineering Innovation is a practical guide that outlines the various processes involved in transitioning a technological innovation from a mere idea to a working prototype, then on to a marketable product. Taking an innovation from ideation to market success involves two separate processes that must occur simultaneously: building a business around the innovation (i.e., **Business Development**) and evolving the innovation from an idea to a viable product or service from a technical standpoint (i.e., engineering the innovation through **Product or Technology Development**). Both the Business Development Process and the Product/Technology Development Process run in parallel and generally correspond to the life cycle stages of the company, which also coincidentally correspond to stages of funding that companies often encounter as they grow from a startup to an established business.

Figure A shows the relationship between the Business Development Process and the Product/Technology Development Process. Figure A also shows the corresponding business life cycle stages of a company and the typical corresponding funding stages. Color has been used to indicate the life cycle stage of the company: the due diligence stage is represented by orange, the startup stage is represented by

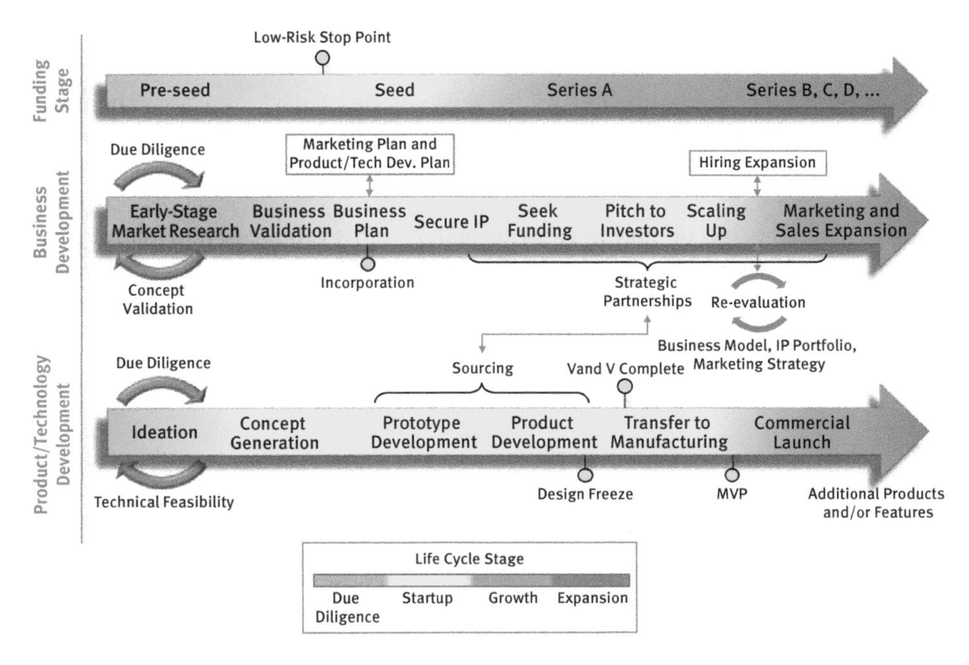

Figure A: Graphic representation of the various processes involved in simultaneously developing a business and bringing a product/technology to market.

yellow, the growth stage is represented by green, and the expansion stage is represented by blue.

The graphics within Figure A will be referred to frequently throughout the text of this book to orient readers as to how the various processes of building a business and bringing a product to market interplay.

Engineering Innovation consists of two parts:

Part A: The Business Side of Innovation

Part B: Engineering the Innovation

Part A: The Business Side of Innovation focuses on the Business Development Process and Funding Stages, which explores the various business aspects of bringing a new technological innovation or product to market. The chapters of Part A are designed to provide readers with guidance to follow, as they work through the Business Development Process.

Part B: Engineering the Innovation focuses on the Product/Technology Development Process and explores the various design, engineering, and manufacturing aspects of bringing a new technological innovation or product to market. The chapters of Part B are designed to provide readers with guidance to follow, as they work through the Product/Technology Development Process.

A. The Business Side of Innovation

The business side of innovation follows the Business Development Process. The Business Development Process begins with Early-Stage Market Research, which involves a feedback loop of Due Diligence (Chapter 1) and Concept Validation (Chapter 9). Chapter 1 focuses on the importance of conducting due diligence and market research (see Figure B) and offers tips on how to gather useful data that can be used to inform your decision as to whether a concept (i.e., the idea that you plan on developing into a product) is viable and potentially marketable. To say this another way, data collected from your due diligence and market research activities can be used to validate your concept.

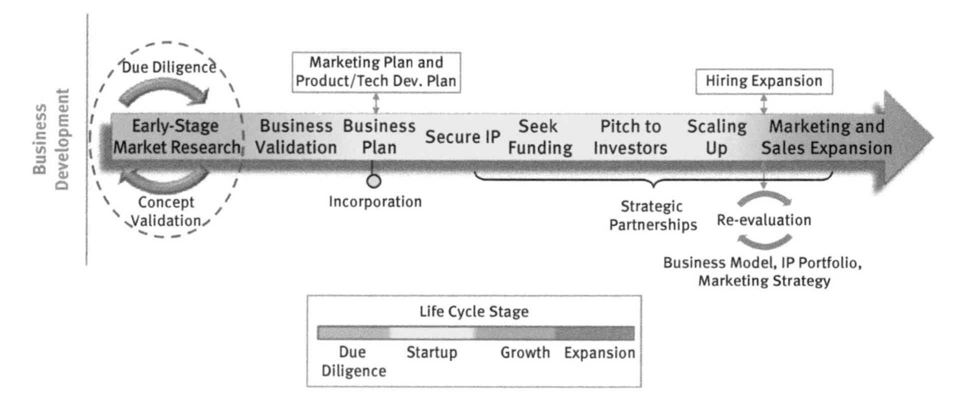

Figure B: Chapter 1 focuses on Due Diligence and Early-Market Research.

The next step of the Business Development Process involves Business Validation, which is discussed in Chapter 2 (see Figure C). Validating your business model involves evaluating whether the market will support the business model by performing market situation analysis; conducting Strengths, Weaknesses, Opportunities, Threats (SWOT) analysis; and holding conversations with various stakeholders. The first portion of Chapter 2 focuses on the Business Validation stage of the Business Development Process.

While Business Validation is an important, early step in the Business Development Process, it is also important to simultaneously validate the underlying concept for the product, technology, or service that you plan to bring to market. Concept Validation, which is part of the Product/Technology Development Process, can be influenced by information gathered during your Business Validation efforts. The same mechanisms used to validate your business model can also be used to validate your concept. For instance, Concept Validation involves defining user requirements through conversations with stakeholders and conducting ongoing market research.

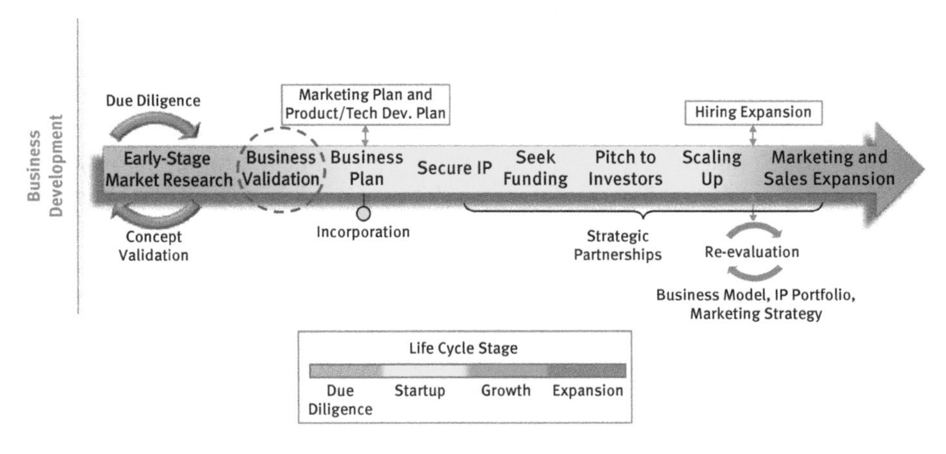

Figure C: The first portion of Chapter 2 focuses on Business Validation.

The second portion of Chapter 2 is focused on techniques that can be used to inform Concept Validation throughout the Product/Technology Development Process.

Chapter 3 focuses on how to develop and build a business. Building a business involves making a lot of decisions and requires a lot of careful planning. For instance, choosing a business structure, establishing a smart team of people to help the business achieve success, laying the groundwork for a business plan, and identifying support structures and opportunities that the business can use to form and grow are all important steps of the Business Development Process. In reality, none of these steps for starting a business occurs sequentially. Rather, they all seem to occur in parallel. The process is a whirlwind, but rewarding. Figure D shows the step of Business Incorporation relative to the Business Development Process.

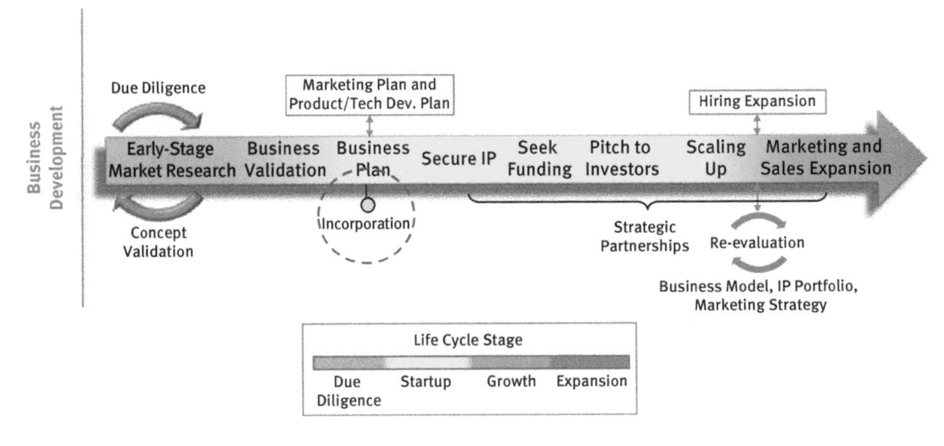

Figure D: Chapter 3 addresses business incorporation and establishing a team.

The next step in the Business Development Process is to prepare a Business Plan. Chapter 4 discusses the various different strategies that underlie how a business works, such as a marketing strategy, sales strategy, expansion strategy, intellectual property (IP) strategy, and, in the case of technology-based businesses, a technology development and implementation strategy. But a strategy is only the idea or plan of action for how a company achieves its goals. Companies must go one step further and capture these various strategies in a corresponding number of plans, such as a marketing plan, sales plan, expansion plan, IP plan, and, in the case of technology-based companies, a technology development and implementation plan. The highlights of each plan are summarized in an overarching, high-level Business Plan. The end of Chapter 4 provides additional guidance on how to prepare a written business plan. Figure E highlights the step of preparing a Business Plan in the Business Development Process.

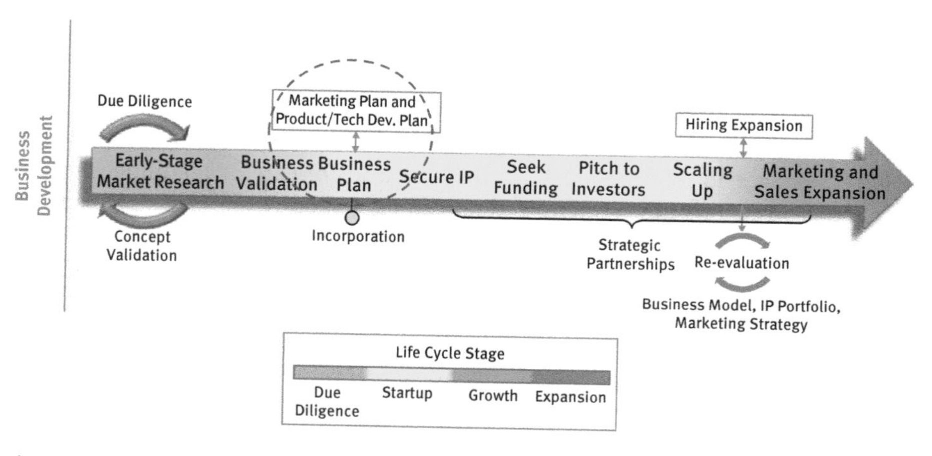

Figure E: Chapter 4 focuses on the various business strategies and plans involved in preparing a Business Plan.

Chapter 5 explores IP and explains the value to be gained by investing in various forms of IP protection related to the innovation, such as patents, trademarks, copyrights, and trade secret protection. Not only does a comprehensive IP portfolio offer legal protections against infringement (i.e., others making, using, selling, or importing the innovation and its associated IP rights), but it can also be monetized (for instance, through the licensing of IP rights) to produce additional sources of capital for the company. Securing IP rights involves careful planning. Leveraging those rights involves the cunning use of strategy. Entrepreneurs and innovators simply cannot develop and market an innovative new technology or product without securing some form of IP. IP protection is a significant source of competitive advantage and value for a company. Figure F shows the step of securing IP as part of the Business Development Process.

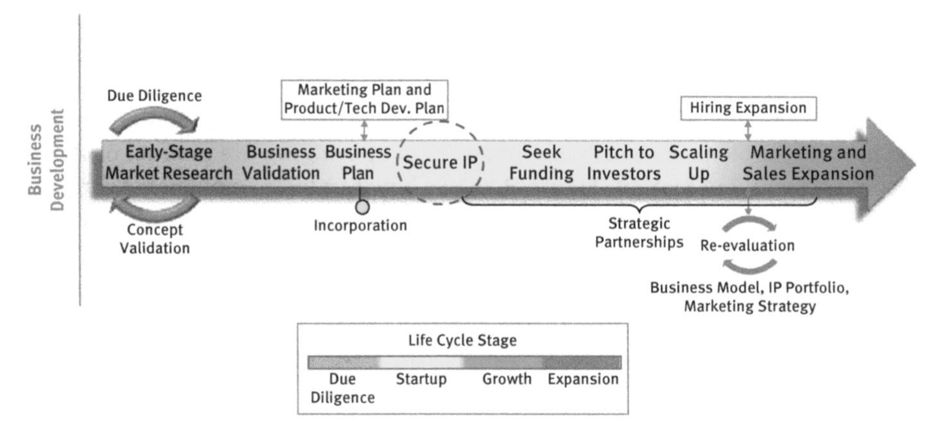

Figure F: Chapter 5 discusses the types of IP rights that are available and IP strategy during the Business Development Process.

Chapter 6 introduces sources of funding and explains the processes for seeking or obtaining funding. Many companies must search externally for sources of funding to build and grow their company and to finance the development of their product or technology, but knowing where to start looking can be challenging. In addition to providing a guide for the various sources of funding, Chapter 6 also offers guidance and practical tips on how to go about applying or asking for funding, including a reflection on lessons learned by the authors concerning pitching to prospective investors. Figure G highlights the steps of seeking funding and pitching to investors as part of the Business Development Process.

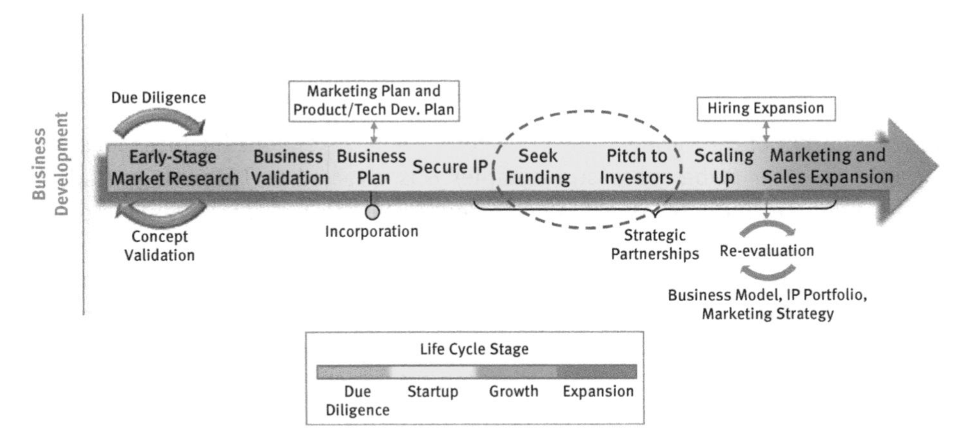

Figure G: Chapter 6 discusses the types of funding that are available and provides tips on how to pitch to prospective investors during the Business Development Process.

Once plans are made for the business, they must be strategically implemented. Strategic Implementation is the focus of Chapter 7. A strategic implementation plan (SIP) takes the components of a business plan (such as the Business Plan, Marketing Plan, and Product/Technology Development Plan) and organizes them into actionable tasks with the ultimate goal of bringing the product or technology successfully to market. Figure H shows how Strategic Implementation fits into the Business Development Process. The planning associated with a SIP occurs simultaneously, while developing the Marketing Plan, Sales Plan, and Product/Technology Development Plan during the Business Plan stage of the Business Development Process. Execution of the SIP usually occurs once funding is secured and concludes with the successful launch of the product or technology to market.

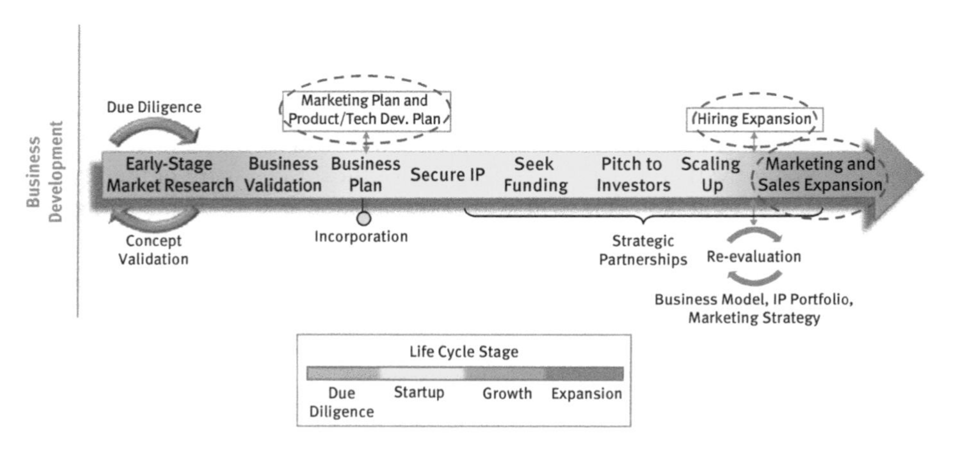

Figure H: Chapter 7 discusses strategic implementation during the Business Development Process.

Implementation of the various plans that form the basis of your business must be managed in order to be successful. Chapter 8 is directed to Technology Project Management. Technology Project Management is the process of managing and delivering successful projects, and it takes place throughout the Business Development Process. By defining strategy, managing scope, mitigating risk, executing proven technology management processes, and applying project management tools and lessons learned from previous projects, companies will have a greater chance of being successful. Chapter 8 explores a variety of useful tools that can be used to effectively manage the critical aspects of technology development projects.

B. Engineering the Innovation

In Part B of this book, the process of Engineering the Innovation will be discussed, in terms of the Product or Technology Development Process. The chapters of this section are designed to provide readers with methodologies to follow, as they work through various technical and manufacturing aspects of the Product or Technology Development Process.

Any product or technology that goes to market starts as a concept. A concept is the idea behind the product or technology that you plan on developing. When a concept is chosen that satisfies an unmet need in the market, there is a likelihood that the product will be a success. The start of the Product/Technology Development Process, as shown in Figure I, begins with the generation of a concept. Concepts are then tested using early-stage, physical prototypes, such as proof-of-concept prototypes, and α-prototypes. Chapter 9 focuses on the beginning of the Product/Technology Development Process.

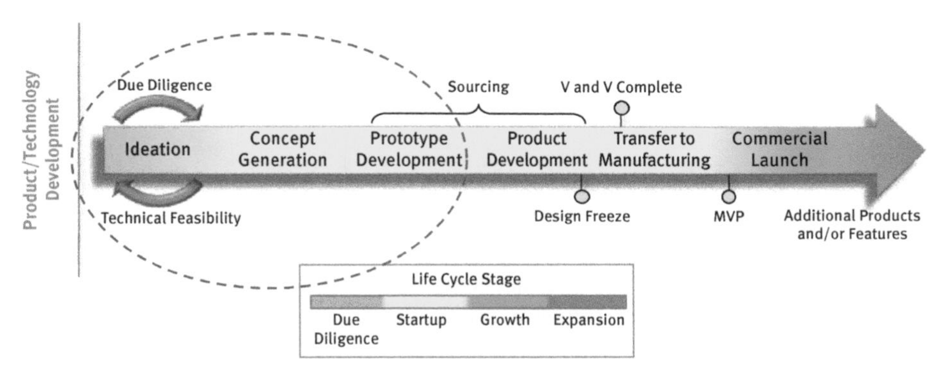

Figure I: Chapter 9 focuses on the ideation, concept generation, and early-stage prototyping stages of the Product/Technology Development Process.

Proof-of-Concepts (POC) prototypes are used to assess the feasibility of top concepts. Once the top concept is selected based on the POC, then early-stage prototypes (α-prototype) are developed, using iterative prototype development and design modification. A β-prototype, or a system stable enough to be tested by users, is developed from the α-prototype and is used to obtain user feedback. Feedback from user testing can be adapted into future iterations of the product design. A finalized product design goes to Design Freeze, which is a stage in the Product/Technology Development Process where a binding decision is made concerning the final design of the product. The Design Freeze product design is subjected to Verification and Validation (V&V) processes, which verify that your product meets the design specifications and validates that your product satisfies the needs of your intended users. The verified and validated product design leads to a minimum viable product (MVP). The MVP is the simplest form of the product design that meets the intended user needs and is economical to produce, and it is the design version that is intended for Transfer to Manufacturing and Commercialization.

Figure J illustrates the Product Development, Design Freeze, Verification and Validation, Minimum Viable Product, and Transfer to Manufacturing stages of the Product/Technology Development Process. The first half of Chapter 10 focuses on the development of the physical product, and the second half of Chapter 10 explores the documentation that is needed for Product Development and Manufacturing.

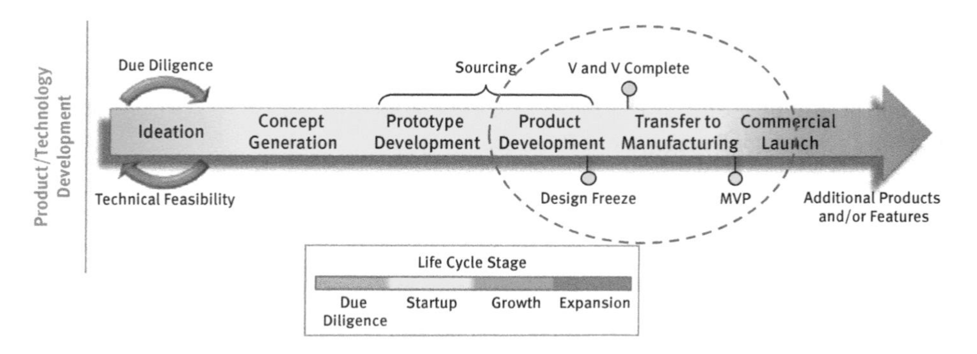

Figure J: Chapter 10 focuses on the Product Development and Transfer to Manufacturing stages of the Product/Technology Development Process.

Bringing a product to market requires that companies focus on ensuring the quality of the product. Ensuring consistent quality of a product involves developing and maintaining a quality management system (QMS) to identify, track, and monitor every component of a product at every stage of manufacturing. Companies must develop a QMS that provides a framework for executing processes, storing and archiving information, and executing various projects according to company quality policies and quality controls. Chapter 11 reviews the types of quality management systems, the design of quality management systems, the purpose of quality management system audits, and the checks and balances that exist within an effective quality management system. Chapter 11 concludes with a discussion focused on Total Quality Management (TQM), which is the methodology that uses statistical tools to analyze the data that is collected and stored within quality management systems. TQM tools such as Lean, Six Sigma, and the combination of Lean and Six Sigma are discussed with examples.

Chapter 12 serves as a case study that brings together the concepts discussed in earlier chapters. Chapter 12 looks specifically at the unique management, technical, clinical, regulatory, and other aspects of bringing medical device products and services to market and focuses on applicable regulations and examples in the United States.

In closing, the Business Development Process and the Product/Technology Development Process typically occur simultaneously. At the outset, it can seem overwhelming. However, early-stage companies have plenty of resources available to them to help guide their growth and development. For instance, companies can model their Business Development and Product/Technology Development Processes on the processes used by more established companies. Early-stage companies can also use industry or regulatory standards to set a framework for how to bring together the various processes involved in building the business and bringing their product through the development process. Early implementation of these standards within a company, or even simply taking the time to contemplate the implications that industry and regulatory standards could have on how the company carries out the Business Development and Product/Technology Development Processes, can be hugely beneficial.

Part A: **The Business Side of Innovation**

1 Starting Out with Due Diligence and Early-Stage Market Research

Abstract: At the outset of any business, it is important to conduct due diligence and early-stage market research. Due diligence is a form of research that involves looking to the relevant field of endeavor and reviewing the literature, market, and intellectual property (IP) landscape surrounding the area of technology of interest for the purpose of making a determination as to whether efforts to develop and commercialize an innovation are worth pursuing. The goal of due diligence is to identify and assess any potential obstacles and potential ways for overcoming these obstacles, and to evaluate the likelihood of long-term success. The value of due diligence is that, in many cases, it can be done with little to no monetary cost; only time. One important step in conducting due diligence is conducting early-stage market research, which involves collecting data and useful information about the needs and preferences of the consumers that make up the target market. The information that is gathered during market research will give you a better idea of whether your innovation is worth developing and bringing to market, that is, the information gathered during the market research portion of due diligence will inform the validation of your innovation.

Keywords: Due diligence, life cycle, startup, growth, expansion, maturity, exit, market research, market segments, value proposition, market reports, consulting firms, data acquisition firms, target consumer, competitors.

1.1 The Life Cycle of a Company

Before delving into the business aspects of innovation, it would be good to start with a fundamental understanding of the typical life cycle that a business goes through. Businesses have a beginning; a middle, in which they grow and develop; and an end. Figure 1.1 shows the typical life cycle stages of a business.

- **The Due Diligence Stage (also referred to as the seed and development stage).** The due diligence stage is the gestation period in which a business is conceived and born. It is when the business is still just an idea that is taking form. During this stage in the business's life cycle important assessments are made using due diligence investigations (see Section 1.2), initial market research (see Section 1.4), and validation techniques (see Chapter 2), to determine whether the idea behind the business is worth pursuing. If so, then the business progresses to the next stage in the life cycle.
- **The startup stage.** The startup stage is when the team is assembled and a company is formed and officially launched. This stage is often very exciting, but

https://doi.org/10.1515/9783110521900-002

Life Cycle Stages of a Business

Due Diligence	Startup	Growth	Expansion	Maturity/Exit
Gestation period of a business. Due diligence, initial market research, and validation techniques are used to decide if business should go forward.	Company is officially formed and launched. Very exciting, but also most risky stage of life cycle. The search for funding starts here.	Business grows into an established company. Stability improves, revenues rise, and the company scales up to meet demand.	Company makes expansion moves, such as expanding the company's workforce, product lines, moving into new regions, and/or markets.	Company either continues business as usual (as a lifestyle business) or makes an exit.

Figure 1.1: Typical life cycle stages of a business.

also very risky. It is the exciting period where the company prototypes, seeks early-stage funding sources, and initially goes to market, but it is also the period in the business's life cycle where the company is most likely to fail. If a company is successful in obtaining funding and launching to market, the business will move to the next stage of its life cycle.

- **The growth stage**. Successful startups advance to a period of growth and development, during which the company transitions into an established company. In this stage, revenues are consistently being generated by the company and profits increase. As success is incrementally earned at this stage in the life cycle, the business must grow in order to meet demand. Growth can involve increasing the company's workforce, scaling up production, managing increased revenues, developing new strategies to remain competitive, and reevaluating how the company is managed.
- **The expansion stage**. After establishing the company in the growth stage, it might make smart business sense for the company to go through a period of expansion. This could include expanding the company's workforce, expanding product lines, expanding into new geographical regions, expanding into new markets, and participating in mergers and acquisitions.
- **The maturity and possible exit stage.** Owners of established companies must decide how to proceed beyond the growth and expansion stages – the owners can either carry on routinely as they have, making a lifestyle out of their business, or the owners can opt to exit the business.

Just as a business has a life cycle, the development of an innovation from an idea to a prototype and then to a marketable product also follows a life cycle, known as the Product/Technology Development Process. At the same time, a business must also grow and develop through a series of phases known as the Business Development Process. Figure 1.2 illustrates how the Business Development Process and the Product/Technology Development Process run in parallel to one another, and also correspond to the progressive life cycle stages and funding stages of a business.

All innovation starts with an idea, and passion with a purpose drives the transformation of an idea into an innovative technology. Many scientists, engineers,

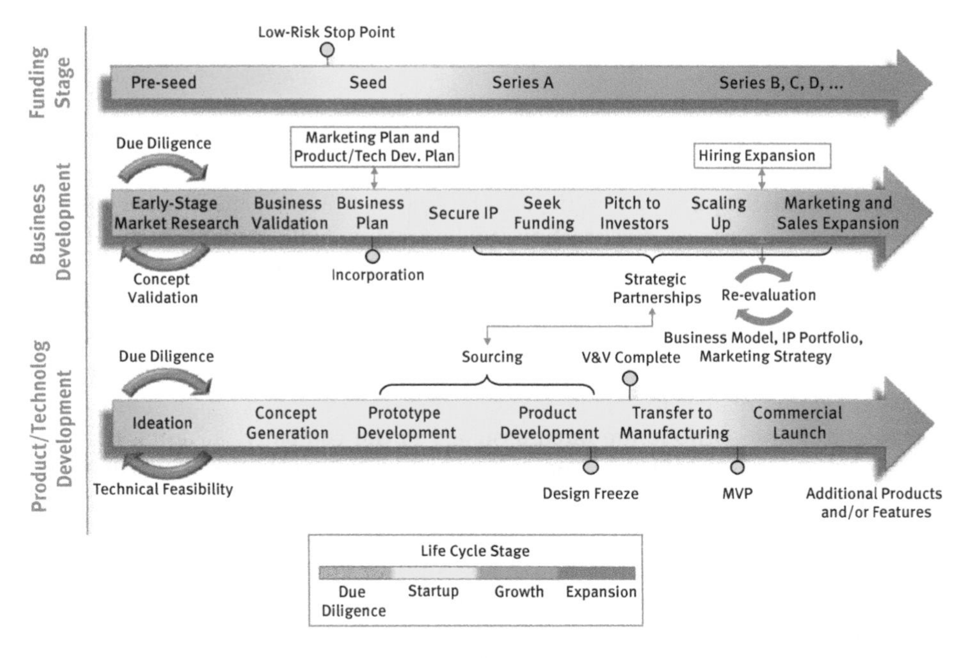

Figure 1.2: Engineering innovation life cycle: funding, business, and technology development.

researchers, inventors, and entrepreneurs are struck by an idea, like the proverbial flash of genius, and then are motivated to develop the idea into a viable technology or product. Not only is there a sense of intellectual accomplishment associated with bringing an idea to fruition, there is also commercial success to be had, if the innovative technology is marketable.

But how does an innovative technology develop from an idea to a prototype then on to a marketable product? This is a question that many intellectual people ask themselves. Conducting research and development, building a functional prototype, and testing the prototype can consume significant amounts of time, resources, and energy. It would be a shame and a waste to discover that a similar technology or product exists in the market after investing significant time and money into efforts to turn the idea into a working prototype.

Fortunately, there are steps that innovators can take to help assess whether development of the idea should move forward. With a little investigative know-how and diligence, innovators can understand what problems the innovative concept (which can be a technology, product, or service) solves; what value the concept has; where the concept fits in the market; and whether developing the concept is worth the effort it will take to create a commercial product. One of the first things innovators should do before investing a lot of time, money, and resources into developing a new concept into a marketable technology or product is to conduct thorough due diligence.

1.1.1 Life Cycle Deviations

It is important to note that the above-mentioned stages represent the typical life cycle of a business. However, not every business's life cycle is the same. Alternative paths can include the following:

- **Intellectual Property Portfolio Monetization.** Some businesses are formed solely for the purpose of generating and leveraging intellectual property (IP) assets (these businesses are often referred to as nonpracticing entities, meaning they hold IP assets, but do not produce a product). Nonpracticing entities often never go through growth or expansion phases. Nonpracticing entities may be serial inventors or a business. The nonpracticing entity acquires a collection of patent rights related to an invention in what is called a patent portfolio. The patents in the portfolio are often assigned to the company, meaning that the company owns the patents – not the inventor(s). By assigning the patents in the portfolio to the company, the company can easily monetize the patents by transferring, selling, or licensing the patents to others. This can be a successful business model for people who do not want to grow or expand a business. Rather, the business can focus its efforts on strategically developing, selling, and licensing patents.
- **Early-Stage Funding.** A business that receives significant investor funding during the due diligence phase may quickly rocket to the growth phase, spending little time in the startup phase of the life cycle.
- **Early Acquisition.** A business that is acquired during the startup stage will be accelerated to the exit stage, skipping over the growth or expansion phases entirely. Planning for growth and expansion is key for getting acquired early, but in certain industries, the current system of big industry players encourage acquisition of smaller, startup entities. Therefore, the exit stage of the life cycle can arrive much earlier than expected.
- **Initial Public Offering (IPO or "going public").** Many businesses opt to do an IPO, in which shares of the company are sold to investors (after an underwriter – usually an institutional investor(s) – places the shares for sale on an exchange). These investors can include entities such as banks, insurance companies, pension funds, hedge funds, and individuals.

1.2 What is Due Diligence?

In essence, due diligence is doing your homework and making sure that any risks associated with developing a concept into a marketable technology or product have been carefully considered. Due diligence is a form of research, or an investigation, that involves looking to the relevant field of endeavor and reviewing the literature, market, and IP landscape surrounding the area of interest for the purpose of making a determination as to whether the innovative concept is worth pursuing. The purpose of due diligence is to identify and assess any potential obstacles, potential ways for overcoming these obstacles, and to evaluate the likelihood of long-term success. The value of due diligence is that, (1) it can help an innovator avoid wasting time and resources on an unmarketable product, and (2) it can encourage the innovator to revise or refine the concept to better reflect market realities. In many cases, the incentive of performing due diligence is that it can be done with little to no monetary cost; only time.

Due diligence takes time, and – as the name implies – diligence. However, there is no specific formula when it comes to how due diligence should be conducted. Due diligence is not going to be the same for every business. The scope and breadth of due diligence will be different at each stage of the business life cycle, and while there is a general formula for how to conduct due diligence, there is plenty of flexibility in how due diligence can be performed (Figure 1.3).

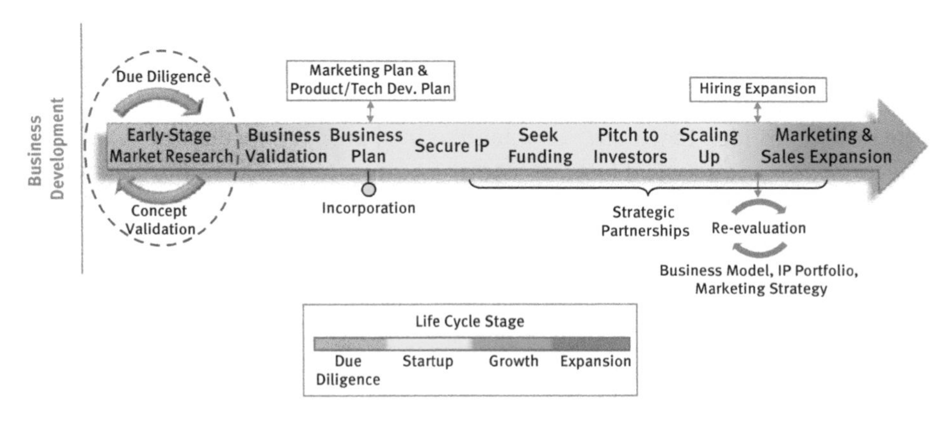

Figure 1.3: Business Development timeline.

In short, due diligence depends on the specific circumstances surrounding the innovative technology or product. For instance, a technological invention in a field where there is a lot of innovation, that is, a crowded market such as online social media platforms, may require more careful due diligence to ensure that there is a place in the market for the new technology.

Due diligence for developmental stages:

1. Determining if a concept is viable.
2. Developing a business model.
3. Determining strategies for market expansion.

Different types of due diligence include:

- Analysis of market trends.
- Evaluation of technological feasibility of developing the product.

- Conducting cost analysis (for instance, evaluating what it costs to develop the prototype, what it costs to secure IP protections, what it costs to scale up production, etc.).
- Conducting policy analysis (such as conducting an evaluation of the local, state, and federal laws and rules (and international considerations) associated with bringing the product to market).
- Analyzing the IP landscape for the relevant market.

Not all businesses will need to conduct all types of due diligence, especially considering the type of business that the company is involved in and the life cycle stage of the company. For instance, a local furniture manufacturing company will most likely not need to conduct federal policy analysis, whereas a large-scale furniture production company that plans to export furniture to foreign countries will be concerned about tariffs and international trade agreements.

It can be prudent to have an independent third party (third parties that can conduct due diligence include consulting firms, risk assessment analyzers, business lawyers, etc.) conduct the due diligence, because all too often innovators are biased by the greatness of their own idea. Innovators sometimes are so excited and passionate about an idea that they overvalue the idea and undervalue the work and innovations being introduced into the market by competitors. Thus, they frequently do not give their innovative idea a proper due diligence investigation. Inadequate due diligence only hurts the innovator in the long run.

But that is not to say that there is no such thing as too much due diligence. Overdoing due diligence can overwhelm an innovator out of their excitement for developing the technology. Too much risk analysis can trigger a fear response or can generate a sense of apprehension, which can ultimately kill the project before it ever really gets started. Due diligence is a balancing act, and it is more of an art form than a science. You must be able to find facts and figures that help tell an effective story. Sometimes the ideal statistics you need are not available, so you will need to adopt an alternative storytelling strategy.

1.2.1 Why Conduct Due Diligence?

When an idea is new, it is exciting. For innovators who are excited and passionate about a new idea, it is easy to accidentally overlook or downplay the significance of potential obstacles in the development and marketing process for the new innovation. Some of the biggest obstacles to innovation include:
- Competitors in the relevant field of innovation.
- The size of the total addressable market (TAM) is too small to justify moving forward with development of the product.
- The inability to raise funding capital.

- Difficulties replicating technology that is highly complicated or intricate.
- Intellectual property exists that serves as a barrier to entry into the relevant field of endeavor.

Due diligence is a tool for assessing these obstacles. Due diligence helps evaluate whether bringing the innovation to market will be worthwhile or a bust. It is critical to know at the outset whether there is a place in the market for the innovation, whether someone else holds IP rights that prevent you from entering the relevant market, and what will be involved in developing, prototyping, testing, scaling up, and selling your product.

Good due diligence provides a solid foundation upon which to develop a winning strategy for the development and marketing of your innovative technology or product. Due diligence must be reasonably sufficient in breadth and scope to ensure that developing the technology is worth the effort. An example of due diligence is shown in the following case study.

Case Study

To demonstrate the value of due diligence, consider the concept of a hand-held ultraviolet (UV) dermatoscope. A dermatoscope is a device that dermatologists use to look at skin to determine whether skin cells may be cancerous. It is often used in conjunction with an UV black light that typically sits in the corner of an examination room.

A forward-thinking dermatologist was frustrated that he needed to wait 10 min after turning on the UV lamp before he could use the device (this type of customer frustration is known as a "pain point"). LED UV lights had become readily available and did not require a warm-up period. Therefore, he came up with an idea to merge the two items, a dermatoscope and a UV black light, into a single, easily portable device that would not need to warm up. The doctor's thought was that this innovation would optimize clinical workflow and reduce the amount of equipment in the exam room. With a tool like this, a physician would no longer be tied to a specific exam room and would be able to get better images of the skin, ultimately making the process of looking for certain types of skin cancer faster and more convenient. In his own practice the dermatologist saw a clinical need:

> *Clinicians need a better tool to assess skin cancer in order to improve clinical workflow.*

At the time of conception, there was nothing on the market like the handheld UV dermatoscope. In order to figure out whether the handheld UV dermatoscope had market potential, the dermatologist first performed due diligence, starting with the question:

> *"Why is there no product on the market like this already?"*

To answer this question, an assessment of the clinical need and market validation for the potential product were conducted. As a beginning step, the dermatologist assessed the clinical need by interviewing other dermatologists, other types of physicians, physician assistants, nurses, and technicians to see if anyone would use the invention. The next step was to conduct formal observations of the daily activities within his own practice and that of a few of his colleagues in different clinical settings (such as private practice, small clinic, or department in a hospital). On average, the typical procedure in the clinic was that a physician would greet a patient in the exam room, turn on a UV lamp, and then go to another exam room to treat another patient while the UV lamp warmed up. The average physician would then return after 10–15 min to examine the first patient under the UV lamp. The result of his clinical observations and interviews concluded that only dermatologists would be interested in the tool. The need was then refined to:

> *Dermatologists need a better tool to assess skin cancer in order to improve clinical workflow, provide more accurate diagnosis, and shorten patient wait-times.*

Next, due diligence was conducted for market validation. Looking at the market segments, as of 2010, there were approximately 10,000 dermatologists in the United States, with approximately 250 new dermatologists graduating medical school each year (a compounded average growth rate [CAGR] of 2.5%).

Research determined that the average cost of a standard dermatoscope was $40, UV lamps typically cost around $80, and replacement UV bulbs $40. The product usage life cycle required that clinics only needed to purchase a single dermatoscope and a single UV lamp and a replacement bulb every four years.

Turning to the costs associated with developing the handheld UV dermatoscope, a prototype of the handheld UV dermatoscope innovation would cost approximately $2,000 to make, and a minimum viable product (MVP) was estimated at $125 in material costs (or cost of goods sold, COGS) per unit.[5] Additionally, since the handheld UV dermatoscope would be used to make clinical decisions and diagnoses, the handheld UV dermatoscope device would require clearance from the U.S. Food and Drug Administration (FDA), meaning that the sales price per unit would be upwards of $700–$1,000.[6] At this cost to the dermatologist customer (i.e., roughly $800 per unit), the total TAM for the handheld UV black light dermatoscope device would compute to just under $10,000,000 in the United States (derived from the assumption that *every* dermatologist needed to purchase a handheld UV black light dermatoscope for their practice at a cost of $800 per dermatologist).

By comparing the estimated costs, it was concluded that the new product would require convincing 10,000 dermatologists to purchase the handheld UV dermatoscope product, even when the existing technology serves the same purpose and is one-tenth the cost. Though there was a convenience of using the new device, the clinical need was not significant enough to justify the additional expense for clinicians.

The dermatologist inventor concluded that the small market size, low TAM, and the cost barriers for clinical adoption did not support a viable business model.

In the dermatoscope case study discussed previously, it costs nothing more than the time required for the doctor to gather the information needed to determine that there was not a viable business model around the new technology. If the dermatologist

5 Minimum viable product (MVP), sometimes also referred to as a minimally viable product, is the simplest form of a product design that meets the intended user needs and is produced very economically. A MVP is used to collect user data to validate the design and functionality of the product and to inform the innovation team if changes are needed for the next iteration of the product.

6 The cost would be dependent upon a combination of factors: costs of goods sold (COGS or the costs attributable, both direct and indirect, to produce or manufacture the dermatoscope), the time, labor, and costs associated with getting FDA clearance or approval, and overhead. The FDA's classification of the handheld UV black light device would depend whether there were any predicate devices on the market (the UV lamp being a predicate device for the handheld UV black light

decided to build the prototype, it would have cost up to a couple thousand dollars. If the dermatologist decided to build and manufacture the device, then it would have cost hundreds of thousands of dollars, not including the costs (time and money) associated with getting clearance or approval from the FDA. The value of due diligence is not only determining if a single market is viable, but polar to the above-mentioned case study, one can often find additional market segments or customer needs.

1.3 Types of Due Diligence

Due diligence is an umbrella term for the research that one conducts in order to evaluate a business opportunity. The research underlying due diligence is multifaceted, and a good way to think about it is to break it down into various categories, or types, of due diligence. A few examples of the various types of due diligence that might be relevant include:

- **Technical feasibility.** A determination will need to be made early on whether development of the innovative technology or product is technically feasible, meaning the innovation is technologically and physically possible to create and can be built, prototyped, and manufactured at a reasonable cost. If an innovation cannot be developed in a cost-effective way, the resulting product will have to be priced high and there will be no room for profits. Technical feasibility can also include determining if development of the technology or product will require certifications, approvals, permits, or licenses, and whether it is possible to secure such in a cost-effective manner.
- **Business model viability.** Most technology and product innovations spawn a company in order to conduct business. But running a business and successfully marketing a new technology or product must be possible and likely to succeed, or else the business will fail. A product without a market is not viable and is doomed to fail (unless some other clever market approach can be worked out). Business viability must consider the following:
  - **Market Saturation.** Market saturation exists when there are plenty of options available on the market to satisfy the needs of consumers.
  - **The Competition.** When a market is crowded with competitors, it can be difficult to penetrate the market with a new product.
  - **Market Segmentation.** The market can be broken down into segments or groups based on specific characteristics of each group.

---

device). For example, if the regulatory path for the handheld UV black light dermatoscope device is a Class II device, then the above-mentioned cost range of $700–$1,000 is a realistic range. If this device was determined to be a Class III device, then the costs would increase dramatically.

  - **Customer Analysis.** Customer analysis and creating specific customer profiles for each target customer group helps identify the needs of each customer group and how the company intends to satisfy the needs of each customer group.
- **Intellectual property landscape.** In all likelihood, a new technology or product will face competition in the market. Since other competing products exist, it is important to review the IP rights that exist for any and all competing products. Knowing what IP rights competitors hold, and understanding whether there is room in the relevant IP landscape for you to assert and secure some of your own IP rights, is another aspect of due diligence.

1.4 Performing Market Research

In today's modern curricula, entrepreneurship courses are often integrated into engineering programs. Still, without having gone through the process of commercializing an innovation, it is very difficult to understand the nuances of what it takes to get a product to the marketplace. The most important part to the innovator is the idea. It is also important to keep in mind the needs of the potential investors, customers, and any other stakeholders. You have to be passionate about your idea, whether it is a product or a service. But your product or service is only a small part of the company that needs to be built around your innovation.

An important part of due diligence is conducting market research. A market is the group of consumers who will buy the product, and market research is the process of collecting data and useful information about the needs and preferences of the consumers that make up the market. Market research requires taking the time to properly understand:

- **The Need (Chapters 2 and 9)** – What is currently lacking? Are there not enough suppliers? Are only poor quality products available? Are people not getting the care, attention, or services they need, etc.?
- **The Value Proposition (Section 1.5 and Chapter 4)** – Why would customers be interested in using your product?
- **Stakeholders (Chapter 2)** – A stakeholder is anyone who can affect or be affected by a product, technology development project, company strategy, or even a company itself. Market research of stakeholders involves considering who (for instance, customers, investors, distributors, competitors, etc.) is interested in the success of your product and what are their demographics?
- **Competitors (Section 1.4.4 and Chapters 2 and 4)** – Who are the main players, and what are their positions in the market, their strengths, their weaknesses?
- **Market, Market Segments, and Market Trends (Section 1.4.3 and Chapter 2)** – What industry or space will your product or service exist in? How can that

space be broken down to make customers more approachable? Is that space growing or shrinking?
- **Implementation Strategy (Chapter 7)** – How is your team going to take a concept from ideation to commercialization?
- **Strategic Partnerships/Alliances (Chapter 2)** – Are there any strategic alliances or partnerships that you can forge that will help increase your business success? Strategic partnerships can potentially reduce costs, streamline development, improve market share, and increase profits.
- **Operations (Chapter 4)** – What are the day-to-day operations of the business going to be, and what are the overarching, larger-view operations of the business?
- **Resource Needs (Chapter 7)** – What resources (such as funds, people, time, and energy) will be required to get your product from the design phase all the way to market? What resources will be needed at each step of the product development timeline?
- **Expenditures (Chapters 4 and 5)** – What are the anticipated costs associated with salaries, rent, fluctuating expenditures (e.g., fuel prices), COGS, licensing agreements, IP, legal fees, computers/hardware, etc.?
- **Revenue Streams (Chapter 5)** – What are the streams of revenue produced by the business? Revenue streams could include sales from selling the product, revenues from licensing IP rights, etc.

Market research is a critical part of due diligence. The information that is gathered during market research will give you a better idea of whether your innovation is worth developing and bringing to market, that is, the information gathered during the market research portion of due diligence will inform the validation of your product (discussed more in Chapters 9 and 10).

1.4.1 Utilizing Market Reports and Other Resources

A good source of information for trying to understand the market landscape are market reports. Market reports are formal reports developed by industry analysts that contain relevant information (typical pricing, typical cost expenditures, etc.), data (demand, supply, trends in the industry, etc.), and statistics concerning a particular type of business in industry – effectively a report on the status of a particular market situation.

A market report will help you better understand the market you are planning on entering. The information in the market report should tell you:
- How large the potential market is, that is, the size of the TAM.
- What supply and demand is in the target market.
- Information about the consumers that make up the target market, such as demographic and economic information.

- Whether the potential market is segmented (segments are specific subgroupings of customers in your target market) (discussed more in Section 1.4.3 and Chapter 2).

Market reports are typically published by reputable sources that are well known and respected in their relevant field. In most cases, not all information is in one place; therefore, it is important to find multiple credible sources that provide statistics and trend data. In the case of an innovative product, it may be difficult to find targeted information, so an initial broad market search can help guide the research to more targeted information downstream.

Market reports are available from a number of different sources: some market reports are available online and accessible for free, some market reports are made available by organizations that provide support to small businesses, and there are many businesses dedicated to producing market reports for a fee. The best place to look for a market report will depend on the type of technology or product you are planning to bring to market, and what industry you are trying to enter. There are broad industry reports and much more specialized reports specific to specific industry segments.

A number of other resources are publicly available to businesses that are just starting out, and these resources can be immensely useful when conducting market research. For instance, various departments of the United States and state governments (such as the Department of Commerce, Department of Health and Human Services, US Census Bureau, and the Small Business Administration at the state level) offer useful resources and market reports.

1.4.1.1 University Business Schools and Libraries

Many colleges and universities that have a business school typically have a Master in Business Administration (MBA) program where students can earn an MBA degree. Sometimes MBA programs partner with local small businesses to provide the MBA students with the opportunity to conduct market research on behalf of real, local small businesses. The small businesses that participate in these partnerships can request market research and market report information that are tailored to their business (and usually for free) in exchange for allowing the MBA students to perform the research as a real-world educational experience.

Additionally, any college or university that has a business school has other resources that small businesses can use to conduct market research. *As a general rule, the public can use university and college libraries* (although access may require identity verification, just like accessing any public library). Universities and colleges that have a business school usually subscribe to a number of business databases (such as Bloomberg BNA, Business & Industry, and Dun & Bradstreet) and publications that furnish market reports. These schools usually also have a dedicated librarian who is trained to help people access and navigate these databases.

If your innovative technology or product relates to a specific industry, such as biotechnology, the library at a college or university that has a department, school, or college dedicated to that technology will most likely have access to market reports specific to that industry. For instance, Drexel University in Philadelphia, Pennsylvania, has a dedicated School of Biomedical Engineering, Science and Health Systems, and a College of Medicine. Due to the university's interrelated fields of study, Drexel University's library staff is trained to help those who need assistance accessing information from medical and health information databases.

1.4.1.2 Consulting Firms and Third-Party Data Acquisition Firms

There are a number of businesses dedicated to conducting research for a fee. Consulting firms and third-party data acquisition firms can be hired to conduct market research. The cost of a market report prepared by a consulting firm or a data acquisition firm depends on how comprehensive and detailed the report needs to be.

Market reports prepared by firms are particularly useful when the target market you plan on entering with your innovative technology or product is potentially a high-profit, but small, market. Similarly, it makes sense to hire a consulting firm if your innovation is disruptive to a particular existing industry (as in when drug eluting stents disrupted the interventional cardiology space) or if your technology or product falls at the bleeding-edge of innovation, meaning the technology is so innovative and novel that no clear market currently exists for the product (e.g., when desktop computers were first introduced to consumers, there was no existing market for desktop computer products). The consulting firm can offer the best market report data available for your particular situation, and can offer good advice about how to create a new market (where none exists), and/or how to convert potential customers from an existing market into your customers.

Market research can be very useful, but the output data from market research conducted by consulting firms and third-party acquisition firms is dependent on the specificity and quality of requests for information made by the clients or customers. If the questions to be answered by the consulting firm are not the right questions, then the data generated will not be useful. In other words, be very clear and specific when working with consultants, or you may end up with a deliverable that does not meet your expectations.

1.4.2 Bringing It All Together with an Example

Collecting the information needed to conduct due diligence can be difficult. In most cases, the use of indirect facts and statistics will be necessary to help validate your innovative concept. This information can be collected from many sources. Examples of reports you can consider reviewing include:

- Industry market reports.
- Reports produced by academic entities.
- Reports produced by professional associations.
- Reports produced by foundations.
- Reports produced by government agencies (Census Bureau, Department of Commerce, Department of Labor, Education Department, Centers for Disease Control and Prevention, Environmental Protection Agency, etc.).
- Peer-reviewed journal articles.
- Other credible sources.

Several useful tools for finding reports and information:
- The Internet.
- University Business Schools.
- Libraries.
- Third-party data acquisition firms.

Example

Samantha is a biomedical engineer. Through her research, she notices every academic laboratory she visits has bulky, poorly assembled, and poorly monitored systems for conducting research on growing three-dimensional (3D) organs for transplant. She has an idea to develop a system for growing human organs that will automate the process while allowing precise fluidic control, delivery of nutrients, and data acquisition to occur. She knows she can build the system, but it would take some time to develop and will cost a lot of money. Before getting started, Samantha decides to conduct due diligence on the market and technical feasibility of the system. She thinks to herself:
- Has someone already made this?
- If so, how many competitors are there?
- Do competitors make money off of it?
- If it does exist, why is this not being used?
- If it does not exist, why not?
- Are there patents for things like this?

When looking at the market, her first step is to conduct an internet search. She queries keywords together and individually like:

> "tissue organ growth system flow rate apparatus sterile"

She conducts online internet searches:
- In Google and Bing (different search engines give different query results).
- Within each search engine she searches under the "All" tab as well as "Shopping" tab.
- Google Scholar for peer-reviewed publications.
- In Google Patent for similar or existing patents.
- Online retailers for existing products available for sale.

In each search, she finds different words to help refine her search. She finds terms like "climatic simulator" and "tissue growth chamber" and "organ incubator." She repeats her search and finds the name of a few companies that have similar products. Samantha adds these companies to her list of competitors that she will look into much more closely (Figure 1.4).

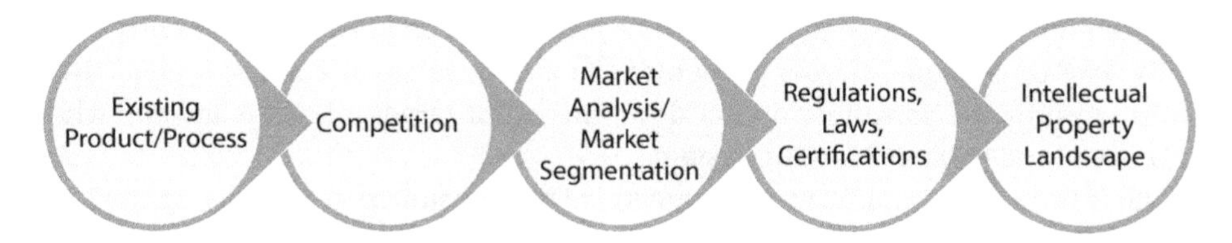

Figure 1.4: Samantha's technical due diligence process.

Samantha takes her list of competitors and tries to look up their DUNS,[7] NAICS,[8] annual earnings, if the competitors are private or publicly traded. To do this, she goes to her local Small Business Development Center (or University's Business School). The staff help her look up market reports in Hoovers, IBISWorld, and several others. She also finds a "for pay" industry report, but it is out of her price range. She is able to assess from this information that her idea has the potential of being a $1 billion-dollar market in the United States.

Once Samantha obtains this information, she sees that she only has one significant competitor. Ever so curious, she asks herself, "Why?" After conducting stakeholder interviews (such as interviews with laboratory managers), she finds that (a) the costs are too high for academia, and (b) there is an obscure federal law that restricts the type of work that can be used coming from a system like this.

Using this information, she can now make a decision whether to spend more time, or her own money, on developing this technology.

1.4.3 Identifying Target Consumers

The next step in conducting market research is to identify the potential consumers, groups, or segments that make up the target market. From your review of market reports, you should have a good sense of what the target market is, and what type of person or organization buys the product you plan on making and selling versus who the end-user of your product or service is (see the stakeholder mapping healthcare product example in Chapter 2, Section 2.3). Deeper analysis to identify the target consumer or consumers can help you better plan your marketing strategy to reach the most potential customers.

7 The DUNS Number (Dun & Bradstreet) is primarily used by prospective customers, partners, and creditors to verify the credit history and financial reliability of a business.

8 The North American Industry Classification System (NAICS) is used to classify businesses by the US Federal Government for assessing economy-related data.

Oftentimes, you will be able to identify your target market as a whole fairly easily. By way of example, if your technology is a new personal cooling system that can be worn by the consumer on hot days, then your target market is anyone who has a need for a personal cooling system.

But it is an oversimplification to categorize the consumers that make up the target market all –together into a single category – the target market could comprise a wide variety of customers that have a wide range of needs, price points, and so on. A one-size-fits-all marketing approach may not work on all of the customers that make up the target market, so it often makes sense to further break a target market down into market segments, which are specific subgroupings of customers in your target market. Once market segments of the target market are identified, specific marketing strategies can be tailored to have the biggest impact on the customers in the specific market segments.

Various market segments exist in the personal cooling system target market. For instance, one market segment would be outdoor laborers, who have to work in hot weather conditions. Another market segment would be customers who want to use a personal cooling device for sports and leisure activities (for instance, golfers, campers, fishermen, etc.). A third market segment could be based on customers who reside in a hot and humid geographical location. A fourth market segment could be people who are more susceptible to heat stroke or heat exhaustion.

To get a better understanding of the target consumer, think about who they are, what they want from the product, why they would buy it, and how much they would likely pay for the product. To help you come up with realistic answers to these questions about the target consumer, try to paint a picture of who the target consumer is. Be specific.

For instance, a target consumer from the outdoor laborer market segment might be Mike. Mike has worked in construction for his whole adult life. He doesn't mind hard work, but it can get tough when the weather is too hot. Since Mike works in construction, he makes around $20 per hour and works long shifts. He could use a personal cooling device that is durable, lightweight, easy to use, and affordable. When marketing to Mike, it might be a good strategy to emphasize low price and suitability for use of the product, while he is on the job.

Similarly, a target consumer from the sports and leisure market segment might be Charles. Charles is retired and loves to spend his hard-earned retirement doing things he loves, such as fishing and golfing. But when the weather is too hot, much of the joy Charles experiences from his hobbies is quickly sapped. Charles could use a product like a personal cooling device, especially if it is lightweight, so that it does not impinge on his mobility. Marketing to consumers like Charles might involve using print advertisements in age-targeted magazines (targeted at retirees), and could emphasize the lightness of the product and how long the cooling capabilities last (such as for a whole fishing trip).

Another target consumer could be fire department personnel or emergency medical response teams that need a compact, lightweight, inexpensive, and disposable cooling system that does not have cables and is easy to put on and take off patients quickly. Marketing to consumers like a fire department or emergency medical response unit may require exhibiting the personal cooling system at emergency response trade shows, talking to medical associations, and publishing data in peer-reviewed journals demonstrating the system's safety and effectiveness.

Also, to better plan your marketing strategy, think about all the ways your company can interact with your target consumer. Know your target consumers' needs in each market segment.

- How technologically savvy are your target consumers?
- How would your target consumers prefer to interact with your company: digitally and/or non-digitally? Prior to purchasing? While purchasing? After purchasing?
- Would your target consumers prefer more or less customer support?
- Which types of advertisements and services would earn the attention and loyalty of your target consumers?

1.4.4 Closely Scrutinize Competitors

Good market research, and thus good due diligence, requires careful analysis of other groups or businesses that are developing similar products to your innovation. Carefully analyze these competing products, and determine how your innovation is different. Look for distinctions between the two, identify shortcomings in the competing product, and decide if your innovation addresses these shortcomings. You are not only looking for differences in the engineering and design of your innovation compared to close competitors, but you are also trying to determine if your innovation provides a valuable contribution over the competing product.

Once you have identified potential competing products, it is prudent to identify the groups or entities that are working on these competing products, so that you can determine the IP landscape for the relevant field of innovation you are working in (this concept will be discussed in more detail in Chapter 5).

As a side note, keep the information you collect on your closest competitors handy, because it will be useful later down the road when you go to prepare a competitive analysis as part of your business plan (competitive analysis is discussed in more detail in Chapter 4).

1.5 At the Conclusion of Due Diligence and Market Research, Prepare A Value Proposition

After conducting due diligence and market research, you should have a good idea whether it is worthwhile to go forward with development and commercialization of your product. Additionally, from the information you gathered during due diligence and market research, you should also be well-situated to develop a value proposition for your product and the business you plan to build around your product. A value proposition is a short summation of your company, your product, your position in the market, and your target market that explains why your product is attractive to customers. To think of it another way, a company's value proposition is a promise of the value to be delivered by the company or by the product that is purchased.

There is no single "right way" to prepare a value proposition. Some companies include an explanation of how their product addresses customers' problems or satisfies the customers' unmet need. Other companies focus on the specific benefits that customers enjoy when using their product. Your company's value proposition should be tailored to the message you want to send to your prospective customers. You could even have multiple value propositions tailored for specific market segments or groups.

Treat your company's value proposition statement as if it is the first impression you are going to make on your prospective customers. Engage them using language that they understand, and let them know that your product is just what they need, without being pushy or too verbose. Also, it is a good idea to pair your value proposition statement with a relevant visual, since a picture can convey volumes.

In Figure 1.5 is an example of a value proposition statement for a fictional company: PetSpec.

Check Up on Your Pet Any Time You Are Away from Home

Free mobile app lets you connect via webcam with your furry friend anytime–your pet misses you, just like youmiss them.

- Visit with your pet in real time!
- Audio lets you speak to your pet!
- Receive alerts when you pet is active!

Figure 1.5: Value proposition for PetSpec.

Knowing nothing about the company PetSpec prior to seeing the above-mentioned value proposition, you now have a clear understanding of what products and services the company offers, who the target customer is, and why the target customer would want the product. All of the following information can be gleaned by consumers from a 54-word value proposition:

- The company PetSpec sells webcams that can be set up by pet owners in their homes to provide a means for checking in on the pet during the day while the owner is at work.
- The owner only needs to have a smartphone or tablet to access the free mobile app that enables the owner to see and speak with their pet in real time.
- Additionally, to avoid wasting time checking up on a sleeping pet, owners can receive an alert on their phone through the mobile app when their pet is awake and active in the home.

With a well-crafted value proposition, less is more. Creating a value proposition statement takes practice, and you may find that over time your company's value proposition may change. Value propositions do not have to be set in stone – they can be iteratively revised to reflect the current state of the company or product.

Keep in mind that your value proposition statement is very important. Your company's value proposition should be on your company's homepage of its website and should be one of the first sections covered in your company's business plan (business plans will be discussed in more detail in Chapter 4). Additional strategies for formulating a value proposition statement can be found in Chapter 4, Section 4.2.4.1.

2 Validating Your Business Model

Abstract: Once sufficient due diligence has been performed, a business model needs to be determined and validated – this involves validating the business (commercial) assumptions and the conceptual (technical) assumptions regarding the new idea or technology. This chapter discusses both business and concept validation. Validating a business model often involves analyzing the market by looking at the relevant market segments, market growth, market share, competition, and by profiling the target customer. It also involves identifying the Strengths, Weaknesses, Opportunities and Threats (i.e., conducting SWOT analysis) on the business model and consulting with important stakeholders (i.e., those who are interested in the success of your product, such as customers, investors, distributors, and competitors). Concept validation focuses on concept validation refers to identifying, qualifying, and quantifying a concept (i.e., the underlying idea that will be developed into a marketable technology or product) by assessing the user requirements of the product for each of the relevant stakeholders. Oftentimes, the lines separating business validation and concept validation are blurry, due to the overlap of the validation activities on the development timeline. Concept validation will be discussed in more detail in Chapter 9. Finally, a highly useful tool for gathering data concerning business and concept validation is Voice of the Consumer surveys.

Keywords: Business validation, business model, market segments, market growth, market expansion, market share, competition, customers, SWOT analysis, stakeholders, user requirements, concept validation, Key Opinion Leaders (KOLs), partnerships, stakeholder mapping, human factors, voice of the consumer (VOC), surveys.

2.1 Business Validation

Even if your team has developed an amazing problem-solving product or technology, unless the business model surrounding the product is valid, then there is a good chance no one will ever experience your novel innovation. A product or technology does not sell itself. Commercial success of a product requires an understanding of customers, how and where your product fits into the market, and how to make money from your endeavors. Any number of business considerations need to be assessed. Customers may be hard to reach, your solution may not be unique enough, or provide enough value to differentiate itself from the goods of your competitors, or the market – swamped with competition – may not react. The end result is that there are not enough prospective customers and partners, customers lose interest, sales targets are not met, and the project is a waste of the company's time and money. Once the team completes Business Validation, then

https://doi.org/10.1515/9783110521900-003

the company is aware of most of the external factors that could influence how the business model will grow to thrive in the current market (the known) and future market (the unknown).

As indicated in the Business Development timeline in Figure 2.1, Business Validation begins at the tail-end of the Due Diligence stage and involves discussing and "testing" your proposed business model with various stakeholders, to get insight on possible scenarios as to how your business model would be received by consumers. Business models can be tested by poking holes in the assumptions that underlie your business model and discussing "what-if" scenarios with stakeholders, representative users, customers, investors, and business partners.

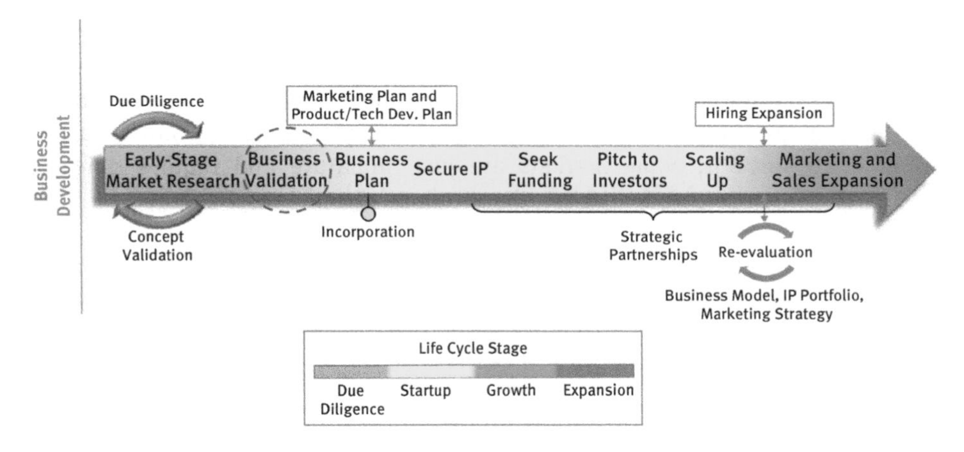

Figure 2.1: Business Development timeline.

Business Validation mainly comprises *risk identification* and *risk mitigation planning* (for more information about Risk Management, refer to Chapter 8). It is important to educate yourself about any external factors that could impact your business model and how the business model could change. The business environment is never constant – there will be disruptive technologies that impact your business model, emerging technologies and competitors will change the demand for your solution, economic downturns and upturns will influence purchasing decisions of customers, and changes in international trade regulations will impact your business's supply chain. Executive leadership of a company needs to understand how external factors (such as political, socioeconomic, environmental, regulatory, and legal factors) impact the market and each market segment. If you do not know much about these external factors, then join associations, and get involved by engaging in efforts to influence policy.

Uncertainty in the business environment is given, but use that uncertainty to your advantage and discuss the different types, and drivers, of uncertainties and constraints with stakeholders, to ensure your business model is robust. Business drivers and influencers, or variables that can shift the business environment, include, and are not limited to, the following types:

- Technological advances.
- Regulatory.
- Legal.
- Economic (macro and micro).
- Competition.
- Substitute or alternative solutions.
- Logistics (e.g., commodities pricing and availability).
- Changing user needs, preferences, and market demands.
- Availability of funding, both private and public, for entrepreneurs.

Business Validation addresses the risk that the market does not have a strong need for your product or that there is no problem in the market that requires a product to solve it; some markets are just immature and require time to develop the justification for using the product. Therefore, innovate for the market you are in.

For example, if you are developing a healthcare solution for an emerging market (for instance, in rural communities), your solution may not require all the user options and technical features that users would expect in an established healthcare market, such as in the metropolises of the United States, Japan, or Europe. Your solution can be developed in a shorter amount of time and can be manufactured for a lower cost, and the market potential is still there because in emerging markets customer needs are not being met and customers cannot afford the products that are being produced for more established markets.

Finally, Business Validation is not complete if you have not utilized the resources that are discussed in Chapter 1 to determine whether there is an overlap between the needs in the market and the capabilities of your solution.

2.1.1 Will the Market Support the Business Model?

Regarding markets, there are fundamental market variables that will impact your business model. For example, the total addressable market (TAM), or total market opportunity, may be growing or shrinking, and understanding the reasons for these trends will drive development planning. Breaking down the TAM into individual market segments will make it easier to see how the market is poised to accept and/or reject your solution.

Market Segments

Market segments[9] can be organized and analyzed according to:
- Geography (Should the solution be launched in one country or another?).
- Customer demographics (such as gender, age, and race/ethnicity).
- Customer psychology/personalities (Do customers want long-lasting relationships with the brand or are customers brand agnostic?).
- Customer actions and behavioral patterns (Are customers looking for the best-quality solution or the cheapest?).
- Organization type.
- Industry focus.

For development planning, study which market segments are rising, which are falling, and which are poised to rise. Identify the barriers to entry that exist in your market, and assess how these barriers differ across market segments. Identify any potential ways for overcoming these barriers to entry. Once this is understood, refine the technology design and development strategy (such as establishing a project schedule and plan that outlines the resources and partnerships and funds that are required for bringing the solution to market), based on the size and accessibility of the target market (i.e., how intense is the competition?), the compound annual growth rate (CAGR) of the target market, the price for your solution that customers are willing to pay in the target market, and the profits you aim to achieve. How you frame your business's position in the market needs to align with the market segmentation strategy – know how your business and how your competitors' businesses would be positioned in the market landscape.

Market Growth/Market Share

A fun analogy for clarifying market growth versus market share can be found in a pizza pie. The size of the pizza pie is analogous to the size of the market. To say this differently, the bigger the pie, the bigger the market. Each slice of the pizza pie represents a share of the market. The more slices that a pizza pie has (meaning the more competitors there are in the market), the less market share there is for each competitor. Studying market growth and market share will determine if the technology development project is worth pursuing and how attractive the profit margins can be. For example, if the market is not growing but is consistent and the business's market share is high, is it worth it for the company to develop and market products for anticipated returns at low risk, such as to make and sell consumables? Or does your

9 Alternatively, if you decide to not go after a particular market segment and you want to launch your solution to a broad market, is there opportunity to differentiate based on price or service?

business only want to go after a growing market? If the intent is to tackle a growing market, then your business will need to define the target market segment, be a leader in that market, and control the market segment.

Note: Be careful to not spend too much time and money on products with low market share in a growing market, if the market is flooded with competitors. It may be better to license your product to a competitor.

Competition

Regarding competition,[10] understand how your competitors are currently positioned and how they are competing with one another for market share. Competitor rivalry is a good thing for market growth because competition drives an increase of advertising and promotional activities, and this increased awareness of product availability and consumer choice increases customer demand for all products that are being marketed in a particular sector, including your product.

Additionally, study the merger and acquisition trends among your competitors. As competitors consolidate and bolster their portfolios of products and services, this will influence customer demands and expectations for any new ideas that are in development and have yet to be commercialized.

You can prevent new competitor market entrants from taking market share by establishing intellectual property rights, designing and delivering strong branding and customer engagement strategies, and creating other barriers to entry for the new market entrants.

Customers

Know your prospective customers – how loyal are they to competitors, and what will be their pain points (e.g., costs, time, and effort involved) to switch from their current choice to using your solution. How important is brand image to customers? How does advertising influence customer purchasing trends?

Think about how you will communicate your solution in a way that will resonate with your target customers in your target market segment(s). For example, for every feature of your solution and for every function it performs, how can you communicate its value through marketing, that is, how would you describe your solution, and how would this description change depending

---

10 Recall that resources for finding information about competitors, for example, Small Business Development Centers, business school libraries, online resources are discussed in Chapters 1 and 3.

on the audience. Your solution needs to provide *specific and measurable* value to customers.

2.1.2 SWOT Analysis

An exercise worth completing during Business Validation is a Strengths, Weaknesses, Opportunities, Threats (SWOT) evaluation of your business model (refer to Figure 2.2). SWOT analysis is typically conducted as a team brainstorming session. For more information about Brainstorming Techniques, refer to Chapter 8. In a SWOT analysis, strengths and weaknesses are typically categorized as internal to a business (such as customer loyalty to a brand, a product's share of the market) and opportunities and threats are typically categorized as external to a business (for instance, substitute products are decreasing customer demand, buyers of your product are trying to reduce prices, and supplier diversification). The goal of the SWOT analysis is to match a business's internal strengths with external market opportunities. For example, by utilizing your business's current market share, you can increase the technology/product portfolio (for instance, via internal or outsourced innovation projects) to take advantage of additional potential revenue streams. Example SWOT analyses are discussed in Chapter 4.

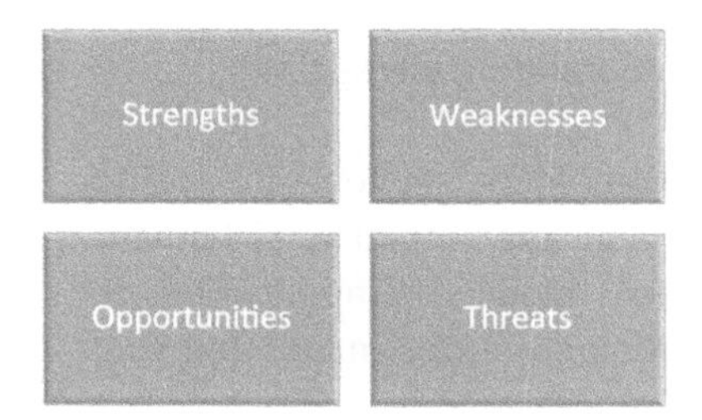

Figure 2.2: SWOT model.

Through market research, business model validation and SWOT analysis utilize the knowledge learned to define an innovation and technology development strategy for the company that will generate the optimal portfolio of products and solutions for realizing the highest profits. This strategy should take advantage of market opportunities and should match those opportunities to a company's strengths in design, development, engineering, testing, project management, supply chain management and marketing, and so on. The data gathered during business model validation informs the decision-making process and should be in alignment with the company's mission, vision, and business goals.

2.1.3 Validate Business Models with Stakeholders

Validating business model assumptions with stakeholders (i.e., those who are interested in the success of your product, such as customers, investors, distributors, and competitors) creates value because stakeholders' feedback and reactions to the business model confirm whether the value proposition (i.e., the intended benefits and features of the solution) needs to be refined or improved, and stakeholders can help businesses answer important questions, such as:

- Does the solution fit the market?
- Will the solution solve the problem that is creating the need for the solution in the first place?
- How willing is the market to utilize the solution?
  - If the market is not willing to use your solution, there could be a number of contributing factors, such as the supporting infrastructure or supply chain may not be in place, the customers may not be ready to adopt the solution, and so on.
- How can your solution reach customers faster and easier?
- How should commercial launch be planned?
  - What should be the beachhead market (i.e., the initial target market for commercial launch)? Beachhead markets are discussed in Chapters 4 and 7.
  - Who should be the solution's early adopters (i.e., the customers that are most likely to adopt or buy your solution the most quickly after you launch your solution)?
  - How will customers react to advertising and branding messaging?

Then, based on the information collected from stakeholders, set goals for positioning your solution with target customers. Prior to and after commercial launch, it will be necessary to refine payment/reimbursement models (if applicable), projected demand and sales cycles, customer acquisition strategies, marketing strategies, and distribution strategies.

2.2 Concept Validation

Concept validation identifies, qualifies, and quantifies the user requirements of each of the stakeholders surrounding a concept.[11] It is an ongoing, value-added process that is conducted throughout the life cycle of a new idea, from initial concept

11 Concept validation is not to be confused with user validation. User validation activities that occur during the technology development timeline confirm that the user needs or requirements, which are identified during the concept validation process, are being met. User validation activities are especially important for validating that the solution meets the intended use of the design.

to final product. This type of consumer-oriented market research occurs during business development and technology development and validates the concept through assessment of the technical design, use cases (i.e., ways that customers use products or services), pricing, sales strategy, and downstream customer retention strategies during product development.

Concept validation evaluates whether a new product, service, or process will meet the user requirements of all of the stakeholders. Concept validation is dependent on the business model strategy. For example, if a company determines that it needs to make a choice between developing a capital piece of equipment to sell as a stand-alone product or developing a capital piece of equipment to give away with the intent that customers will purchase the equipment's associated disposables over the long-term, then the user requirements will be different during the development of the product. Defining business strategy and assessing design can be performed on innovation and technology development projects in parallel.

2.2.1 Defining User Requirements

User requirements define the specific user needs that the design of the solution must meet. User requirements can be documented in a specification. Examples of user requirements can include things like the product packaging must be durable, the product must be made available in the specific sizes, and/or the product must adhere to applicable labeling requirements. User requirements can also be generally implied without being expressly stated. For instance, the product will not cause harm to the user. User requirements should be written at the system and subsystem or component levels. User requirements (also known as design inputs) are discussed in more detail in Chapter 10.

Stakeholder engagement is important to understanding user requirements. During concept ideation and exploration of a new development project, seek to clarify and validate the problem(s) that exist with the current solution on the market through interviews with the stakeholders. Try to understand the "pain points," or the problems that arise with existing products on the market or existing processes that the potential consumer wants to see changed and that they are willing to pay for. When conducting interviews or surveys, define the problem in terms that prospective customers, users, and investors will understand, and show how the problem influences the market. Your thought process must be communicated clearly and concisely to stakeholders, even though the problem may be as complex and jumbled as the word cloud in Figure 2.3.

Designing a set of comprehensive, mutually exclusive user requirements that address the frustrations of the stakeholders will ensure that the design and development process will not need to be repeated or overhauled, once a product goes to market. Additionally, after your minimal viable product (MVP) goes to market, the

Figure 2.3: Word Cloud – What problems exist with the current solution on the market?.

insight obtained from the continued concept validation will help determine the next iteration of product to be developed and will help to ensure customer retention. Stakeholder frustrations can be described as user requirements in a number of ways, such as a need for:

- Better information/data, such as:
  - Defined measurements.
  - Screening tools.
  - Integration and collaboration tools.
- Improved processes, such as:
  - Time savings.
  - Precision and accuracy.
  - Transparency.
  - Accountability.
  - Optimization.
    - Automation.
    - Customization.
    - Standardization.
- Better results, such as:
  - Customer engagement.
  - Customer satisfaction.
  - Procedural outcomes.
  - Quality compliance.
- Automated documentation needed for regulatory approval (if applicable):
  - Process completion.
  - Quality control testing.

Customers should react strongly and positively when discussing the user requirements that are being addressed by your new innovation. If the stakeholders do not care enough to react, then there may not be a true need for the product, the proposed price point is too high, or the design does not improve the user's experience.

2.2.2 Early-Stage Market Research Overlaps with Concept Validation

As discussed in Chapter 1, market research is a critical part of due diligence. The information that is gathered during market research will not only assess market segmentation and competition, but it gives you a better idea of whether your innovation is worth developing and bringing to market, that is, the information gathered during the market research portion of due diligence will inform your company's business strategies and the technical design of your product.

As related to concept validation, market research seeks to understand:

- What product(s) or solution(s) are the gold standard(s)?
- With existing technologies, what are the major flaws (pain points) and is there an opportunity to penetrate the market?
- Why is no one else in the market currently solving this problem?
  - Example: Is the problem too technically challenging to address or is the solution too costly to develop or produce by your prospective competitors?
- Which stakeholders will and will not benefit from your innovation, and/or which stakeholders will lose money?
- How much can you realistically charge customers for use of your solution, in order to make enough profit after cost of goods sold (COGS)? What do the financing/pricing models indicate?
- How are you going to entice and acquire customers with your solution? You are going to need to know how the buyers of your solution operate. How are these buyers structured? How can you get access to the key individuals with buying power (specifically, those involved with procurement)? Which collaborations and partnerships would be most fruitful?
- How much effort is required to secure each sale from each customer (i.e., a microeconomic analysis) versus what percentage of the TAM is realistically attainable within the targeted market segment (i.e., a macroeconomic analysis)?

2.2.3 Validate Throughout Development

Referring to the Product Development timeline shown in Figure 2.4, many of the concept validation activities can be performed throughout the Product Development

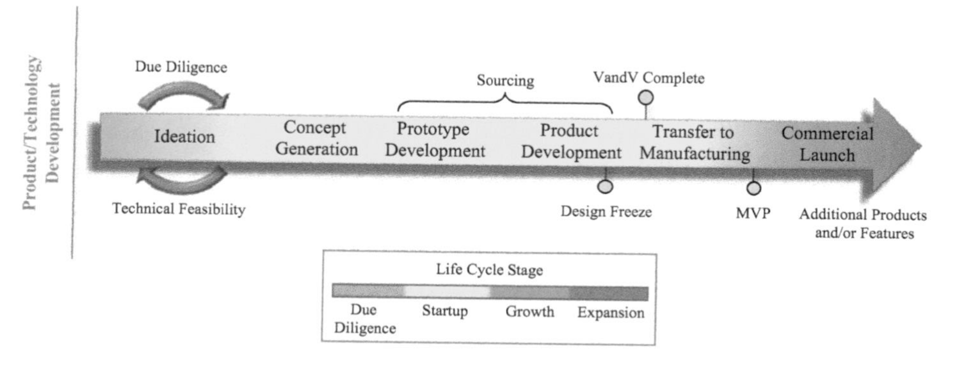

Figure 2.4: Product Development timeline.

cycle to maximize the value that your solution brings to customers. As more validation information is gathered, the final solution and go-to-market strategy can be adjusted and optimized accordingly.

Concept validation is continuous throughout all Product/Technology Development phases (discussed more in Chapters 9 and 10). The goal is to learn everything you can about how the stakeholders feel about your concept, then the prototype, then the minimum viable product (also referred to as minimally viable product or MVP), and even after you introduce the product into the market. This is accomplished through research, such as:

- Reading trade journals.
- Analyzing customer databases.
- Attending academic and industry conferences, trade shows, and expos.
- Talking directly with prospective stakeholders, especially customers.

Stakeholders who are familiar with the pain points of the existing solution can clarify potential new ways customers can use your product and identify potential applications for your solution. These direct sources of feedback cannot be overlooked. Do not undervalue them. They will help guide product development, prior to commercial launch and after the product is on the market. Establishing open lines of communication with stakeholders can drive customer retention.

During the **Concept Generation, Prototype Development,** and **Product Development Phases** of the Product Development timeline, as you seek to understand the needs of every stakeholder, infuse that user-centered perspective into the design of the solution, supporting platform, and tools, and so on. Continue to survey and speak with prospective users and customers. The value proposition for your concept may evolve, as additional solution capabilities and use cases or new ways customers can use your concept become apparent during these interviews. Engage with Key Opinion Leaders (KOLs) look to establish partnerships.

Key Opinion Leaders

Additionally, by seeking input from advisors and involving KOLs at important stages of design and development, you can pave the way for these influential individuals to advocate on your behalf in the future. KOLs are respected members of the industry in which the concept will be used, and KOLs are experts in the field. An example of a KOL could be the chairperson of an industry or trade association, a political figure, or a prominent industry executive. They are known for their innovation, skills, publications, accolades, and acumen. These individuals can help guide product development by identifying technical pros and cons with early and late-stage prototypes. Or these individuals can help generate market interest in your product or service by representing the product (or company) at events or at key meetings, speaking at symposia and conferences on your behalf, or by authoring or publishing peer-reviewed research papers or favorable articles about your product or service, and so on. Use KOLs as strategic publicists for your startup!

Establishing Partnerships

To maximize opportunities for collecting data about your technology, the establishment of a single key industry partner or multiple industry partnerships (e.g., manufacturing facilities, academic institutions, research labs, or hospitals) can facilitate access to stakeholders. In these cases, sponsored research agreements or licensing agreements will need to be implemented to clarify terms, if additional intellectual property is developed during the relationship. Partnerships should provide a place to conduct pilot studies, while testing different iterations of prototypes with different groups of stakeholders. In the case of medical devices, pharmaceuticals, and/or biologics, partnerships can be utilized to conduct clinical trials with the final design of a product, to validate safety and effectiveness.

2.2.4 Final Thoughts on Concept Validation

When competitors in the market finally figure out how to design and develop a similar solution, how will you differentiate your solution and the service offering to keep customers from switching sides? A promising development project can be quickly killed if you stop:

- Learning about your solution.
- Thinking about how to optimize the delivery of your solution.
- Meeting with prospective customers, buyers, and investors.
- Developing strong networks and partnerships throughout the supply chain.
- Collecting data about your solution – how it is being used, why it is being used, and why it is not being used.

If you stop trying to make your solution the best it can be, you won't make it. A competitor, whether entrenched or new to the market, will always be there to try and decrease your potential share of the market.

2.3 Stakeholder Analysis

"Stakeholder" is a general term used to refer to any person who can affect or be affected by a product, technology development project, company strategy, or even a company itself. Stakeholders are those with any interest in your project or company's outcome. They can be internal or external to a targeted customer's organization. Stakeholders that are within the customer's organization can be at any level (e.g., C-levels, directors, top-level managers, mid-level managers, employees, various departments, or work groups). Every product has a ripple effect that will be valued differently by each stakeholder.

As part of your continued due diligence, the activity of mapping the stakeholders and determining their involvement and interest in your product will help determine what kind of information you want to collect from the stakeholders and who is the target audience (e.g., marketing department, investors, and product development team) for the data that is collected. Before speaking with your stakeholders, decide what type of information you would like to extract and how you would use it. For instance, interviewing the product's end user might result in user requirements and usability information for your product development team to assess, or alternatively, interviewing an end-customer survey group may give you acceptable price points and inform your marketing department of effective marketing approaches.

Mapping stakeholders is the process of looking "upstream" and "downstream" from where you think your product will be used, to identify who is affected by the product. An example is shown in Figure 2.5. Multiple stakeholder maps can be developed for a given product. A map can be completely internal to a targeted customer's organization or encompass a whole market or even a supply chain. When targeting a market segment, the stakeholder mapping will determine not only who is using the product, but everyone else who is affected by the product (e.g., who is benefiting, who makes purchasing decisions, who the product disrupts, and who it saves time for), or who needs to justify why the product is being used. Often mapping stakeholders follows the chain of command within a customer's organization, but does not encompass all of the customer's organization. Several stakeholder mapping examples are described in the following sections.

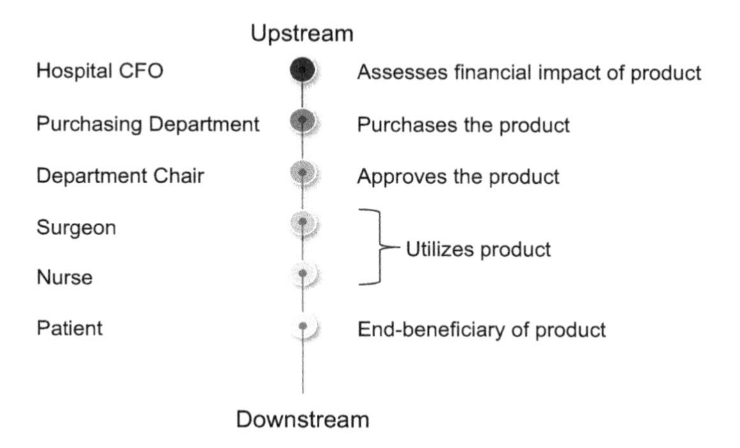

Figure 2.5: Example of upstream and downstream gauze bandage stakeholders.

Stakeholder Mapping IT Example
If you have an accounting software program for use by a hospital, the stakeholders would include the Chief Financial Officer (CFO), Purchasing Department, IT Manager, Accounting Department Chair/Manager, and Administrative Staff. The CFO would want to see cost savings and a return on investment (ROI) from purchasing your accounting software program. The Purchasing Department might assess the ROI or an increase in workflow efficiency. The IT Manager will need to understand how the program will be deployed, what resources are needed, and what security measures will need to be implemented in order to use the software. The Accounting Department Chair would need to know how much training and disruption will occur for the staff during the adoption and implementation of the new accounting software, which would have short-term costs affecting his/her budget. Finally, the Administrative Staff will be the actual end users of the program. Notice that all of the stakeholders tangential to the new accounting software have different interests and/or needs regarding the platform.

Stakeholder Mapping Healthcare Product Example
Often, the people who use a service, product, or process are not the same people who actually make the purchase. When developing a new gauze bandage, the focus would most likely be on the way the bandage interacts with the patient and the person applying the bandage to the patient. When assessing a concept, one should look upstream and downstream from who will be using the product and how the product will be used (see Figure 2.5).

As an example, consider a nurse applying a novel gauze bandage to a wound. Though the nurse is a stakeholder for the gauze bandage product, he or she is not necessarily the person making the decision to purchase that specific gauze bandage, nor is the nurse the end user, or end beneficiary, of the bandage. In the case of a new gauze bandage, to conduct stakeholder mapping, you must look upstream and downstream from where the bandage is being used. Start with the nurse. The nurse applies the gauze bandage product. Moving downstream, the patient is the downstream stakeholder and end user, or end beneficiary, of the gauze bandage product. Looking upstream, the hospital purchasing department and the CFO are also stakeholders.

Each stakeholder will have different interests regarding the gauze bandage product. Starting from the furthest downstream point, the patient would want comfort and sterility. The nurse will want the gauze product packaging to be easy to open and the gauze bandage product to be easy to apply to the patient. The purchasing department will want cost-effectiveness, low risk of complications due to use of the gauze bandage product, and reliability of the distribution. The CFO will want so see lower costs and better outcomes. These key observations and feedback from the various prospective upstream and downstream stakeholders offer subtle, but important, insights for product development and business strategy.

When speaking with stakeholders who are in direct contact with your product, such as the surgeons or nurses from the gauze bandage stakeholder example, ask pointed questions to ensure that the needs of these direct users are being met.

2.3.1 Stakeholder Feedback on Human Factors

Human factors involve the study of how people interact physically and psychologically with a new product or technology. Accounting for human factors, such as mitigating a fear of ominous-looking examination equipment, can be as simple as painting a Magnetic Resonance Imaging (MRI) machine to look like a train in a children's hospital, so that the pediatric patients won't be afraid of the machine, or as complex (from an engineering and design perspective) as designing a surgical tool that can be used ambidextrously. In relation to the gauze bandage example mentioned earlier, what would happen if the bandage was an unappealing color or the sterile packaging was difficult to open?

Since designing a product to address human factors involves predicting the potential user's capabilities and limitations, it is a good idea to design for human factors iteratively and to talk to the stakeholders who will be using the product. With each iteration of design and redesign, develop a survey for a well-rounded stakeholder sampling. The survey does not need to be complicated, but it must capture the subtle suggestions and complaints offered by the stakeholders participating in the survey, which can make a difference between producing a good product and producing a great product.

The physical component of addressing human factors encompasses designing for safety, usability, and function, while the psychological component of addressing human factors is much more nuanced and encompasses designing to reduce human error and to address human capability, as well as predicting how the users will interact with the system.

Examples of Physical Human Factors include:
- How many clicks of a mouse does it take the user to accomplish the task in a graphical user interface?
- Can the device be used by left-handed and right-handed users?

- Does the product fit the user comfortably?
- Is the cycle time appropriate for the task?
- Is the data that is collected or distributed by the system accurate?

Examples of Psychological Human Factors include:
- How many clicks of a mouse does it take before the user becomes frustrated with the user interface?
- Is the graphical user interface aesthetically pleasing, easy to understand, and easy to use?
- Does the user find the system fun or engaging?
- Is the color scheme pleasant or offensive?
- Are there any potential cultural or religious concerns that may need to be accounted for in the overall design?

2.4 Carefully Analyze the Voice of Your Prospective Customer – Survey and Focus Group Design

To establish lines of communication with influential and/or interested stakeholders, find a sampling of prospective users and establish a professional relationship with them. They can provide invaluable feedback during the product development process. Conducting Voice of the Customer (VOC) surveys can formalize the information collection process.

This sampling of prospective users is representative; it does not need to be a large number, but the perceived usability of a service, product, or process can be very different from one user to another user. For example, if you are developing a software tool for employees of a given industry or sector, a person who works by himself/herself in a startup organization versus a team of employees in a large organization can have very different experiences interacting with your software. Conducting VOC surveys offers the ability to incorporate user feedback to confirm the utility of and/or refine a new concept or prototype.

If there is already a solution on the market, VOC surveys can help identify competitors and which attributes define the industry "gold standard." Technically oriented VOC questions that are used within concept validation studies (refer to Chapters 9 and 10) can help optimize the design of the next generation product: this is a great source of information for iterative innovation. VOC surveys should be conducted ***before*** concept and prototype development, ***during*** concept and prototype development, while the idea and design are being finalized, and ***after*** the product, service, or process is released to the market.

VOC questions can be open-ended (used for qualitative data analysis and to obtain customer insight, which may not be readily known) or closed-ended (used for quantitative data analysis to obtain statistical insights from the data).

Example questions to ask prospective users are:
- Would this concept make your life easier?
- Would you spend money to use this?
- Do you want this solution?
- If not, what would make you want this solution?
- What features would you like to see in this concept?
- Is it comfortable to use?
- Does it work the way you would want it to?
- What is frustrating about using it?

Things to keep in mind when speaking to prospective users are:
- Keep questions simple and to the point.
- Do not ask too many questions.
- Try not to use technical terms that may be overwhelming.
- Learn to read between the lines in responses; ask open-ended, follow-up questions for clarity.

2.4.1 Getting Started

A good way to start preparing a VOC survey is to identify the motivation behind the survey – what data do you intend to gather from the survey and how do you intend to use the information that is gathered. Start by specifying a research question that you want to answer. Following are a few examples of motivations behind preparing VOC surveys.
- How do patients feel about using a new, noninvasive healthcare technology, while visiting their doctor?
- Would customers be more inclined to purchase and use a new product, if that product could be customized by the customer to suit the customer's personal preferences?
- What customizable features would customer's most like to have in a new product?
- Would using an exclusively social media advertisement campaign on the three biggest social media platforms (i.e., Facebook, Twitter, and Instagram) for the launch of a new product be more effective than a product launch campaign utilizing a more traditional approach, such as an advertising campaign that is half print advertisement and half television commercial?

Information from VOC surveys can be used to optimize business models, identify new sources of customers or revenue streams, clarify marketing messaging, refine the design of your solution, confirm labeling and packaging choices for your product, and so on. Prospective investors are keen to learn what "gems" your team uncovers during these VOC exercises, so taking the time to design and execute an effective VOC survey is a good investment for the team.

2.4.2 Psychology of VOC

There are two main challenges associated with conducting VOC surveys: getting survey takers to participate in the survey in the first place, and getting survey takers to finish the survey. Since survey takers are usually volunteers, and they do not want to waste their time, it is important to keep surveys short and to the point. Every question needs to count. Every question must have a purpose, and no question should be redundant, that is, two questions in a survey should not result in the same answer data.

Depending on the credentials that are required of the VOC survey participants, that is, medical device manufacturers may need to survey clinicians and prospective users for their medical device, and pharmaceuticals may need to survey healthcare specialists who could prescribe the use of their medications, VOC sponsors may offer to compensate credentialed VOC survey takers as a token of appreciation for their time and invaluable input. However, this compensation must be legal and should not pose any ethical concerns.

2.4.3 Designing the Survey/Focus Group

Preparing questions for a VOC survey takes a lot of thought and deliberation. You must have a clear idea of what type of information you want to collect, and then must create questions that will get you the data you need. Survey questions can be broken down into different types, based on the type of data that they are designed to produce. Survey question categories including:

- **Categorical (i.e., self-characterization) questions.** Categorical survey questions enable survey takers to categorize themselves according to the categories defined in the question. In essence, survey takers are empowered to evaluate themselves and categorize or describe themselves according to predefined categories.
- **Behavioral questions.** Behavioral survey questions are questions focused on consumer behaviors. Answers to behavioral survey questions paint a picture of how survey takers behave or would behave in given situations.
- **Perception questions.** Perception survey questions are designed to extract information from survey takers about their perceptions of things that they are being asked about.
- **Demographic data collection.** It is usually a good idea to collect demographic data about survey takers, since this information can be useful in identifying

whether specific categories of potential consumers (based on age, gender, and other demographics) have specific attitudes about the product. Survey questions about demographic information should be tailored for relevancy to the particular market.[12]

One of the biggest challenges to conducting VOC surveys is that survey takers are generally unwilling to participate in surveys that are too long or confusing. Survey questions need to be crafted in such a way that you can extract information and data from survey participants with the shortest possible number of questions. One way to extract more data from a limited set of questions is to make the survey questions contain a number of subparts. For example, take a look at the following sample survey question:

Q1. How many hours a week do you (please fill in the blank):
Use all of your mobile devices (e.g., a tablet, smartphone, or laptop connected to the internet): ______ hours
Use a tablet device: ______ hours
Use a smartphone device: ______ hours
Use a laptop computer with internet: ______ hours

Q1 asks one question "How many hours a week do you... ?" but provides four subparts (i.e., use a mobile device, use a tablet device, use a smartphone device, use a laptop computer with internet) where survey takers can enter in answers in terms of hours spent engaging in the specified activity per week. By designing the survey question in this way, survey takers feel like they are answering one question, but in reality, they are providing answers to four separate questions.

It is prudent when preparing survey questions to include a brief definition of any key terms used in the survey that may be confusing or unclear to the survey takers. Providing a short, clear definition on the survey questionnaire will indicate to the survey takers what exactly is being asked in the survey questions.

Focus Groups

Interviews that are conducted with focus groups enable your company to gather information from many individuals who fall under predefined market/customer segments. Because you can identify and specify the particular product and service needs for a market segment, you can then choose to invite individuals from the segment to participate in a focus group, so that your company can confirm or refute its assumptions about that market segment. For example, if you are developing a medical device, then meet with clinicians who practice and have experience

12 Examples of the various types of survey questions can be found in the second example of Section 2.4.7.

with the types of procedures for which the medical device would be used. Do not conduct focus groups with orthopedic specialists, if your company is developing a cardiac medical device. Select cardiologists who conduct a minimum number of procedures annually in which the device will be indicated, and select cardiologists with a certain minimum number of years of experience. If there are enough financial resources in place, it is imperative to conduct multiple focus groups, so that you can gather information from multiple market segments. Conversely, you can choose to conduct several focus groups with a diverse group of participants who represent multiple market/customer segments.

With focus groups, ensure that enough participants are present so that ideas can be shared and developed amongst the participants, that is, one idea leads to another and another, but do not invite too large a number of participants, which would prevent all participants from sharing their ideas and opinions. It is also good to invite participants who have an interest in the outcome(s) of the focus group, and therefore will be more inclined to participate with valid and honest feedback.

Focus groups tend to collect qualitative data versus quantitative data. Interviewers ask open-ended questions to collect feedback from the focus group participants about current products and services on the market or future products or services that your company is developing, for example, "How do you feel about our customer service policies and practices?" "How would you improve a particular feature of our product?" "Why do you feel the way you do?" "What aspects of our product are not meeting your needs?" After the focus group is conducted, participant responses are analyzed and grouped according to common themes or trends. This information can be used to define and prioritize customer requirements that address the needs that were shared during the focus groups.

If there is no one in your company who has prior experience with designing and conducting focus group interviews, then you may want to consider hiring an outside firm that specializes in that type of VOC research, to ensure you receive the most valuable information that can be used to make informed decisions about the development pathway for your product or service.

2.4.4 Drawing Conclusions About Human Factors (Regarding the Concept/Product) from VOC Results

Human factors data is invaluable for determining whether the product or service, based on its current design, is a scalable product that will find success in the market. Human factors data will indicate whether your solution will be understood and used easily by a majority of the target customers.

For VOC feedback on human factors to be credible, however, the population that is surveyed needs to be diverse yet representative of the intended group of users or customers. And if, as a development team, you decide to not heed VOC feedback regarding human factors, then the team's rationale for that decision needs to be documented as part of the VOC findings.

Example

A product development team receives VOC feedback on the product's operating instructions, legibility of the product labeling, and the packaging prior to commercialization. VOC participants voice concerns about the pictures in the operating instructions and lack of instruction for how to remove the product from its packaging. The product development team, therefore, decides to update the operating instructions prior to commercialization. VOC participants then voice their concerns that the symbols and dimensions that are included on the product labeling are too small and too close together. The product development team, therefore, decides to update the product labeling, prior to commercialization. VOC participants lastly voice suggestions for changing the design of the inserts and trays that are part of the product packaging. The product development team decides to not change the design of those components, because the benefits of the packaging design change do not outweigh the costs of making that design change this late into the product development cycle. The product development team is confident that by not making the change, there is no added risk to the customer. The product development team recognizes that the packaging change can be made to the next version or generation of the product that is introduced into the market.

2.4.5 Conduct Customer Surveys

If you are conducting live, face-to-face surveys with preselected participants, send out invitations in advance with all the necessary information (e.g., date, time, place, and reservation details) and follow-up with phone calls or emails as required. Surveys give your company a chance to interact with potential customers, so it is important to portray the company as polished and organized, and it is important to practice good customer service techniques, for example, welcoming participants to the survey and thanking participants for their time.

Prior to conducting a survey, preparation is key. Before executing a live, face-to-face VOC survey, write a comprehensive script which can be used by anyone who is tasked to conduct the survey. This script will complement the body of the survey and it should include:

- An introduction to the survey.
- A summary of what types of questions will be asked to the survey participants.
- Any actions participants will be asked to complete as part of the survey.
- How long the survey will take?
- How survey responses will be scored?
- What each of the survey score responses indicate, for example, a score range of 1 through 5 represents XYZ?

- Opportunities for the survey participants to provide additional feedback and/or to ask questions.
- How their responses will be used to make design and development decisions?
- How their responses will remain anonymous and how their personal information will be kept confidential?

If there is time, scripts can even include a brief overview of the product or service that is being featured in the survey. Then, practice and rehearse the execution of the survey with your team. Many survey improvements can be made after a survey "dry run" or "dress rehearsal" uncovers small issues with the survey. For example:
- Is it easy for your team to complete the survey?
- Are any of the survey questions unclear?
- As you read the script out loud, is there a consistent flow to the script?
- Are transitions between sections of the survey seamless?
- Are additional pauses needed in the delivery of the script?
- Is the survey flawed in a way that leads survey participants to provide bad data, for example, survey participants are unable to select a "Not Applicable" option?

Think about what materials will be needed to execute the study, for example, writing materials, printouts, models, or prototypes. Are these materials readily available and easy to use?

Companies can use surveys as an opportunity to think about product use cases. Does the location where your survey is conducted need to simulate the conditions in which your product will be used? For example, if you are designing a medical device that will be utilized in an operating room at a hospital, the operating room will have varying noise and light levels, which could make it harder for a user to see, hear, or operate your device. Therefore, consider altering the light and noise levels within your survey location, especially if participants will be asked questions during the survey regarding the look and feel of the device or whether the user can see and hear visual and audio notifications on the device.

While participants complete the survey, survey facilitators should document any observation that can be considered a variance from normal survey execution, for example, survey discrepancies, errors, or deviations, and general observations, such as survey participants asking clarification or follow-up questions.

If live, face-to-face surveys are not possible or not desired, then consider distributing surveys online. Digital platforms, like SurveyMonkey, are widely used for distributing electronic surveys to many individuals who represent your target market/customer segment. For example, while designing and uploading the survey onto the digital platform, you can select for the survey to be sent to individuals with a specific career type or educational background, individuals who reach a threshold household income, individuals of a particular age range or gender, and so on. Digital platforms are of great benefit to companies because they can distribute multiple surveys to many participants and collect and analyze responses for you, all in a short amount of time for a small fee.

2.4.6 Extracting Useful Information from VOC Data Through Statistical Analysis

VOC data does a company little good unless the company can extract meaningful information from the collected VOC data. The company needs to be able to extrapolate the attitudes of the target market based on a sample of customers surveyed. Statistical analysis is a great tool for extracting useful information from VOC data. When it comes to statistical analysis, the complexity of the data analysis can run the gamut. For instance, some statistics are simple to derive and understand, such as the:

- **Mean** – The average of all the data points that make up a data set. The mean can tell you what the average response was in the VOC survey for a particular survey question.
- **Median** – The middle data point in a sequential set of data points. The median of VOC survey data tells you the central data point of a data set, and when compared to the mean, the median can tell you whether the data is normally distributed about the mean or is skewed. Identifying the cause of a skew in the VOC data can be useful information.[13]
- **Mode** – The most frequently recurring data point or points (in the case of bimodal or trimodal data). In a data set, if a high percentage of VOC survey participants (e.g., 80%) gave the same response (e.g., gave a rating of 8 on a scale of 1 to 10), creating a mode of 8, this information can be predictive of how other customers would likely also respond.
- **Range** – The range associated with the data set, bounded by the highest data point value of the data set and the lowest data point value of the data set. Range can be useful in judging whether certain survey questions have value or should be dropped.
- **Variance** – An assessment of how much variance from the mean there is in a data set. Each data point in a data set is subtracted from the mean to provide the difference between each data point and the mean, and then the differences are squared. Next, the squares are averaged together to yield the variance of the data set. Variance can be a useful indication of where other data points

13 If your VOC dataset follows a normal distribution, then mean and median will be the same value. If, however, your VOC dataset follows a non-normal distribution, then mean and median will not be the same value. Mean is typically utilized more often than median, but median may be a more representative value of the dataset, if there are outliers in your VOC data.

might fall relative to the mean. Histograms or frequency plots are also useful for graphing variance data.
- **Standard Deviation** – An indication of how spread out the data points in the data set are from the mean. The square root of the variance produces the standard deviation. Standard deviations are taken from the mean to indicate how far out a particular data point falls from the mean.

More complicated statistical analysis is necessary in order to draw conclusions from multiple sets of data that are comparable to one another. An important aspect of statistical analysis is to improve confidence in the accuracy of the data, and thus the information that the data conveys. A confidence interval in statistics is a range of data values that indicates that the data is accurate with a certain level of confidence. Confidence intervals indicate statistical significance, that is, the reliability of the result. Usually, confidence intervals of 95% or 99% are used when analyzing statistics, but confidence intervals can be higher (e.g., 99.9%) or lower (e.g., 80%) as necessary for the analysis. By using a high confidence interval, there is a higher likelihood that the VOC data is accurate (as representative of the larger target market's attitudes), and thus a higher likelihood that the conclusions that are drawn from the data are accurate.

VOC survey data is often drawn from customers who can be grouped in different ways. Groupings of VOC survey participants, and subsequent analysis of the data in terms of groups, can be revealing. For instance, perhaps you might find useful information when you look at the VOC survey data after grouping participants by age: children under the age of 12, teenagers ranging from 13–18, and adults 18 years and older. There are three groups of data, and determining the differences between the three groups can provide insight into the attitudes of the target market age segments.

Analysis of variance, also known as ANOVA analysis, is a statistical tool used to determine if three or more groups of data have similar or divergent means.[14] ANOVA analysis can only indicate whether one of the data sets has a mean that is different from the others, but it cannot indicate which mean is different. In relation to determining whether the means of the groups of data are different with ANOVA analysis, when the means of the groups are different from one another then it becomes important to assess which group mean(s) are different. Tukey analysis (another statistical analysis comparison test) is often conducted in conjunction with ANOVA analysis to determine which means are different in a group of means. The

14 T-tests can be used to determine if the means of two groups are significantly different.

details of how to conduct ANOVA and Tukey analyses are beyond the scope of this book, but it is important to know what tools are available for your use, as you interpret VOC survey data.

2.4.7 Two Practical Examples

Two example VOC Surveys are provided in the following sections. In the first example, notice how the survey begins with a brief introduction and defines the scores for all possible responses. The survey design makes it easy for the survey to be taken: the customer can easily circle his or her chosen answer, the customer can select from the same set of responses for each question, and the survey also lets the customer know how many questions are included in the survey.

Example

Voice of the Customer Questionnaire

Thank you for participating in this survey, which is designed to assess how you would feel about using a new, noninvasive healthcare technology at your doctor's office. We are currently developing a medical device that detects early signs of orthopedic injuries. This survey will ask ten questions. We look forward to your feedback.

ID	Question	Score			
		Disagree	Somewhat disagree	Somewhat agree	Agree
1	Would you be willing to prevent more long-term damage after an injury by using a new technology at your doctor's office?	1	2	3	4
2	Would you like to be able to track your recovery after surgery?	1	2	3	4
3	Would you be interested in early detection of a health problem by using a new technology at your doctor's office?	1	2	3	4
4	While waiting to see the doctor, would you be willing to spend 10–15 min using a new, noninvasive technology that could detect a health problem?	1	2	3	4
5	Would you be willing to use this new technology before and after surgery?	1	2	3	4

(continued)

ID	Question	Score			
		Disagree	Somewhat disagree	Somewhat agree	Agree
6	Would you be willing to use this new technology before and after a period of physical therapy?	1	2	3	4
7	Would you be willing to use this new technology after experiencing an injury?	1	2	3	4
8	Would you be willing to use this new technology as part of your yearly physical exam?	1	2	3	4
9	Would you be willing to pay extra (on top of your co-pay) to use this new technology?	1	2	3	4
10	Would you be willing to pay an average of $X to use this new technology?	1	2	3	4

In this above example, notice how the first three questions engage the customer – most people would probably respond with a "4" for these first three questions. By agreeing with the survey initially, customers subconsciously relate to the survey and are more willing to finish the survey.

The second example below includes both the survey questions, as well as an analysis of data that was collected from the survey takers. In this second example, a marketing team wanted to analyze whether using a traditional marketing approach consisting of a combination of print advertising and television commercials, or launching a strictly social media-driven advertising campaign targeting the three biggest social media platforms (Facebook, Twitter, and Instagram), would be a better marketing strategy.

Example

The survey questionnaire below contains a definition as to what a social network is and is limited to ten questions, which are broken down according to type: categorical, behavioral, perception, and demographic information.

Survey Questions

Definitions of "social media platforms" – Social media is an aggregation of internet-based platforms where people socially interact with others. Popular social media platforms include Facebook, Twitter, Instagram, and LinkedIn.

Categorical Questions

1. In a typical week, how many hours do you spend:
   Engaging with social media platforms: ______ hours
   Viewing TV programming: ______ hours
   Reading magazines: ______ hours
   Reading a newspaper: ______ hours

2. I have a profile on (check all that apply):
   Facebook
   Twitter
   Instagram
   LinkedIn
   I do not have a social media account/profile

3. Rate the following statements based on how applicable they are to you:
   Scale: Does not describe my use of social media 1–10 Describes my use of social media well
   a. Social media is how I stay in touch with people I know.
   b. Social media is the primary way that I make plans with people I know.
   c. I use social media to keep my friends posted on what I am doing.
   d. I use social media to keep track of what my friends are doing.

Behavioral Questions

4. I usually buy products
   Scale: Poorly describes me 1–10 Describes well
   a. Over the internet
   b. Over the phone
   c. From a retailer/store
   d. Secondhand or used

5. Do you prefer to make an impulse purchase decision...
   Scale: Do not prefer 1–10 Highly prefer
   a. Online
   b. Over the telephone
   c. At a retailer or store
   d. At a secondhand store or retailer

6. What is the likelihood that you would purchase a new product based on an advertisement seen on one of the following forms of media?
   Scale: Very unlikely 1–10 Very likely
   a. Print advertisement in a magazine
   b. Commercial on television
   c. Advertisements on billboards
   d. Nonsocial media webpage as a banner or sidebar advertisement
   e. Social network platform as a banner or sidebar advertisement

Perception Questions

7. Please rate how applicable the following statements are to you:
   Scale: Poorly describes me 1–10 Describes well
   a. Social media is a waste of time.
   b. Social media gives me a sense of belonging to a community.
   c. Social media platforms let me control the narrative of what is going on in my life.

8. Please rate how applicable the following statements are to you:
   Scale: Strongly disagree 1–10 Strongly agree
   a. Social media platforms do a good job of targeting advertisements to me based on my profile, likes, and interests.
   b. TV commercials are good for breaking up regular programming.
   c. Advertisements in the magazines that I purchase are targeted to my interests.

9. Please rate how applicable the following statements are to you:
   Scale: Strongly disagree 1–10 Strongly agree
   a. I believe that social media platforms are critical to keeping me up to date on current events.
   b. I make all of my product purchases based on an advertisement I have seen.

10. Demographic Information:
   a. Age   b. Gender

The survey was distributed to 400 individuals. Two hundred of the surveys were distributed via email. The other 200 surveys were distributed to survey participants as a private message through their social media accounts. Once the marketing team received the survey results from the participants, statistical analysis was conducted to produce the following data.

Survey Answers Data

Categorical Questions

1. In a typical week, how many hours do you spend:
   Use social media platforms: Average: 5.3425 h Std. Dev.: 6.335 h
   View TV programming: Average: 14.875 h Std. Dev.: 13.49 h
   Read magazines: Average: 2 h Std. Dev.: 2.03 h
   Read a newspaper: Average: 1.85 h Std. Dev.: 2.03 h

2. I have a profile on (check all that apply):
   80% of Survey Participants reported having a profile on Facebook
   55% of Survey Participants reported having a profile on Twitter
   40% of Survey Participants reported having a profile on LinkedIn
   35% of Survey Participants reported having a profile on Instagram
   10% of Survey Participants reported “I do not have a social media account/profile”

3. Rate the following statements based on how applicable they are to you:
   Scale: Does not describe my use of social media 1–10 Describes my use of social media well
   a. Social media is how I stay in touch with people who I know.
      Mean: 2.95 Std. Dev.: 1.96

b. Social media is the primary way that I make plans with people who I know.
Mean: 3.9 Std. Dev.: 2.9
c. I use social media to keep my friends posted on what I am doing.
Mean: 4.25 Std. Dev.: 2.3
d. I use social media to keep track of what my friends are doing.
Mean: 3.10 Std. Dev.: 2.12

Behavioral Questions

4. I usually buy products
Scale: Poorly describes me 1–10 Describes well
a. Over the internet
Mean: 6.70 Std. Dev.: 1.49
b. Over the telephone
Mean: 4.40 Std. Dev.: 2.18
c. From a retailer or store
Mean: 3.05 Std. Dev.: 2.24
d. Secondhand or used
Mean: 1.05 Std. Dev.: 1.23

5. Do you prefer to make an impulse purchase decision...
Scale: Do not prefer 1–10 Highly prefer
a. Online
Mean: 3.80 Std. Dev.: 2.23
b. Over the phone
Mean: 1.85 Std. Dev.: 1.65
c. At a retailer/store
Mean: 1.85 Std. Dev.: 1.45
d. At a secondary market
Mean: 1.10 Std. Dev.: 1.23

6. What is the likelihood that you would purchase a new product based on an advertisement from one of the following forms of media?
Scale: Very unlikely 1–10 Very likely
a. Print advertisement in a magazine
Mean: 4.65 Std. Dev.: 1.82
b. Commercial on television
Mean: 4.95 Std. Dev.: 1.95
c. Advertisement on billboards
Mean: 3 Std. Dev.: 1.96
d. Nonsocial media webpage as a banner or sidebar advertisement
Mean: 5.95 Std. Dev.: 1.75
e. Social network platform as a banner or sidebar advertisement
Mean: 6.45 Std. Dev.: 2.04

Perception Questions

7. Please rate how applicable the following statements are to you:
Scale: Poorly describes me 1–10 Describes well

a. Social media is a waste of time.
Mean: 3.15 Std. Dev.: 1.78
b. Social media gives me a sense of belonging to a community.
Mean: 5.75 Std. Dev.: 1.94
c. Social media platforms let me control the narrative of what is going on in my life.
Mean: 4.75 Std. Dev.: 2.36

8. Please rate how applicable the following statements are to you:
Scale: Strongly disagree 1–10 Strongly agree
a. Social media platforms do a good job of targeting advertisements based on my profile, likes, and interests.
Mean: 5.95 Std. Dev.: 2.08
b. TV commercials are good for breaking up regular programming.
Mean: 3.85 Std. Dev.: 1.21
c. Advertisements in the magazines that I purchase are targeted to my interests.
Mean: 4.25 Std. Dev.: 1.7

9. To what extent do you agree with the following statements?
Scale: Strongly disagree 1–10 Strongly agree
a. I believe that reading magazines and newspapers keeps me well informed.
Mean: 5.70 Std. Dev.: 1.54
b. I make all of my product purchases based on an advertisement I have seen.
Mean: 3.25 Std. Dev.: 1.62

10. Demographic Information:
a. Age: Range: 16–59,
Mean: 32.5 years Std. Dev.: 9.1 years
b. Gender: Female: 52% Male: 48%

Based on this data, the market researchers wanted to know (1) whether survey participants were more likely to purchase new products based on advertisements seen on one form of media over another, and (2) which advertising vehicle survey participants prefer (i.e., do they prefer ads on social media, on TV, or in print).

One-way analysis of variance statistical analysis (ANOVA) was conducted to compare the means of the data collected for Question 6a, b, and e. This involved a comparison between data concerning how likely the survey participants are to purchase new products based on an advertisement that is in print (Question 6a), a TV commercial (Question 6b), or on a social media platform (Question 6e). A second one-way analysis of variance analysis was conducted to compare the means of the data collected for Question 8a, b, and c. The second analysis compared whether there was a statistical difference in the survey participants' preference for how they view advertisements – on social media platforms (Question 8a), on TV (Question 8b), or in print (Question 8c).

From this analysis, several important conclusions about whether a strictly social media-based advertising campaign would be better for a new product than a more traditional advertising campaign that is half print ads and half TV ads can be drawn:

- Based on data from Question 6, survey participants on average are more likely to purchase new products after seeing a TV advertisement, followed by seeing a print advertisement, and advertisements featured on nonsocial media-based websites. The results were statistically significant when comparing TV advertising to print advertising or social media web advertising.

- Survey participants' perceptions of the various advertising media formats (Question 8) were inversely related to survey participants' responses to what drove their new product purchasing behavior (Question 6).
  - In Question 8, survey respondents rated social media as the most enjoyable form of advertising, followed by print and TV advertising. Survey respondents disliked TV commercials as an interruption to their favorite shows, but welcomed social media content suggestions in the form of advertisements.
  - Yet based on data from Question 6, respondents indicated feeling as if they are most influenced to make a purchase of a new product by the media format that they enjoy the least (TV ads).
- Online purchasing was the preferred method for purchasing new products (Question 4) among the survey participants. There was no significance found in the purchasing behavior of men and women, or between age ranges.
- The data indicates that launching a social media-based advertising campaign for a new product will be an effective marketing strategy.

3 Forming a Business Around Your Technology

Abstract: Building for success starts by laying a foundation for the development and commercial success of a new product or technology. The product will not be successful in a vacuum – there is a need to develop a business around the product so that the product can be marketed and sold. There is a lot to consider at the outset of building a business and there are a lot of decisions to make. For instance, decisions will need to be made regarding the type of business entity that should be formed, a business plan should be prepared, a team of people to help the business achieve success should be assembled, and a groundwork will need to be set for protecting and securing the investments of time and effort that will be put into the development of the business. In reality, none of these steps for starting a business occurs sequentially. Rather, they all seem to occur in parallel. It is not uncommon to work on a business plan, while simultaneously recruiting talented individuals who would be perfect for your team. And you might be filing the formal paperwork or preparing an operating agreement to make your business structure official, while simultaneously taking measures to protect company assets. The process is a whirlwind, but rewarding. Below we discuss the various aspects you need to consider as you launch your business.

Keywords: Incorporation, business entity, sole proprietorship, partnership, limited liability company (LLC), corporation, Employer Identification Number (EIN), Small Business Administration (SBA), Small Business Development Centers (SBDCs), Economic Development Corporations (EDCs), incubators, accelerators, establishing a team, leadership, mentors, advisors, Board of Directors.

3.1 What Is a Business?

Before getting ahead of ourselves, let's consider the benefits of creating a business out of your innovation. At this stage in the game, you need to start thinking about what you are doing as a business and need to start thinking about dollars and cents. So why is forming a business a good idea at this stage? A business is a legal entity, and its purpose is to provide structure and a degree of certainty to business operations, to protect the business owner(s), and to take advantage of certain tax saving mechanisms (at least in the United States). Every business starts small, but eventually can grow to be quite large. That is why it is important for the business to have an organizational structure afforded by a legal business entity.

As shown in Figure 3.1, incorporation is the step in the Business Development Process that involves organizing a business into a legal entity. When a business is organized into a legal entity, depending on the type of legal entity formed, certain legal benefits can be conferred on the business or can be taken advantage of, and a

https://doi.org/10.1515/9783110521900-004

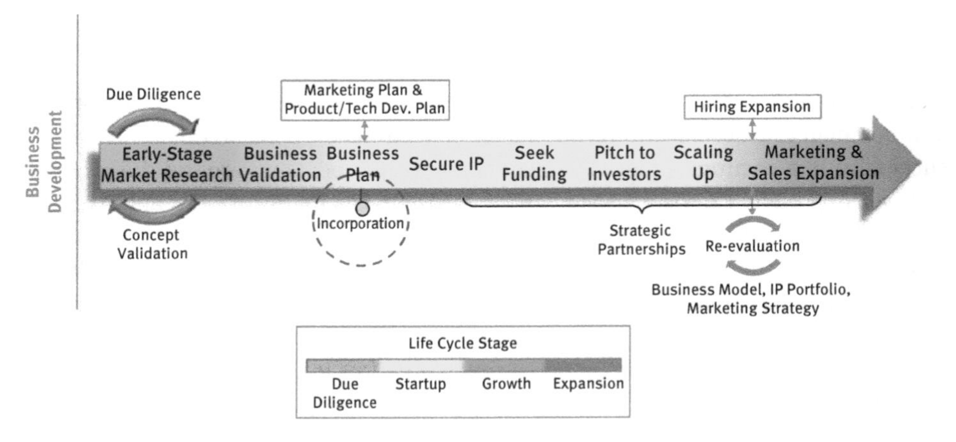

Figure 3.1: Incorporation and establishing a team steps in the Business Development Process.

number of potential problems associated with the growth of the business can be mitigated or avoided entirely. For instance, in the US certain business entities pay taxes on earnings, but at a corporate tax rate (and shareholders are also taxed again later on dividends earned, in what is known as double taxation), while others are "pass through" entities, meaning that income generated by these business entities is passed through without taxation at the business level to the individuals who own the business entity, where it is only taxed at a personal income rate. Other business entity structures legally shield business owners from liability incurred by the operation of the business.

There are a number of different types of legal business entities in the United States, and each type has its own unique qualities. The formation of a corporation is done at the state level. There are pros and cons associated with each structure type, and business owners should weigh each option carefully, in order to determine which business entity type is the best for helping them achieve their business goals and objectives. You should consider consulting with a business lawyer to determine which legal business entity, and what state, is the best for your particular situation.

It is important to point out that once you choose a business entity type, you are not locked into that specific business structure forever. The business structure can be changed, so long as you comply with all of the requirements for converting the business structure. On the other hand, changing the state in which a business entity is formed is a much more complicated task.

It is important for anyone who is starting a business to conduct research in order to better understand what options are available, and to discover what additional requirements will be placed on you, once you form your business. For instance, most businesses will be required to obtain a federal

Employer Identification Number (EIN) in the United States, which is similar to an individual's Social Security Number. An EIN is necessary for businesses to report income for tax purposes, and you cannot open a business bank account without one. Similarly, it is prudent for businesses to secure business liability insurance as a means to protect the business's assets, in the event that obligations are incurred. Table 3.1 gives a brief overview of the most popular US business entities, an indication of whether the entity is a "pass through" entity for taxable income purposes, whether an EIN is required, and whether business liability insurance is strongly recommended.

Table 3.1: Most popular business entities and indication of pass through entity status, EIN requirement, and whether business liability insurance is strongly recommended.

	Pass Through Entity	**Needs EIN**	**Needs Liability Insurance**
Sole Proprietorship	✓	Maybe	✓
Partnership			
General Partnership	✓	✓	✓
Limited Partnership	✓	✓	✓
Limited Liability Partnership	✓	✓	✓
Corporation			
C-Corporation	No	✓	✓
S-Corporation	✓	✓	✓
LLC	✓	✓	✓
Single-Member LLC	✓	Maybe	✓

The type of business entity that is chosen will dictate:
1. What level of liability business owners personally have;
2. what the tax structure is;
3. the means to distribute equity (if applicable).

The type of business structure chosen for a company will determine how it is taxed and what mechanisms are put in place for distributing equity (if applicable). When starting a business, you should take into consideration that the legal entity will fall into the realms of both business law and tax law. Due to the complexity of forming a company, you should consult a business attorney *and* an accountant to determine the best structure for your particular business situation. What follows is a high-level overview of the most common types of US business entities that are available in most states.

3.1.1 Sole Proprietorship

The easiest business entity to form is a sole proprietorship. In a sole proprietorship, a single individual owns all of the business's assets and is responsible for all of the business's liabilities or debts. Basically, the business owner is the business itself and he or she operates the business in a personal capacity.

Many people choose the sole proprietorship business structure, at least at the early stages of their business's life, because it is the simplest business entity to form. There are no formation formalities, meaning you do not have to file any special paperwork in order to start conducting business as a sole proprietor, or are there any special formalities, once you start running your business.

A sole proprietorship is a "pass through" entity, so any income earned in a sole proprietorship is treated like normal personal income for the business owner. This means that the sole proprietor business owner is taxed on the income that is generated by the business, making the business a "pass through" entity on the owner's personal taxes. Any losses experienced by the sole proprietorship can be deducted from the personal income tax of the sole proprietor.

Sole proprietors generally do not need to secure an EIN, unless they have employees that are paid wages or file excise or pension tax returns. Sole proprietors that do not have to obtain an EIN can use their Social Security Number for tax purposes instead.

A major disadvantage of the sole proprietorship business structure is that the business owner is exposed to liability related to the operations of the business. To say this another way, the business owner of a sole proprietorship can be held personally liable in a legal sense, if something bad happens. Debts, judgments, and any other liabilities fall on the shoulders of the sole proprietor personally.

3.1.2 Partnership

When a business is owned by two or more individuals, a partnership might be an attractive organization structure. The partners must jointly own the business, must voluntarily form the partnership, and must carry on business for profit. A partnership is a "pass through" entity, so any income earned in a partnership is treated like normal personal income for the business owners. Partnerships are required to obtain an EIN for tax purposes.

While there are no formal filing requirements to begin conducting business as a legal partnership business entity, many partnerships are wise enough to know that they need a partnership agreement among all of the partners in the business before their business is fully underway. A partnership agreement is a contract between the partners that sets forth each partner's rights and responsibilities to the business, role in the organization, share of profits and liability, governance for how to exit the partnership, etc. Getting a good partnership agreement in place early can help avoid a lot of headaches among the partners in the future.

There are three types of partnerships:

- **General Partnerships.** The general partnership business structure is popular because each partner is entitled to an equal share of any income generated by the business, unless some other apportionment is agreed to in a partnership agreement. However, a general partnership also makes each partner liable for the acts, omissions, and debts of the business equally. What many people don't like about the general partnership entity is that all the partners are liable for the acts of each other. So, one partner could enter into a business contract with a supplier, and the whole partnership would be on the hook for the contract. Similarly, if one of the general partners is sued, all of the partners, in the general partnership are liable as well.
- **Limited Partnerships.** Limited partnerships are a common partnership business structure that is used when there are some partners who want to run the business on a day-to-day basis (these partners are called general partners) and some partners who only want to act as an investor in the business (these partners are called limited partners). There must be at least one general partner in a limited partnership. The roles of any general partners in the limited partnership are governed by the same principles as those that govern a general partnership discussed above. The limited partners generally do not participate in the day-to-day activities or management of the partnership and thus their liability is limited.
- **Limited Liability Partnerships.** Limited liability partnerships (LLP) are more formal than general partnerships and limited partnerships, sharing many of the same legal features as a limited liability company (LLC) (discussed in Section 3.1.4). LLPs are very common in fields of business that are conducted by professionals, such as lawyers, professional engineers, or accountants. All of the partners in an LLP have limited liability and are protected from the acts, errors, and omissions made by other partners in the LLP. When one of the limited liability partners in an LLP acts in a negligent or wrongful manner, that individual is liable for his or her actions in a personal capacity. LLPs usually require a formal agreement in order to start operating, and LLPs are usually required to make an annual report to the appropriate state governmental entity, but annual reporting depends on the legal jurisdiction of the business.

All partnerships are "pass through" entities for tax purposes, meaning that the profits and losses earned by partners in a partnership are considered personal income or loss for each partner, and are thus taxed at a personal income tax rate.

3.1.3 Corporation

A corporation is a business entity structure that is its own legal entity. A corporation has its own legal name, identity, responsibilities, and obligations under the law. An EIN is required for all corporations. There are several types of corporations, but of particular relevance to you are C corporations and S corporations.

A C corporation is not a "pass through" entity – it is its own legal entity and thus is its own taxable entity. The earnings of the C corporation are taxed at a corporate tax rate, but then shareholders that are paid any dividends by the C corporation will also be taxed at the personal level on their dividend earnings, that is, double taxation. C corporations can decide whether to issue different classes of stock to shareholders, which carry different voting rights.

An S corporation can only be formed if certain qualifications are met. Additionally, some types of businesses are expressly excluded from being able to become an S corporation. S corporations cannot have more than 100 shareholders, can only issue one class of stock to shareholders (so all voting rights are equal), and all of the shareholders must be US citizens or US residents. As opposed to a C corporation, an S corporation is a "pass through" entity, so any income earned in an S corporation is treated like normal income for the shareholders.

Corporations are generally run by shareholders and a Board of Directors. Shareholders purchase an ownership stake in the business by exchanging money for shares in the corporation. Sometimes shareholders can earn their shares in the corporation in exchange for their labor (also known as sweat equity) or by providing professional services to the corporation (the professional is paid in corporate stock). Generally speaking, shares in the corporation entitle each shareholder to voting rights. The Board of Directors often makes decisions regarding the direction of the corporation, but the shareholders often have to vote to approve of the Board's plans or actions. The corporation business structure legally protects shareholders from the debts, judgments, and liabilities of the corporation.

When a business is a corporation, and thus its own legal entity, the business earns its own income for tax purposes. Since the corporation earns its own income, the corporation is taxed on its income at a corporate tax rate. Additionally, when income is distributed to shareholders according to their respective ownership stake in the form of dividends, the shareholders are also then taxed on their dividend earnings. In effect, income that is passed through a corporation to the shareholders is effectively taxed twice.

There are a number of formal filings that must be submitted on a state (meaning that there are state-required formation documents) and federal level (such as obtaining an EIN) before a corporation is officially formed. The process of filing the formation documents is typically referred to as the process of incorporation. The exact filing that is required will depend on the state in which the business is being incorporated. Nearly all states require a corporation to file either a document called

"Articles of Incorporation," a "Certificate of Incorporation," or a "Certificate of Formation," with the Secretary of State or similar state governmental entity.

Additionally, corporations usually are required to have annual shareholder meetings, which require compliance with a number of legal formalities, such as recording of minutes and maintaining records for a certain amount of time.

Some small businesses start out as either a sole proprietorship or a partnership that consists of only the founders of the business. Later the founders convert the business into an LLC or corporation, as part of a fundraising strategy. Other startups start out immediately as an LLC or corporation. Each company will have to choose the right path for itself.

Once formalized as a corporation, businesses can generate funding by selling shares in their corporation. However, while selling shares may generate much needed capital, it's important for founders to remember that every share sold dilutes the founders' control over the business.

3.1.4 Limited Liability Company (LLC)

The LLC business structure is also its own legal entity apart from its business owners. The business owners in an LLC are referred to as "members" and there is no limit on how many members can be part of an LLC. Some states even permit a single-member LLC. The members are governed by a membership agreement.

LLCs that have multiple members generally need to obtain an EIN. Single-member LLCs do not need to obtain an EIN, if the owner of the single-member LLC is the sole owner and the single-member LLC does not have any employees.

An LLC is either considered a member-managed LLC or a manager-managed LLC. In either case, at least some of the members of the LLC are involved in controlling how the LLC is run on a day-to-day basis. If possible, income generated by the LLC is distributed to its members – but not necessarily equally. The membership agreement sets out the terms of how much income from the LLC each member is entitled to, so members that are more involved in the daily operations of the LLC or have more resources invested might get a larger share of the LLC's earnings. Your accountant should ensure that the proper tax forms reflect these distributions.

Like a corporation, the LLC business structure provides some protections to its members and their personal assets from the liabilities of the LLC. But what is nice about the LLC structure is that even the more-involved LLC members, that is, the ones who are tasked with making key business decisions, such as the Board or CEO, are protected from personal liability, unless their actions concerning the business were seriously irresponsible, illegal, unethical, or corrupt. Bad behavior on the part of LLC members can easily pierce the corporate veil of protection offered by the business's LLC status, exposing the bad actors to liability for their misdeeds.

There are a handful of formation requirements when creating an LLC. Specifically, the LLC is required to file an Operating Agreement with the small business administration (SBA) for your state, or similar state government entity. There is usually a form to complete as well, and a filing fee, which varies depending on the state where you file.

Once the LLC is established, there are no formal reporting requirements. The LLC does not have to conduct annual meetings or record minutes, unless doing so is part of the LLC's Operating Agreement or the membership agreement. However, it is highly beneficial to conduct annual meetings and to record minutes, in order to avoid adversaries piercing the corporate veil and gaining access to members' personal assets.

3.2 Support Structures for Startups and Small Companies

Federal, state, and local governments want startups and small companies to succeed. Small companies provide jobs and generate tax revenue. Regional ecosystems exist to give support to startups and small businesses to ensure that there are long-term sources of tax revenues (refer to Figure 3.2).

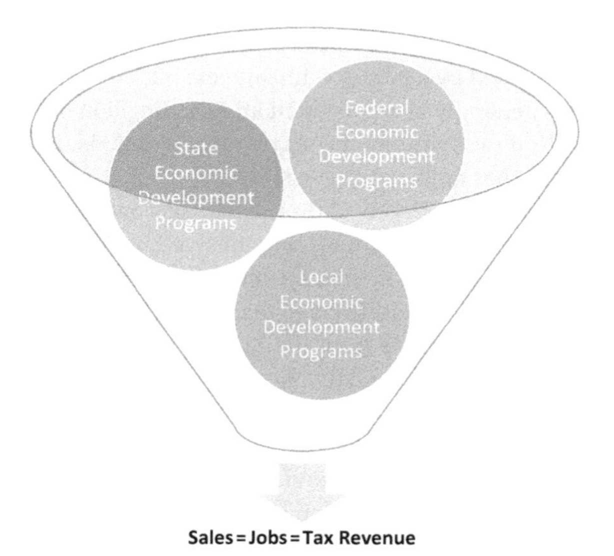

Figure 3.2: Support structures for startups.

Some of the most useful government-run support structures for startups and small companies include the Small Business Association (SBA), state-sponsored support programs, Small Business Development Centers (SBDCs), and Economic

Development Corporations (EDCs). In addition to government ecosystems, there are also incubators, accelerators, and hybrids, which are operated by nongovernmental entities. Below are some of the services that economic development programs can offer startups and small companies:

- **Education** – Structured and regularly offered startup business education (e.g., SBDCs in a particular state).
- **Mentorship** – Local and national mentors across the industry of interest. They may also focus on topics such as operations, finance, and distribution.
- **Networking** – Programs often make introductions to local contract manufacturers, financial services providers, government officials, hospitals, etc.
- **Facilities** – Facilities can include shared spaces, incubators, co-ops, and workshops, but economic development programs also need to be willing to collaborate and provide support for interdisciplinary startups.
- **Funding** – Organized, strategic, tiered funding from sources such as VCs, Angels, institutional entities, state and federal governments.
- **Vetting of Funding Sources** – A good program will vet prospective investors to ensure that the investors will not take advantage of the entrepreneur.

Not all states or programs are comprehensive and not every startup needs each component. The following are additional things to look for to assess the validity of the program or service:

- **Partner organizations** – Most likely will be state and local government, hospitals, universities, and/or large industry players (e.g., Johnson & Johnson, Google, IBM, and Intel).
- **Local leadership buy-in** – Ties into local government and, most importantly, the local leadership's vision, policies, incentives, and funds.
- **Entrepreneur Vetting** – The vetting of the aspiring entrepreneurs for the investors.
- **Talent** – A plan for accessing and/or importing the talent needed for that industry sector. This should include local K-12/college/university support programs. Realistically the plan will need to anticipate publicity and possibly offering incentives for importing talent from different regions.

3.2.1 Small Business Administration (SBA)

The SBA is a US Federal Agency that provides support to entrepreneurs and small businesses. The SBA aims to maintain and strengthen the nation's economy by helping small business become established and viable. The agency principle resources are the ability to provide capital, and to supply contracts and counseling.

The SBA has at least one office in each US state. Since the SBA is a federal agency, it coordinates government-backed loans through banks, credit unions, and other lenders that partner with the SBA. The SBA also provides programs to ensure that contracts reach underrepresented populations, such as minority-owned, woman-owned, and veteran-owned small businesses.

The agency also provides grants to support-providing agencies, such as Women's Business Centers, SBDCs, EDCs, and SCORE, which is a volunteer mentor corps of retired and experienced business leaders. Besides providing mentors, SCORE also offers workshops and has a library of free webinars.

3.2.2 State Programs

Each state has its own system or department for economic development. For example, the State of Pennsylvania has the Department of Community and Economic Development (DCED). The DCED offers grant and incentive programs throughout the state. These programs can range from grant programs for individuals to funding local economic development initiatives, such as additional funding to the SBDCs and EDCs. It also funds the Ben Franklin Technology Partnership, which includes multiple regional nonprofit organizations under the same regulations spread throughout the state that provide high-risk loans to high-tech startup companies.

3.2.3 Small Business Development Centers (SBDCs)

SBDCs are the SBA's "boots on the ground." They usually collaborate with local colleges and universities and are funded by the SBA, state government, and the associated university. Their singular purpose is to help startups and small businesses. Since they are funded by the federal and state government, they satisfy their requirements for funding by showing metrics for how many startups and small businesses they have helped, the number of jobs created, and the revenue those companies generate. SBDCs do not take equity, but they will require that the startup or company participate in multiple surveys for data gathering purposes. They only offer support in the form of consulting services. Examples of SBDC consulting services include:

- How to form a business (including what paperwork to fill out).
- Business training resources (books and manuals).
- International business assistance.
- Market research assistance (they often have access to expensive market databases).
- Technology assistance.
- Government marketing assistance.
- Assistance finding grants or alternative funding.
- Networking.

- Tax credits.
- Student interns.

Not all SBDCs are the same. Different states and regions give different levels of support. The first thing to do is to walk through the door.

3.2.4 Economic Development Corporations (EDCs)

An EDC is usually a nonprofit 501(c)(3), whose purpose is to support a specific type of economic development within a specific geographical area. These organizations are often funded by federal, state, and/or local government. An example of this would be an EDC to promote food co-ops in a tristate area. They often provide programs to give:
- Low interest loans.
- Grants.
- Tax credits.
- Other economic incentives to attract businesses in their local area.

Note: Corporate tax credits are economic development subsidies that reduce a company's taxes by allowing it to deduct expenses from its income tax bill. Tax credits are usually granted for a particular kind of corporate activity a state wants to promote such as job creation tax credits, starting or moving a company to an economic development region, investments spent on new facilities and/or equipment, or research and development (R&D) credits, which are especially lucrative for pharmaceutical, medical device, and technology-based companies.

Tax credits are not the same as tax deductions. A tax deduction is subtracted from a company's income before the amount of tax is calculated, lessening the amount of profit subject to taxation, whereas tax credits are subtracted from the amount of tax owed, rather than from a company's income. In some instances, if a company's credit exceeds its tax bill for the year, the company can carry over the leftover credits for the next tax season. This can be helpful to lessen the tax burden on startup companies.

3.2.5 Incubators

Incubators provide a workspace where a company can grow in exchange for a fee. Incubators can be an office space, a cubicle, or even laboratory space. Incubators are often inexpensive, because they are a part of an economic development initiative from a local and/or state government. Incubators often provide seminars and networking to their tenants. Most incubators are equity free, but some may offer incentives for equity.

3.2.6 Accelerators

Accelerators are educational "boot camps" for startups and small companies with novel products. Startups and small companies must apply for acceptance into an accelerator program. Acceptance is typically accompanied by a funding award. These programs are highly competitive, because they offer curricula and training on how to get the startup company up and running both on a business and technical level. Accelerators provide resources for startup companies, including access to experienced advisors, as well as access to potential investors. Accelerators generally offer support with practicing fundraising pitches, mentoring the team, and connecting a company to vendors, mentors, advisors, and investors. Accelerators can also be industry specific (like digital health or medical devices) or generic (e.g., high-tech or software development). If a company is invited to participate in an accelerator, then most likely angel investors and VCs will be introduced to the company. Accelerator programs explain investor expectations, so there are no surprises. A good example of an accelerator program is the National Science Foundation's Innovation Corp program (NSF I-Corps). NSF I-Corp was developed to prepare scientists and engineers that are skilled in their specialty area to understand the multiple skill sets needed outside of an academic laboratory. This program is of benefit to the U.S. government and the NSF because it trains federally funded scientists on how to commercialize their inventions.

Accelerators and incubators are special ecosystems for startups. The ultimate goal of participating in an incubator or accelerator program is that the company will get to market faster than trying to get to market on its own. Startups will often meet with higher rates of success, because of the support that they receive from the incubator or accelerator program. They can be a win-win for everyone involved, because the purpose of incubators and accelerators is to de-risk innovation and de-risk investment. This means that they serve a dual purpose; one for large corporations and one for investors.

3.3 De-risking Innovation and De-risking Investment

3.3.1 De-risking Innovation

Large corporations like Johnson & Johnson, GE Healthcare, and Philips have found that being "innovative" is risky and expensive. The strength of large corporations lies in streamlining the manufacturing process of a product, distribution, and sales. Innovation itself is very risky for big corporations that need to satisfy shareholders.

Incubator and accelerator programs offer a structure for large corporations to see what innovative technology is on the horizon. They can also see if the new products complement their own product portfolios. In many cases, it is more affordable for a larger corporation to acquire a startup company that has identified a need, started developing a product, and demonstrated early validation of their concept, rather than try to come up with the concept on its own.

3.3.2 De-risking Investment

The risk associated with investing time and money into a startup is that startups have a high rate of failure. Key reasons for business failure include:
- Running out of cash (#1 reason).
- No market need for the product or service that the startup provides.
- Lack of focus, motivation, commitment, and passion of the startup team.
- Unwillingness of founders to learn or listen.
- Taking advice from the wrong people.
- Lack of good mentorship.
- Lack of knowledge in finance, operations, and marketing.
- Not understanding distribution channels or being outcompeted.

Understanding the common reasons why startup companies fail can help you better protect your startup from failure. Notice that the product is not a part of any of the key reasons why startups fail. The most common reasons that startups fail are team-related issues. These issues are usually tied to leadership and its ability to build a strong team, drive a business model, or provide an effective organizational structure. More importantly, if running out of money is attributed to the reason for failure, other factors most likely contributed to the fall of the startup as well. This is where accelerators and incubators de-risk investment. These programs try to assess and mitigate these problems, before significant investment is made.

To do this, an accelerator or incubator (1) invites aspiring companies with novel products into their program or facility, (2) provides educational and mentor support to the startup to help navigate the complicated world of starting a company, and (3) then assesses which companies are suitable for investment or acquisition, after watching the companies grow during the program. Often there is an associated fund or group of funds that will invest in exchange for equity into the top companies in incubator and accelerator programs. The intersection of incubators, accelerators, and investors is illustrated in Figure 3.3.

There are several direct benefits to being in an ecosystem that fosters and supports the growth of a startup. Some direct benefits include external validation of the product and the business model, the ability to refine the fundraising pitch, and

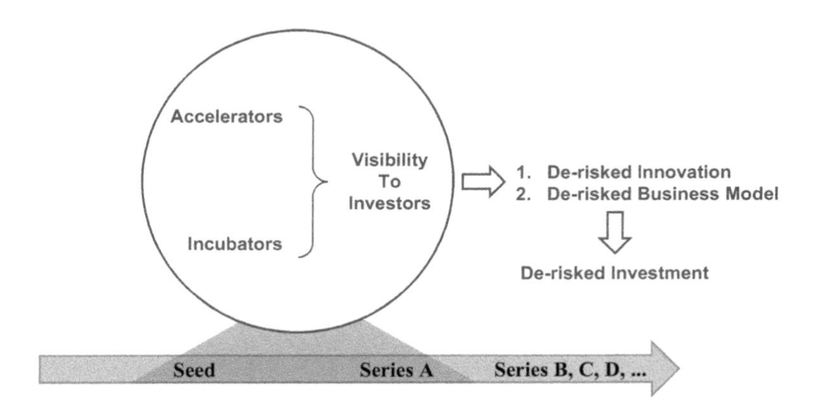

Figure 3.3: Technology ecosystem and investment.

increased visibility to investors. A number of indirect benefits exist as well, such as free marketing and promotion. As a company is getting involved in one of these programs, here are a few things to consider:

1. ***Are all of the components of the ecosystem accounted for?***

As mentioned earlier, incubators and accelerators provide different services to the startup companies that participate in them. When deciding to get involved with an incubator or startup, take into consideration if structures are in place for (a) education, (b) mentorship, (c) networking, (d) facilities, (e) access to funding, and (f) vetting of funding sources (discussed in Section 3.2). Not all programs have all of the above, but this does not mean that they are not good programs. For example, an accelerator program in a rural region may not have access to all types of funding. In this case, though the education and exposure received from the program are of value, a company would still need to reach outside of the region to find the right investment groups.

2. ***What do you want from the program?***

Participation in an incubator or accelerator program sometimes come with strings attached, so if you are considering applying to one of these programs: (1) read the fine print and (2) consider what it is that you want to get out of participating in the program. Many programs offer funding awards associated with acceptance into the program. It is important to determine if these awards are nondilutive, in-kind services, or equity. An assessment should be conducted to determine if the publicity achieved through participation in the program and the award are a fair deal and/or are in alignment with the other investments that are being made into the startup.

3.4 Establishing a Team

You have already determined that your innovation will deliver measurable value to the field, your business, and consumers. So how are you going to get to commercialization? Are you going to do it all by yourself? Unlikely! You are going to need a team of people you trust that will help you reach your business goals. Furthermore, you will need to lead your team, as the team marches forward toward success.

It is important that a company or startup assemble a team of highly motivated, dedicated people who are going to see the company through to maturity. Key personnel can make or break the success of bringing a new technology to market. Start by establishing a team of people who you think have the appropriate skills, dedication, and drive to help your business blossom into a success.

Establishing roles and positions within the company is very important. How you portray the team that forms your company is a key part of marketing your company to potential investors. The founder of a company does not necessarily need to be the CEO. In many cases, the founder is the person that developed the product or technology. It may be best for the founder to be designated as the chief science officer (CSO) or chief medical officer (CMO), while maintaining a majority ownership stake in the company. Appointing the individual who developed the technology to an important role, such as CSO or CMO, shows potential investors that people within the company have been assigned to the roles for which they are best suited.

But how do you find the right people for your team? And how do you decide that they will be a good fit for your team? At what point do you need to bring those people on? How do you incentivize and compensate your team? When it comes to making a team for your business, you are going to have to think long and hard about who you want on board and why. Start by looking at who you know. Chances are good that you already know at least some of the people who should be part of your team, and chances are also quite good that *they* know someone who might be a great fit for your team as well. You will be amazed at just how vast and varied your network of connections is, once you start trying to put those connections to good use.

Your team should have some people with relevant experience to your innovation, as well as some business experience. Legal experience could be beneficial as well, and a team that comprises members with overlapping experience can be a boon for your business. Your particular innovation might involve other considerations that may warrant having a specialist on your team too. For example, if your innovation is a medical device, you may need to be concerned with securing clearance with the Food and Drug Administration (FDA). Obtaining FDA clearance or approval is a lengthy, often complicated and stressful process, and having an FDA specialist on your team could be beneficial, as you are making marketing plans and business decisions.

Be critical when you are considering candidates for your business team. If you have experience working with someone in the past that makes you think that he or she would be a great fit, consider talking to him or her about your business idea to gauge his or her interest. Someone who might seem like a great fit won't provide you with much benefit, if he or she is not enthusiastic about the business and/or the underlying technology. Identify people you think will work well together, and if you have the ability to capitalize on synergy between two people or their skill sets, that would be ideal.

Do not feel rushed to put together a perfect team at the earliest days of your business. Start with just a few people, maybe just the founders, and grow outward from there as you come across new team members. As you talk to more and more people about your product and your business, you might be offered suggestions about individuals to contact or meet. These new contacts could prove incredibly valuable to getting your business off on the right foot. Consider all your options and be patient.

3.4.1 Leadership of the Team

A business leader, such as the CEO or President, is the "voice" and "face" of the company to the outside world – he or she champions and advocates for the company. The fate of a startup largely rests on the character, aptitude, decisiveness, and perseverance of its leaders.

Leadership is sometimes the voice of reason in a chorus of dissident team members with differing views for solving the technical and commercial challenges of the startup. Leadership needs to keep the bigger picture in mind, not get bogged down in the details, and act in the commercial and financial best interest of the startup. For example, if there is a technical fault with a product and the team is finding it difficult to determine the root cause(s) of the issue, a good leader will (1) push the team to stay on track, that is, try to adhere to the schedule, (2) not let the investigation turn into an academic project at the cost of too many resources, and (3) urge the team to pursue the most optimal (long-term) and suboptimal (short-term) solutions in parallel to keep the project and budget on track.

Leadership is responsible for increasing the value of the company, while keeping the company fiscally 'above water.' For example, startups begin with an imaginary "runway" of cash that is sourced from investors, and there is only so much cash available for the startup to use (the length of the runway). Leadership for the startup needs to make sure that runway lasts as long as possible and that valuation milestones (e.g., prototype development, end of the product design phase, verification and validation, regulatory approvals, clearances or accreditations, and commercial launch) get reached along the way. As the startup reaches these critical milestones, the startup's value grows. This is beneficial for the investors, because as more valuation milestones are reached without asking for additional funds, the larger their investment grows.

The rate at which cash is spent in a startup, that is, the "burn" rate, and the amount of cash that is available to the startup determine when the company reaches the "end" of the runway. Once that happens, leadership needs to get approval from their bosses, the Board of Directors, to raise more money, and everyone who has already invested in the startup may see their investments get diluted in value. A CEO needs to make a strong case and should present a thorough implementation strategy to the Board to gain approval. The Board of Directors, who typically are major investors in the startup or represent the major investors, can be very cognizant of and particular when it comes to how the raised funds will be spent.

3.4.2 Mentors, Advisors, and Board Members

Mentors, advisors, and board members operate in unique ways to supplement the skills on your team, facilitate useful introductions into their own professional networks of contacts, provide feedback on strategic moves, and ask the tough questions from an outsider's perspective to keep the startup company on track.

Technology accelerators, entrepreneurial co-working spaces, alumni associations, networking events, recommendations from local leaders, and prior networks of colleagues are great ways to find and get introduced to mentors, advisors, and prospective board members. The same advice for establishing a team applies to finding mentors, advisors, and board members. With any team of mentors, advisors, and board members, diversity of ideas and experiences is invaluable.

Mentors and business leaders for the startup can define a formal, time-bound, goal-driven relationship, or create a causal relationship, for example, meet informally for coffee or lunch while mentoring. In this scenario, business leaders can discuss the "heavy" matters of the present. Mentors can ask clarifying, open-ended questions and make suggestions. Importantly, mentors can share contact information for key individuals in their network who can provide value to the business. Mentors may be asked to sign a nondisclosure agreement (NDA) with the company, but this is not required. Mentors are typically volunteers that want to give back to the business community, and may or may not have invested in the company.

Advisors are also great assets to a startup company, and they are not bound to the company by legal formalities, unlike board members. Advisors are very interested in the success of the company, and may have invested personal funds in the company during early fundraising stages. Business leaders of the startup company meet with advisors to get assistance with charting new territories and defining new strategies, for example, hiring a sales force and establishing a payment and incentives structure. It is a good idea to bounce ideas off of advisors, before presenting strategies and implementation plans with the Board. Since specific details about the company may be discussed with advisors, it is prudent to establish NDAs with advisors. In the early stages of a startup company advisors are generally not compensated, but as the relationships become more formal, sometimes advisors can be compensated with a very small percentage of equity in the company.

Board of Director Members approve business decisions that drive the company in a strategic direction. The CEO reports to the Board. Board members are typically comprised of company founders who are still active with the company, formalized advisors that bring expertise to the company in the fields of technology, finance, marketing, distribution, sales, and so on, investors or representatives of investment rounds, and the CEO. With startups, it is particularly helpful to have board members who have successfully launched other startups. Board members may be compensated and/or have a significant equity stake in the business. Early-stage companies, prior to funding, do not usually have a Board. As a company grows and begins to conduct its funding rounds, the Board will be established. A board of directors is required, if the business wants to take the form of a corporation.

Good leaders will keep their Board well informed of company achievements and difficulties. Board members want their advice to be sought and heeded. The frequency for arranging formal meetings with the Board is dependent on the life cycle stage of the company, the amount of fundraising required prelaunch, or the extent of ongoing business development activities that require the Board's participation. A Board may meet monthly, every 6–8 weeks, or even quarterly.

4 Laying the Groundwork for a Business Plan and Preparing a Business Plan

Abstract: While every company needs a business plan, there is a lot of groundwork and planning that first must be done before a business plan can be created. Entrepreneurs and companies must start by preparing a number of different strategies (e.g., marketing strategy, sales strategy, expansion strategy, intellectual property (IP) strategy, and in the case of technology-based businesses, a technology strategy), and corresponding plans for each strategy (e.g., a marketing plan, sales plan, expansion plan, IP strategy, and in the case of technology-based companies, a technology plan). The highlights of each plan are ultimately captured in an overarching, high-level business plan. Simpler business models can be captured in shorter business plans, while more complicated business models often require lengthy business plans. The first portion of this chapter discusses techniques and tools that companies can use to help prepare the various strategies and plans that form the foundation of the company's business plan, while the second portion of this chapter is dedicated to preparing a business plan.

Keywords: Business Plan, marketing strategy, target market, customer segments, sales or promotion strategy, four Ps of marketing, competitive analysis, marketing plan, promotion plan, market situation analysis, SWOT analysis, value proposition, expansion strategy, intellectual property strategy and plan, technology development plan, technology life cycle, forecasting, financial models.

4.1 The Process of Developing a Business Plan – Preparation for a Business Plan

A business plan is an ever-evolving, written document that lays out, at a high level, your plans for starting and running your business – it is a written codification of your business model that details how the business will operate, how revenues will be generated, identifies who the customer base is, and lays out what the product or service is that will be sold. Every company needs a business plan, but before you can attempt to create a business you will have to do a lot of strategizing and organizing. Figure 4.1 shows the relationship between developing Marketing and Product or Technology Development Plans and the Business Plan in the Business Development Process.

Businesses have many moving parts and there is a lot to be considered about how your company will function, how you will achieve certain milestones, and how you plan on making money. You must start to lay the groundwork for your business plan by formulating a number of different strategies, such as a marketing strategy,

https://doi.org/10.1515/9783110521900-005

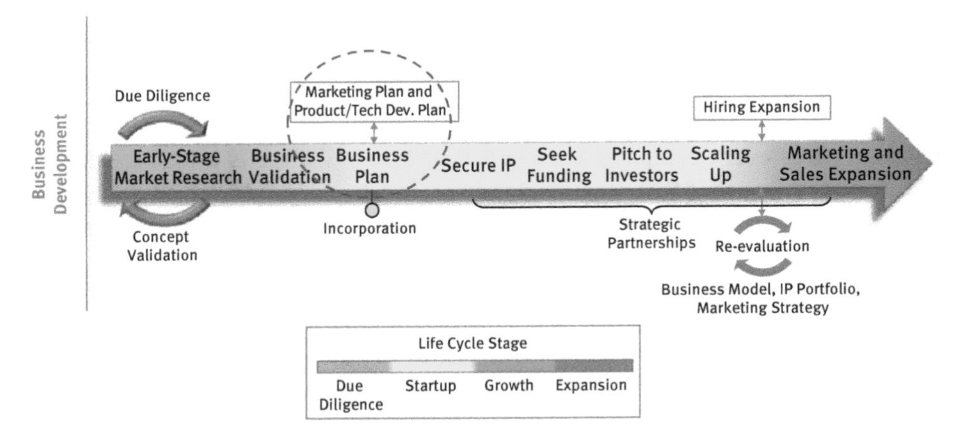

Figure 4.1: A number of Business strategies and Plans must be prepared and then captured in a Business Plan.

sales or promotion strategy, expansion strategy, intellectual property (IP) strategy, and in the case of technology-based businesses, a technology development and implementation strategy. Strategies are the ideas or plans of action that you come up with that will help you and your company achieve your goals. Once these various strategies are worked out, it is time to capture these strategies in a corresponding number of plans, such as a marketing plan, sales or promotion plan, expansion plan, IP strategy, and in the case of technology-based companies, a technology development and implementation plan. Plans are written documents that codify your strategies. The highlights of each plan are captured in an overarching, high-level business plan.

4.2 Developing a Marketing Strategy and a Sales Strategy

Marketing and sales are closely related, but they are different concepts and you must understand the difference between sales and marketing, in order to effectively develop your marketing and sales strategies. Marketing strategy focuses on marketing the product and your company to target consumers over the long-term. This involves understanding and anticipating customer needs, developing techniques for influencing customer perception of your product and company, and determining areas of focus within the market. Sales strategy, on the other hand, is focused on making the sale, that is, what is required to make sales happen. Strategizing for sales purposes is more short term, in that your focus is how to close the deal and retain customers.

At a high level, a marketing strategy needs to identify:

- Target market and customer segments.

- What content will be communicated to those segments and how.
- How customer introductions can be developed into customer relationships, which then convert to product sales and brand loyalty.
- Exceptional customer service and engagement tactics.
- A clear plan of action for achieving marketing objectives and goals (this plan of action will later be written down as the marketing plan).

On the other hand, a sales strategy needs to identify:
- Target market and customer segments.
- Target customer profile(s).
- Effective techniques that will be used to facilitate sales.
- Exceptional customer service and engagement tactics.
- How existing customers can promote and stimulate future sales.
- A clear plan of action for achieving sales and promotion objectives and goals (this plan of action will later be written down as the sales plan).

While a company's marketing strategy (and eventual marketing plan) and sales strategy (and eventual sales plan) are separate aspects of the overarching business plan, there is some overlap in the underlying concepts that go into understanding how to work through a marketing strategy, a sales strategy, and the corresponding marketing and sales plan.

4.2.1 Marketing Strategy: Using the Four Ps of Marketing to Achieve Marketing and Promotion Objectives

Marketing and sales are related and often influence each other. One example of the relationship between marketing and sales is readily apparent in a popular model used in marketing and promotion strategy and planning, known as the four Ps of marketing, also known as the marketing mix. Companies often leverage the four Ps of marketing in order to achieve marketing objectives. The four main factors, or Ps, that play a role in defining that "sweet spot" for any given product or service are Product, Price, Place, and Promotion.
- **Product.** The product is either the good your company sells or the service that your company provides. Many factors play into customer demand for your product. For instance, the particular problem that your product solves for the customer, how easy the product is to use, and the benefits offered by your product to customers all contribute to whether customers will want to buy your product.
- **Price.** Price refers to the amount of money your customers pay for the product or service that your company offers. Customers will perceive your product as having a certain value, and products that are priced in alignment with

customers' perceived value are more likely to sell. Customers do not respond well to products that are priced too high or too low.

- **Place.** Place refers to how the product gets into the hands of customers. Certain products may sell better in a particular distribution channel. For instance, a tech product, like a mobile app, does best when sold in an online app store.
- **Promotion.** Promotion is how the product is communicated to customers. Promotion is largely about advertising, but it also includes other aspects of promoting the product, like managing public relations and making special sales offers (promotion plays a significant role in sales strategy and planning).

To successfully market a product, the product has to hit a "sweet spot" of these four Ps, in order to entice customers to purchase the product. The influence that each of the four Ps has on the marketability and sale of any given product varies with the product that is being sold. That is why the four Ps are also sometimes referred to as the "marketing mix" – the product, price, placement in the market, and promotion of the product all play a role in companies achieving marketing objectives, but to different degrees, based on the specific product. In some cases, the product itself plays the biggest role in whether customers purchase the product (such as what the product does or what the product offers to customers), while the driving force behind purchases of a different product might be founded in the promotion and price of that product. A stellar product at an unjustifiable price will most likely not be successful in the market, unless it is promoted well. Similarly, a mediocre product can do incredibly well, if priced cheaply and promoted effectively. Your marketing strategy should define your company's four Ps and should explain how your company plans to use the four Ps marketing mixture to achieve your marketing objectives.

Example
A good example of a well-known company that has a well-balanced marketing mix of the four Ps is Walmart. Walmart offers a wide variety of everyday goods (Product), and low prices (Price). Walmart places its stores in locations (i.e., cities) where its customers live (Place), and advertises convenience (e.g., open 24 h), low price, and product variety through various marketing platforms (e.g., TV commercials, webpage banners, and mailers; Promotion).

4.2.2 Marketing Strategy: Conduct Competitive Analysis

Another important tool for developing a market strategy is to understand your competition. Using tools such as a strengths, weaknesses, opportunities, and threats (SWOT) analysis (discussed in Chapter 2 and Section 4.2.4.1) will identify the competitors in your industry sector. In-depth competitive analysis of your technology's industry sector identifies opportunities and threats, which can be addressed with business scenario planning and risk mitigation. Opportunities and threats may even emerge from sources other than apparent competitors: the government, policy makers and influencers, standards organizations, investors, and so on.

Competitive analysis needs to assess all types of products or technologies that are on the market or currently in development, including complementary products or technologies and alternatives, which may be preferred by customers in place of the product or technology that you are developing. Alternative products or technologies from other industry sectors may also become competitors. "Breakout" products, technologies, or market disruptors are becoming more and more common in our global ecosystem, as consumer demands for convenience and choice are shifting the way we live, work, and play. For example, the smartphone obviated the need for MP3 players (this example will be discussed further in Section 4.5.2).

The influence that existing competitors have on your identified market and customer segments cannot be overestimated. Will there be a huge switching cost or required change of behavior for customers to switch from the current solution to the product or technology that your company is developing? In other words, will customers need to learn new processes to use your product or will customers require training to use your technology? While assessing existing competitors and their offerings, document reasons why a customer should be willing to switch to your product, for instance, your technology is better, faster, cheaper, and so on.

Example

If your company is developing a diagnostic device that is used in the hospital operating room environment:

- Do existing technologies disrupt the clinical workflow for users of that existing technology, whereas your technology causes no disruptions?
- Do existing technologies increase costs for hospitals and patients, whereas your technology decreases costs, and/or completes diagnostic tasks faster and with fewer mistakes?
- Do existing technologies impose additional harms to users or patients that your technology avoids?
- Does your device provide more information than existing technologies?
- Does your device reduce the risk of patient lawsuits?
- Does your device improve quality outcomes for patients, physicians, and hospitals?

Once questions like these are answered, they can be used to translate the benefits of your product into clear, concise messages for customers, which will be important to differentiate your product from competitor products in the market.

While in development, it is helpful to chart out the commercialization roadmaps for the most likely competitors, including design, development, clinical, regulatory, and market launch plans, and estimated dates for valuation milestone completion, so that your company leadership can be more effective at strategic planning. For example, if you and another company are both developing similar medical devices, then paying close attention to your competitor's IP filings, conference presentations, research publications, and clinical trial activity will allow you to be aware of their competitive advantages, value proposition differentiators, regulatory filing strategy, and estimated market launch dates. Based on that information, as a company you may be willing to take on more risk to finalize design and development activities and expedite regulatory application submission, so that your regulatory approval and estimated market launch date is earlier. Or, you may choose as a company to stick to the current design and development plan and allow your competitor to enter the market first, to see if their device runs into any clinical or reimbursement issues and observe how the average selling price is influenced by the device's perceived clinical utility and value. Since a competitor blazed the trail for you, your management team may feel more confident about the amount of investment money that has been raised to support the project, the chosen regulatory pathway, clinical protocol design, the opportunities for reducing the device's manufacturing costs to drive greater profit margins, commercial launch planning, and so on.

4.2.3 Marketing Strategy: Technology-Based Considerations

In the case of technology companies, the marketing strategy will be dependent on the type of technology that is being developed, whether it is an (a) existing technology, (b) complementary technology, (c) new technology, or (d) breakout (market disruptive) technology. For example:

(a) If the technology that you are developing will compete with an existing technology on the market, then the marketing strategy will need to creatively highlight all the differences between your technology and the existing technology, especially the differences that consumers care about: quality, cost, performance, and so on.

(b) If the technology that you are developing will complement an existing technology on the market, then the marketing strategy should identify market growth opportunities with the fewest and/or lowest market entry barriers. For example, if your company manufactures toothbrushes, then it would be logical for your company to develop toothpaste or mouthwash products as well, because loyal customers are already familiar with your company's brand for hygiene and consumer oral health products.

(c) If the technology that you are developing is a new technology, then the marketing strategy will need to make customers "believers" of the new technology by making a case for change that customer needs are not being met, and their unmet needs are growing in scale and urgency. For example, legal and regulatory requirements change constantly, and the existing technologies on the market may no longer adhere to these requirements.

(d) If the technology that you are developing is a "breakout" technology, then the marketing strategy should leverage word of mouth recognition, acclaim from critics, and market "buzz" that is generated on social media platforms. Advocacy for the breakout technology typically originates from the beachhead market, or initial market, and early market adopters. Let your customers share the work by promoting your technology for you!

Marketing and sales plans differ, depending upon the type of technology that a company is trying to promote. For instance, a mobile app developer will have a marketing and sales strategy with a lot of online promotion for use of the app, while a medical device developer will be promoting its product through demonstrations at hospitals, medical school settings, and conferences.

4.2.4 Documenting Your Marketing Strategy and Sales Strategy in a Marketing Plan and a Sales Plan

Your marketing strategy and sales strategy should be written down in a marketing plan and sales plan. The highlights of your marketing plan and your sales plan will be included as part of your business plan.

A marketing plan is a key document detailing how your company will brand and market itself to potential consumers. The marketing plan provides a roadmap of how your company will market the product, starting with where your company is right now, and detailing your promotion plans and activities in a step-by-step fashion going forward. Marketing plans are prepared with long-term goals in mind, and it is not uncommon to see marketing strategies that have a detailed one-year plan, followed by a less detailed plan covering the steps the company plans to take in 2–5 years.

A sales plan is another key document for your business plan that captures how the company will accomplish making sales and how the company will generate more sales in the future. A sales plan details your company's sales plans and activities in a step-by-step fashion and outlines your company's sales objectives and goals.

There is no set length requirement for a marketing plan or a sales plan. For instance, a fully detailed marketing plan can be many pages long, while simpler business models might need a few pages to capture the critical points of the marketing strategy. In any case, it is a good idea to prepare a marketing plan and sales plan that are as long as your company needs them to be, in order to thoroughly cover the various aspects of your company's marketing and sales strategies. Once you have completed the full marketing plan and sales plan, the highlights can be included in your business plan.

4.2.4.1 Begin with Market Situation Analysis

Market situation analysis can be useful in developing both a marketing plan and a sales plan. As the name implies, market situation analysis looks at the relevant market and assesses your position in it, as well as the position of your competitors in the relevant market. Thorough market situation analysis involves:

- Conducting a SWOT analysis.
- Articulating your company's value proposition.
- Knowing your target market.
- Identifying your marketing and sales objectives.

Conduct a SWOT Analysis

Market situation analysis requires you to take a good, hard look at your company and identify your company's SWOT. This is a SWOT analysis. You can perform a SWOT analysis on both your company's business model and your product.

Internal factors in your company are identified as strengths or weaknesses. For instance, if your company has a great marketing team with years of experience, this would be a strength. On the other hand, being brand new to the market means that customers are unfamiliar with your business and product, which would be a weakness. Due to the internal nature of your company's strengths and weaknesses, you most likely are able to exert some control over these things and influence them over time. For strengths, you can take steps to reinforce your company's competencies, or improve upon what you already do well. For weaknesses, the good news is that you can usually take steps to change or improve upon your identified weaknesses.

External factors can be either an opportunity for your company to take advantage of, or a potential threat to your future success. By way of example, a company that produces a cutting-edge healthcare technology is poised to benefit from the fact that there is an aging baby boomer population about to enter their golden years, when health issues in this population tend to occur at higher rates, that is, there is likely to be an increase in the total addressable market. Conversely, tightening federal regulations that make producing a product more difficult or costlier would be a potential threat to a company's long-term success.

Performing a SWOT analysis is easy, but takes a lot of thought and careful consideration. Many companies use a traditional four-square SWOT analysis template to organize the SWOT that they identify as relevant to their market situation. When adding various SWOT to your SWOT analysis chart, make sure you think deeply about each item you add. A sample SWOT analysis template is provided in Figure 4.2.

Performing a SWOT analysis helps you better understand your company's market situation, and will enable you to brainstorm effective short- and long-term strategies to help your company extract the most benefit going forward. It is ideal and more efficient to devise strategies that address multiple SWOT at once, but this is not

SWOT Analysis	Help Achieve Company Objectives	Work Against Achieving Company Objectives
Internal Factors	**Strengths** Strengths are usually a source of value for a company. Examples of strengths can include having a: • Reasonably priced product • Good company reputation • Industry knowledge • Good business and customer relationships • Comprehensive IP portfolio • Location central to your target market	**Weaknesses** While weaknesses are often viewed as a problem or impediment now, they are usually things that can be improved over time. When you identify weaknesses, also identify a solution for mitigating or overcoming the weakness. Common weaknesses include: • Being new to the market (your company is unknown) • Limited resources or production capacity • Inexperience
External Factors	**Opportunities** Positive external factors that make it likely that your company will be successful are opportunities. Opportunity examples include: • Pre-existing customer base • High demand or unmet needin the market • Capitalizing on the vertical integration of your supply chain, manufacturing, etc. • Being able to take advantage of mergers and acquisitions	**Threats** Common threats include: • Changes in the law • Changes in industry regulation standards • Technology changes • Economic downturn • Reliance on third-party suppliers or distributors • Competition in the market • Social media attacks/negative promotion online.

Figure 4.2: SWOT analysis template/example.

always possible. As your company begins to implement these strategies later on, do not hesitate to reassess the SWOT analysis. Circumstances change over time – opportunities might vanish, threats might eliminate themselves from your concerns, or you might discover new strengths you did not know you previously had.

Articulate Your Value Proposition

Your marketing plan should also clearly articulate your company's value proposition. The value proposition statement is the heart and soul of your marketing plan – it is the promise of the value of the service to be delivered by the company or by the product that is purchased. The value that a company provides is based on the strength of the company, which usually takes one of three forms: (1) the company offers value to customers through operational excellence, (2) the company offers value to customers by being customer-focused, or (3) the company offers value to customers through its strength in product leadership.

1. **Operational Excellence.** Value propositions based on operational excellence need to emphasize that the company offers a high-quality product for a low cost, which is possible through the operational efficiencies of the company. Walmart is a good example of a company that has an operational excellence-based value proposition.
2. **Customer Focus.** Value propositions that are based on customer focus need to stress that the company values its relationship with its customers and treats them accordingly. Coca-Cola has a reputation for having a high level of intimacy with its customers – its products are an integral part in many customers' family occasions, and customers have intense brand loyalty toward Coca-Cola products.
3. **Product Leadership.** Value propositions that are based on product leadership should focus on how the product it offers is top in the industry. Apple has a value proposition that is focused on product leadership.

Know Your Target Market

When developing a new product or technology, it is important to have an overall strategy that addresses all aspects of bringing the product to market. Not only must there be a plan for conducting research and development and prototyping and testing, and a technology plan for how the technology will satisfy the unmet needs of consumers, but there also needs to be a cohesive marketing plan and sales plan as well. What sense is there in developing a product or technology if you have no idea how you will sell it, once it is ready for commercialization?

Deciding how best to bring the product or technology to market involves carefully considering where the new technology will fit in the marketplace, that is, knowing your target market. But how do you identify the appropriate market? Consider the following questions:

- **Where does this new product or technology generally fit in the marketplace?** Try categorizing the product or technology by giving it labels as to what it is and what it does (e.g., a mobile application, a smart device, a new information system management software, a nanotechnology application, a new semiconductive material, etc.).
- **Is there a special niche market that this product or technology fits into well?** Identify how broad or niched the potential market is for the product or technology. Consider if the product or technology could do well in multiple markets/applications. Cross-industry applications for innovative products and technologies can open up multiple markets for your company.
- **What unmet needs does the product or technology satisfy in the market?** Identify any long-felt, unmet need in the market that your product or technology can satisfy.
- **Does the product or technology solve a specific problem?** Are there other solutions to that same problem? When your product or technology solves a

specific problem that no other technology can easily solve, there is a potential market for your product or technology.
- **What does the competition look like?** After conducting some market research and due diligence, you should have a pretty good idea of who your closest competitors are. Think about how your product or technology is different from your competitors', determine whether your product or technology is better than your competitors', and articulate why.

It is important for your company to understand the specific target market you are trying to reach with your product. By identifying the market you want to reach, you can better concentrate your marketing efforts for the target market where you are most likely to be successful. Ask yourself the following questions about your target market:
1. Is your target market based on a geographical area or a particular customer group?
2. How large is the marketing opportunity you are considering by targeting these markets?

It is important to realize that of the total target market you identify, only a portion is realistically reachable/addressable.

Next, identify the target customers by conducting a customer analysis. You should group the customers into customer segments (Chapter 4). To do this, first identify the type of customer in each segment and think about what would drive them to make a purchase of your product, and what might drive them away from purchasing your product. What is the population, demographics, and education level of each target customer segment? How much does the typical person in each customer segment earn? What do they care about?

Go a step further and imagine putting a face to each customer segment by creating customer profiles. Create a representative customer for each customer segment. Give him or her a name and a personality. Explore the representative customer's likes and dislikes, as they pertain to your product.

Example

Vame (a fictional company) is a personal computer gaming platform where users can download video game content created by other users of the platform, that is, independent video games, or indie games. Membership is free, but downloading a game costs between $1.99 and $5.99. The Vame platform allows indie game developers to test mini or beta versions of games they are developing, while players get to see and experience early-stage game content.

The target market for Vame is young to middle-aged males, with a small market of similarly aged female users, who are interested in video games and particularly indie games. Vame's target market is not limited by geographical region because the platform is online, so anyone anywhere in the world can access the platform to download and play game content featured on Vame's platform. Vame understands its target market as largely comprising two customer segments:

- *Content Producers*. They are Vame platform users who want to test beta versions of indie game content they are developing. Uploading their games to Vame's platform provides content producers with multiple benefits. First, content producers can get feedback on their games from players, which can allow the content producers to work out any bugs in the coding for later versions of the game. Second, content producers earn some pay for posting their content to Vame's platform, since Vame and content producers share the profits from downloads.
- *Players*. They are Vame platform users who want to play new, original content and do not mind paying relatively cheap prices to do so. Players generally play video games as a hobby and have a long history of playing all kinds of video games. Players like Vame's platform because while some of the games use familiar and well-known game mechanics, Vame content producers like to try and introduce new elements into game play that have never been done before, which translates to a whole new video gaming experience for players.

Vame is doing quite well in the gaming community. In developing the business around the platform, Vame took the time to carefully figure out who is its exact target market. At the early stages of developing the Vame platform, Vame looked at its two main customer segments and tried to imagine a customer to represent each customer segment.

For the content producer, Vame came up with Matt. Matt is a "tech guy" – he's been interested in technology, the internet, gaming, coding, and so on, since he was young. Now that he is older, he has a job in information technology (IT), but is still just as passionate as ever about gaming. For Matt, gaming is a hobby that he has relied on as a source of fun, camaraderie, and stress relief for practically half his lifetime. Matt is drawn to the Vame platform because he likes to code for fun in his free time. He likes the challenge of making an indie game that other people want to play – it gives him a sense of pride. He likes that he can make money from his game content, if he uploads it for sale on the Vame platform. Matt also likes the ability to have other people try out his game and the chance to get feedback from players, because players will always try to do things Matt never considered doing, when he was developing the game. Player feedback gives Matt new ideas for future game content. Matt does not like having to pay a subscription fee to have someone host his game content. Matt would rather have hosting fees taken out of any earnings he makes from downloads of his content.

Regarding the representative customer for the player segment, Vame created Jeff. Jeff loves video games – they are a long-time hobby that he loves dearly. Jeff has a dedicated game play station in his home where he has handpicked his favorite hardware (e.g., keyboard, controller, headset, etc.) that he uses to play his games. Jeff has always been fascinated by video games. Specifically, Jeff likes how he can play as a character who has abilities that he does not have in real life – Jeff likes the fantasy or role playing aspect of playing video games. Jeff has played all of the major titles, and likes to explore the indie game space from time to time to see what is new. Jeff does not like having to pay a monthly subscription fee to access a gaming platform. Jeff would much rather "pay-to-play."

Vame worked very hard on identifying each target market customer segment, and prepared a value proposition statement for each group.

Value Proposition for Content Producers

Want to test your indie game but need access to players? Upload your indie game content on the Vame Gaming Platform. Vame makes your content available to players who can "pay-to-play" your game content with a download. While Vame gets a small hosting fee, the profits from the downloads go directly to you. Get feedback on your game content from hundreds of players.

Value Proposition for Players
Vame Gaming Platform lets players demo new indie game content. Check out cool new game mechanics, sample some original titles, and test out new mind-bending physics engines. Whether you are into Role Playing Games, adventure style, or puzzle solving, you will find it all on the Vame Gaming Platform. Accessing the Vame platform is free – no monthly subscription fees. Games are "pay-to-play" and cost less than an energy drink and a bag of chips. Only download content that interests you.

Another important consideration your company should think about as you try to understand your market is whether there are any other applicable market drivers (i.e., factors that influence the market) or risks (i.e., potential threats to the success of your business). There are a number of external factors that might impact your target market. For instance, there could be political, religious, or cultural considerations you need to be aware of concerning the product you are trying to sell, or concerning the target customer market you intend to sell to. There could also be social market drivers, or technological market drivers that could impact your success in the market as well. Identification of market drivers allows you to strategize ways to maximize your marketing efforts by capitalizing on the market drivers that are applicable to your business.

On the flip side of this analysis, you need to look at your product, your company, and your target market under a microscope to evaluate whether there are any major risks to your success in the target market. If there are risks, you should also identify ways to mitigate the risks you face in the market. It would be detrimental to your business, if you inadvertently were to offend or alienate your target market. For instance, culturally insensitive advertising, or a product that is misaligned with the target market's political ideals, could be devastating to your bottom line. Take the time to deeply explore the various cultural, political, religious, and social aspects of your target market.

Understanding your market can make or break your business. Marketing that backfires could severely impact your ability to enter a new market, to enter a new geographical region, or to introduce a new product or service. Having a thorough understanding of your target market will help you and your company make the biggest impact possible.

Identifying Your Marketing and Sales Objectives

If you are going to take the time to devise a plan for your marketing efforts, shouldn't you have some marketing goals to strive for? Absolutely! A portion of your marketing plan should be dedicated to identifying your company's marketing objectives and setting a timeline for achieving those goals. Most companies have similar marketing objectives, such as:

- Encouraging potential customers to consider purchasing the product.
- Raising brand awareness among the target customer base.

- Generating new leads and potential new customers.
- Making and increasing sales.

Companies only have so much time and a set amount of resources to dedicate to marketing, so it is wise to try and maximize the return on any marketing investment. This is why companies typically limit their marketing objectives to only a handful of the most important and realistic ones. Companies often also limit the scope of their marketing objectives to one year at a time. By being selective and limiting the timeframe for reaching the marketing objectives, companies can focus their marketing resources into promotional efforts that are most likely to produce the greatest result in a reasonable amount of time.

Companies rarely attempt to take on all of their marketing objectives at the same time. It makes more sense, and is often more practical, to take steps toward achieving marketing objectives in phases. This is particularly true for startup technology companies that have limited resources. Marketing objectives should be quantified and should have a set timeframe in which they are to be achieved. They should also feature the specific marketing tactic that will be used.

Example

Restivas is a small technology startup. The company developed a wearable fitness tracker device with online tracking tools that has the potential to be very popular with customers who are athletic and health conscious. Restivas wants to sell not only the wearable fitness tracker, but also downloadable content through a web-based platform and a mobile app. Restivas's marketing objectives (in order of ease of implementation) for the upcoming year include:

- **Getting 50 potential new customers per month to consider purchasing the product.** Restivas's marketing plan aims to achieve this goal through a promotional offer for a free-trial version of its mobile app and web-based platform to new customers who purchase Restivas's wearable fitness tracker. A free trial does not cost Restivas anything, and allows potential customers to sample the product risk free.
- **Generating 25 new leads and potential new customers per month.** Restivas has entered into arrangements with several local gyms and health food stores over the upcoming year to conduct several "demo days," where Restivas sales representatives will demonstrate the wearable fitness tracker and the mobile app to prospective customers.
- **Raising brand awareness by 5% per quarter among the target customer base.** Restivas has created a radio advertisement that will be aired on local radio stations. Restivas has paid for strategic advertising time slots (e.g., (1) at the start of the new year when people have New Year's resolutions to work out more, (2) heavier advertising during the spring, when the weather warms and people want to be more physical outdoors, and (3) during the weekday 4–6 p.m. rush hour, when the target customer is stuck in traffic daydreaming about being at the gym instead).
- **Restivas is striving for a 7% increase in sales for the year.** Restivas has conducted analysis and believes that approximately 7% of potential customers who are introduced to the product will be converted into sales. This customer conversion rate translates to a 7% increase in sales for Restivas for the year.

Tailor your marketing objectives to focus your promotional efforts on marketing strategies that will yield the most fruit for the effort that you put into them. You want to set marketing goals that require your company to reach for them, but at the same time, those goals should be realistic.

4.3 Expansion Strategy: Planning for Entry into Future Markets

Sometimes it makes sense to include the anticipated path of entry into future markets within a marketing plan. While initial marketing efforts will be focused on winning over the initial target market (also known as the focus market or beachhead market, which is discussed in more detail in Chapter 7), careful forward-looking thought should be given to expansion of the company and expansion into new markets.

Example of a Market Expansion Plan

Consider the following example of a milk additive product produced by a small startup called LactLuster. At first, LactLuster's target or beachhead market may be dairy farmers located in a geographically limited region near the company's headquarters where the additive is manufactured. For instance, if LactLuster is headquartered in Pennsylvania (USA), then LactLuster's initial market may be limited to dairy farmers in the state of Pennsylvania. The initial, or beachhead, market is represented by the innermost circle on the diagram in Figure 4.3.

Once the milk additive product has successfully launched in Pennsylvania, the next step in LactLuster's marketing plan is to expand regionally into the dairy communities in the nearby states of Ohio, West Virginia, New York, and New Jersey. This marketing plan expansion step is represented by the next largest circle on the diagram in Figure 4.3.

Figure 4.3: Example market expansion plan.

From here, LactLuster plans to expand further to the total addressable market in the United States, that is, by marketing nationally to all dairy farmers in the US. This marketing plan expansion step is represented by the next largest circle on the diagram in Figure 4.3.

Finally, LactLuster's last phase of market expansion is to introduce its milk additive product into other milk- and dairy-related products, such as powdered milk, cream products, cheeses, and so on. This marketing plan expansion step is represented by the largest circle on the diagram in Figure 4.3.

4.4 IP Strategy and Plan

Another important piece of groundwork for a business plan involves formulating the company's IP strategy and corresponding IP plan. Most technology companies are teeming with IP, ranging from the company name, logo, and website design to the underlying invention behind the product or technology that the company sells. An IP strategy needs to identify the company's major IP assets, focus on what forms of IP (e.g., patents, copyrights, trademarks, trade secrets) will be most beneficial for protecting the company's IP assets, how the company will get the IP it needs, and what will be done with the IP once it is secured (e.g., monetized through licensing agreements, leveraged in litigation, etc.).

The IP plan will focus on when and how the necessary IP rights will be procured. Some forms of IP require registration, while others must be acquired through an application process. In some cases, a company might be using IP that belongs to another company and may need to secure a licensing agreement in order to use the IP without risk of infringement. Obtaining IP rights takes time, and certain actions must be taken by the company to secure IP rights within specific time frames, or else legal protections could be jeopardized or lost. That is why it is so important to have an IP plan. Guidance on how to develop an IP strategy and document an IP plan can be found in Chapter 5.

4.5 Technology Plan

Developing an innovative technology or product requires extensive strategizing and planning. A strategic technology plan – not to be confused with the Technology Development Plan (discussed in Chapter 10) – is an overarching plan that identifies at a high level the goals and objectives for the technology, for example, how the technology will be used, how many consumers will need the technology (e.g., the quantity of units that need to be produced, whether scaling up will be needed, etc.), and pairs these goals and objectives with a technique, tactic, or plan for how to achieve them (e.g., a promotion plan for getting first-generation products into the hands of influential consumers, a marketing technique that will be used to tap into a desired target customer base, a branding strategy to get consumers interested and introduced to the product, etc.).

Technology, as it is developed, goes through a life cycle known as the Technology Life Cycle. Your technology plan should highlight each stage of the technology's life cycle and identify what steps the company plans to take at each stage. The stages, which are discussed in Section 4.5.1, include Concept, Design, Development, Commercialization, Continuous Improvement, and End of Life.

Next, consider whether the product or technology is cutting-edge. Innovation is a spectrum that includes cutting-edge innovation and bleeding-edge innovation. Technologies and products that are at the cutting-edge of innovation are often easily and readily adopted by consumers, because the innovation is based on a technology that has been around for a while, even if it has not been marketed to consumers en masse yet. Since the technology is new, but not completely unfamiliar, consumers are likely to be intrigued by cutting-edge innovations, especially when it is easy to teach users how to use the technology. Technologies and products that are at the cutting-edge of innovation tend to be lower cost to develop and produce, compared to technologies that are at the bleeding-edge of innovation.

Bleeding-edge innovations are more difficult to get consumers to accept, because they are so new or difficult to understand that consumers often turn away from them – consumers are often reluctant to be early adopters of bleeding-edge innovations. This is because bleeding-edge innovations often are untested or have many kinks in how they operate or function that need to be worked out more carefully, before consumers will become interested in them. Another way to think of bleeding-edge technology is that it is technology that is too advanced for everyday consumers to use effectively or in a way that is valuable to them. Highly specialized consumers, however, may be interested in bleeding-edge technology for research and development purposes, but the market for bleeding-edge technology is often small, if one exists at all.

It is also important to include technology forecasting in your technology plan. Technology forecasting helps assess whether the technology has a place in the market by predicting market demand. Poor technology forecasting can result in bringing a technology to market that misses the mark.

4.5.1 Technology Life Cycle

The duration of a technology life cycle is dependent on the type of product, process, or technology that is being developed, its complexity, and its scale. Despite this variation in duration, the phases or stages of a technology life cycle are consistent at a high level. The stages of a technology life cycle, as shown in Figure 4.4, include:
- Concept
- Design

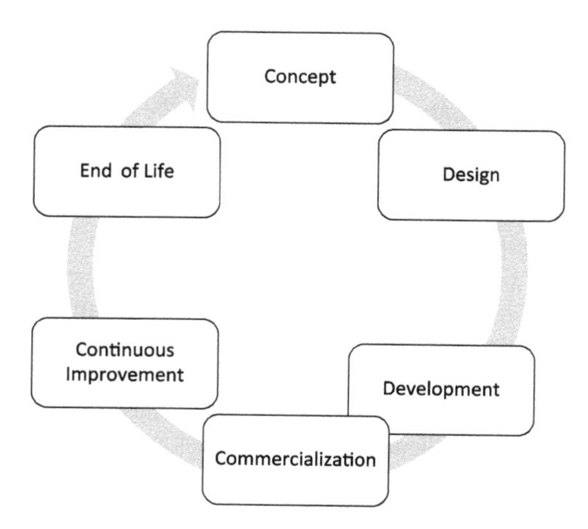

Figure 4.4: Technology life cycle stages.

- Development
- Commercialization
- Continuous Improvement
- End of Life

Information learned in each stage is used as feedback, while project teams develop the next iteration of the product, process, or technology. To progress from one stage to another, project teams hold Design Review meetings (Design Reviews are discussed more in Chapter 10) in which the outputs and milestones achieved during the current life cycle stage are documented and presented to a cross-functional team comprised of management, engineering, operations, regulatory and quality roles, and an independent reviewer. This team reviews the evidence and quality of the work accomplished to date and determines whether the project is ready to move to the next life cycle stage and the company is ready to expend additional resources, or if additional deliverables are needed to meet the requirements of the current life cycle stage. Technology Design Reviews are an important aspect of maintaining project design controls, which are discussed more in Chapter 10.

The *concept life cycle stage* comprises concept origination (or ideation), concept refinement, and concept selection, and is the birthplace of a new product or technology. The various aspects of the concept life cycle stage are discussed in more detail in Chapter 9. This life cycle stage relies on concepts (or the underlying ideas for what will become new products or technologies) being generated from a good understanding of fundamental scientific and engineering principles. The goal of the concept stage is to agree on a single concept or group of concepts to progress into the design stage. Concept ideation and refinement can be achieved a number of ways, including via:

- Market research and market needs assessment (refer to Chapters 1 and 9).
- Team brainstorming (brainstorming techniques are discussed in Chapter 8).
- Stakeholder observation (stakeholder activities are discussed in Chapter 2).
- Key Opinion Leader interviews (refer to Chapter 2).
- Utilizing sketching, storyboarding, and process mapping techniques to explore various aspects about each concept (refer to Chapter 8 for more information about storyboarding).

Raw ideas and concepts are refined by defining decision or selection criteria and then applying these criteria to the concepts that were created during concept ideation. To select a single concept, additional subject matter expertise may be sought and simple benchtop tests may be performed to increase the project team's confidence in the decision. During this life cycle stage, project, design, and process risks should be explored and documented.

The *design life cycle stage* comprises defining the requirements, building, and testing the proof-of-concept (POC) prototype (product or process), and finalizing requirements and specifications. POC is discussed in more detail in Chapter 9. The goal of the design stage is to de-risk the technical design of the product or process through iterations of prototype development and refinement of user requirements (design inputs) and technical requirements (design outputs). Design inputs and outputs are discussed more in Chapters 9 and 10. To decrease the complexity of the development project, user requirements need to be analyzed for whether they are mandatory (must have requirements) or nonmandatory (should have requirements). This way, if there is a technical challenge or burden that cannot be overcome, and the technical challenge is associated with a nonmandatory requirement, then it is not a project showstopper, that is, it will not kill or end the project. To ensure the technical requirements and specifications are accurate, benchmark testing on similar products or processes is conducted during this life cycle stage as well. During this life cycle stage, project, design, and process risks should be explored further and mitigated.

The *development life cycle stage* comprises building and testing of product or modeling-up of the process to ensure performance and safety, gathering evidence to prove the product or process is compliant with applicable quality and regulatory standards and laws, and mitigating outstanding design and process risks. Various aspects of the development life cycle stage are discussed more in Chapter 10. The goals of this stage are to complete the design of the product or process (sometimes referred to as Design Freeze) and complete verification and validation testing to ensure design inputs (user requirements) are being met through the testing of design outputs (technical requirements). Verification and validation testing are discussed in Chapter 10.

The *commercialization life cycle stage* comprises preparation for scale-up or ramp-up of production, finalizing supply chain procedures and agreements, and implementing marketing launch campaigns (Chapter 7 provides more information about strategic implementation of a business plan for getting to commercial launch and Chapter 10 covers more information about scale-up of production/manufacturing). The goals of this stage are to ensure that there is enough supply or product or process to meet customer demands, and that the product or process is successfully penetrating the marketing segments via marketing channels that were identified in the marketing strategy and plan.

Prior to marketing launch, manufacturing preparations include establishing process quality controls, and verifying that equipment and resources are ready and operating at sufficient manufacturing capacity to meet projected customer demands. Scale-up or ramp-up activities may include the following:
- Streamlining manufacturing workflows.
- Investing in manufacturing automated technologies.
- Reducing Cost of Goods Sold (COGS) by reducing the time to manufacture or the labor hours required for manufacturing.
- Expanding manufacturing sites.
- Expanding manufacturing operations.

Regarding supply chain preparation, due diligence is required to select distributors and importers that will form key partnerships in target markets that will help your company define a commercialization strategy. These partners are familiar with the market needs, competitors, compliance requirements, product registrations, and marketing authorization processes, so do not overlook their knowledge and contributions.

The *continuous improvement life cycle stage* comprises continued marketing and customer surveillance (which is discussed further in Chapter 8), sustainable engineering and operational performance improvements, maintaining quality, and eliminating the impacts and effects of commercialization issues (e.g., production inefficiencies, product or process failures, etc.). The goal of this stage is to meet changing market demands by being flexible and ready to capitalize on new opportunities, for example, expanding customer bases, launching into new markets, marketing new ways of using the product or process, and finding ways to increase profit margins. A company cannot survive by utilizing just one way to use a single product or process.

For this stage, continuous improvement looks to improve all areas of operation, including the placing of manufacturing orders to meet customer demands, the scheduling of supply chain activities, customer service processes, including product or process installation and servicing, returns and complaint handling processes, postmarket surveillance, and so on.

The *end of life* stage comprises a gradual loss of competitive advantage, repurposing of existing infrastructure, and transitioning of functions (e.g., manufacturing production), work processes, and property to prepare for the next technology project. The goal of this life cycle stage is to transition in a way that does not compromise customer loyalty or create excessive waste, while maintaining quality and regulatory compliance (if applicable), and protecting IP.

In this stage, a loss of competitive advantage may indicate the following:

- Your product or process does not have a distinguishing competitive advantage (with regard to competitors or your own company's technology portfolio), and therefore, market share is diminishing.
- Your product or process no longer provides an advantage over your competition (e.g., cost to produce your product or technology is higher than the competition and your profit margins are shrinking).
- Your product or process utilizes a technology that is viewed by the industry as a commodity.
- Your company decides that it is cost prohibitive to keep the technology in-house and seeks to license the technology to competitors.

The stages of a technology life cycle should be discussed at a high level within the technology plan portion of the business plan. Additional detail for each stage can be discussed within the Product or Technology Development Plan (discussed more in Chapter 10).

4.5.2 When Cutting Edge Innovation Satisfies Unmet Consumer Need, Commercial Success Isn't Far Behind

Generally speaking, consumers are inclined to adopt cutting-edge innovations and are resistant to adopting bleeding-edge innovations, because they are untested and carry too much risk. More technologically advanced consumers, or specialized consumers, may be capable of understanding and using technology that is at the bleeding-edge of innovation, but these consumers are limited in number. An innovative technology that is too advanced will fail, because consumers cannot put the technology to use in a way that benefits them. When an innovative technology or product is at the cutting-edge of innovation and addresses unmet needs of consumers, it is a promising combination.

Case Study

Consider a real-world example of an innovative product that both satisfied an unmet need (that consumers did not even know they had!) and was cutting-edge at the time of its introduction to the market. One of the most innovative products that enjoyed immense commercial success at the start of the twenty-first century was a portable music/media player that practically everyone is familiar with now – the iPod.

The iPod was introduced in October of 2001. At the time, the potential market of consumers for this innovative portable music player was vast – literally anyone with ears and a love of music could be a potential customer. And after the commercial successes of the portable cassette player in the 80's and the portable CD player in the 1990s, the iPod was bound to have a huge potential market in the 2000s and beyond.

There is no disputing that the iPod was cutting-edge, at the time. When Apple introduced the iPod, consumers had never seen anything like it. The device itself was small and portable with a sleek design and shape. The controls were futuristic, but intuitive after a small amount of experimentation, which felt to the consumer more like discovery. The earbuds (a novel concept in their own right) were small and inconspicuous, compared to headphones commonly used with portable tapes and CD players, and had easily accessible volume controls on the cord (in later versions). Furthermore, the consumer was empowered to pick and choose what songs would go into the iPod's music library, making the device highly customizable to the consumer's musical tastes.

Every technological aspect of the iPod was carefully strategized based on consumers' unmet needs, despite consumers not even knowing they had unmet needs. Up until the introduction of the iPod, consumers had gotten by well enough using portable tape and CD players. While these older technologies had shortcomings, consumers still managed to use these devices. For instance, a tape or CD player was often a bulky and awkward device that was only capable of playing a handful of songs, making the variety of music fairly limited, unless the consumer also carried around additional tape cartridges or CDs. Cassette tapes were a dying technology at the turn of the century anyway, and portable CD players tended to skip, if the player suffered even the slightest impact or bump.

The iPod was specifically designed to overcome all the problems portable tape and CD players posed to consumers and to capture the consumer's imagination with a novel space-aged design. For instance:

- Unlike bulky and awkward portable music players of the past, iPods were small and could easily fit in the consumer's pocket.
- iPods were mini-computers that played songs that were in MP3 format, eliminating the consumer's need to carry around extra CDs or tape cartridges. Consumers could listen to music without the song skipping a beat, even if the iPods were used while jogging.
- The MP3 format for music storage allowed for up to 1,000 songs to be stored in the iPod's 5 GB of memory, increasing the consumers' nearly instant access to much more of the music they preferred than conventional portable music devices.
- The earbuds were small, tight-fitting earphones that settled into the ear cavity, making them inconspicuous and comfortable for most consumers.
- Volume controls were incorporated into the cord of the earbuds in later iterations, so that consumers did not have to remove the iPod from their pocket to control the volume. Volume controls on traditional portable music players were on the body of the player.

But Apple did not stop there. Apple went on to create multiple iterations of the iPod: Multiple generations of the Classic model, the iPod Mini, the iPod Nano, the iPod Shuffle, and the iPod Touch. Each iteration offered more storage space and new features.

Then Apple opted to enter the smartphone market by introducing the iPhone. When Apple developed the iPhone, it saw an opportunity to integrate its successful iPod technology into a mobile phone device. The end result was a smartphone and personal music player combo device that was so successful that at the time of this writing, Apple has produced more than a dozen different iPhone models.

Strategy is the key to success, when developing an innovative technology or product. Apple had a specific technology strategy in mind, while developing the iPod portable media player, that specifically focused on extracting as much value out of the innovative technology as possible for the consumer's benefit by addressing the unmet needs of the consumer. The iPod's massive success as an innovative technology was meticulously planned, and the technology strategy for the iPod has been continuously updated and revamped over the years, in order to keep the iPod (now iPhone) on top as one of the most successful portable music players of all time.

4.5.3 The Importance of Technology Forecasting

It is also prudent to conduct technology forecasting to help determine whether the technology has a place in the market. Technology forecasting involves predicting market demand for a technological innovation. A new technology must bring value to the marketplace, or no one will be interested in buying it.

Case Study

For more than 100 years, photographers counted on Kodak as the supplier of photo film and as a provider of photo development services. However, when digital photography technology started to emerge, Kodak was not an early adopter of the new digital technology, since consumers traditionally liked physical photographs that could be held, touched, and framed. Kodak's technology forecasting regarding digital photography as a fad, or commercial novelty that would wear off over time, was inaccurate. Kodak had the potential to be a major player in the digital photography sector, but poor technology forecasting made Kodak arrive late to the game.

Taking into consideration the performance and technical details of the technology, the economic characteristics of the technology, and how many potential – yet realistic – applications the technology can be used for, a technology forecast can be made as to whether your innovative technology is ripe for the market.

Case Study

Consider technology forecasting for Schoeller Textil AG's NanoSphere technology. NanoSphere is a nanoscale textile finishing technology that utilizes nanoscale properties to keep fabrics and materials clean and dry. Textile surfaces incorporating the nanoscale finishing technology are made to permanently repel water, stains, and dirt through the application of a nanostructure surface on the fabric. Dirt, water, and debris adheres only to a reduced area of the fabric surface, due to the nanostructured surface of the fabric.

In addition to being useful in the home and commercial textile industries, this innovative nanotechnology-based textile finishing could have a number of other applications in different markets. For instance, this technology could be useful in:

- Carpet and upholstery applications.
- The development of pet products.
- Luggage, bags, and other accessories applications.

Furthermore, if the nanotechnology could be adapted for use on other materials besides fabrics (e.g., wood surfaces, plastic surfaces, natural stone surfaces, etc.), more potential markets become accessible. Be creative when you are looking for potential cross-industry applications for your innovative technology.

It is possible that the technology is forecasted as being on the bleeding-edge of innovation, that is, the technology is too advanced to appeal to the market. Rather than be discouraged by this finding, consider making an opportunity out of the situation. For instance, developing a beta version of the technology or product with less than optimum specifications could make for a marketable product that consumers will be interested in presently. This leaves the optimized version of your technology or product for future use and commercialization and gives you more time to strategize about how to bring the optimized version to market.

4.5.4 Developing a Technology Portfolio

Many technological developments are not isolated to a single technology area – most technological advancements involve technologies from multiple fields. In other words, most technological innovations involve a suite of related technologies, and the technology in development needs to be compatible with and demonstrate interoperability with each enabling technology.

Companies involved in technological innovation need to develop a technology portfolio around their innovative product. Developing a technology portfolio helps your company understand the various aspects of technology involved in making and using your innovative product.

Case Study

A digital assistant for use in the home, like the Amazon Echo, Apple HomePod, or Google Home, involves a number of different but related technologies. For starters, there is the physical hardware. The main speaker/station for the home is formed from hardware such as circuitry, a housing, speakers, a microphone, just to name a few critical physical components that are key to the digital assistant's operation. Additionally, the main speaker/station must have internet connectivity and Wi-Fi capabilities. Next, there are a number of smaller satellite devices (e.g., Echo Dots, Google Home minis, etc.) that can be placed in different rooms throughout the home. These satellite devices are physical hardware as well, but they also involve relay technology to communicate with their home base.

These digital home assistant devices also utilize loads of software technologies. For instance, Alexa is the voice-activated software that controls the Amazon Echo. Users can access additional Skills for Alexa (i.e., additional software programs) through a mobile application (e.g., software delivery is facilitated through a mobile app platform).

In any given digital assistant device, there are at least several different technologies that work together to create the final product. The hardware of the main home station/speaker is one, and the satellite devices that are scattered about the home are another. How the home station/speaker communicates with the smaller satellite devices is a third technology, while the voice-activated software that controls these devices is a fourth. Downloading new capabilities in the form of software is a fifth technology, while being able to access new software through a mobile platform is a sixth type of technology. And these are just the technologies that are readily apparent on the surface – there is no doubt that additional technologies exist in these digital assistant devices, such as digital security technology, data logging technology, and so on.

One aspect of developing a technology portfolio is to develop a corresponding IP portfolio (Chapter 5) around the technology portfolio. An IP portfolio is a collection of IP rights (e.g., patents, software licenses, licenses to use technology not owned by the company, trademarks, copyrights, etc.) related to the various technologies that form the backbone of your innovative product. IP portfolios are useful for companies because an IP portfolio protects the aggregate of related IP rights associated with the technology, and can be used offensively to stop competitors from infringing on rights that you hold, or can be used defensively to scare away potential copy cats and competitors.

Once you have developed a technology portfolio, it is important to take the time to carefully prepare a corresponding IP portfolio that documents the rights and licenses that your company holds to make your innovative product possible.

4.5.5 Developing a Realistic Timeline

Realistic development timeline planning is mainly discussed in Chapter 8, but there are also several key points to keep in mind at a high-level regarding technology development timelines and longer term (next generation) development and commercialization plans that are associated with the strategic implementation plan (SIP), which is discussed in Chapter 7, and the Technology Development SIP, which is discussed in Chapter 10.

A technology development project is initiated to solve a problem, and sometimes this problem is so complex that it may take 5, 10, or even 20 years to solve, so the company and project team need to prepare themselves for the short, medium, and long term to solve the problem. Project timelines are dependent on the type of technology being developed and the industry that governs how the technology is implemented and used, for example, to solve a problem in immunology may require 10–15 years, to solve a problem in spaceflight may require more than 30 years.

At this early stage of business formation planning, technology development timelines should be high level, meaning that tasks should be organized by quarters, not by months or weeks. For example, an example of a high-level technology development timeline can be found in Figure 4.5.

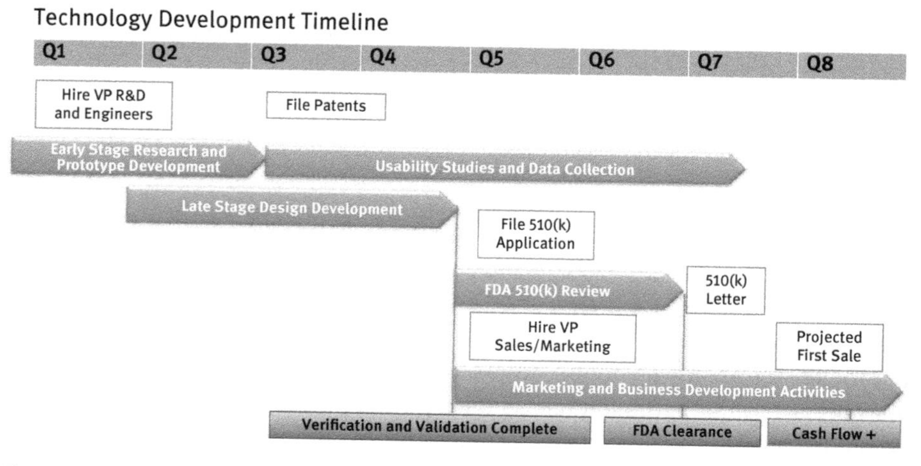

Figure 4.5: High-level technology development timeline.

A technology development timeline will also depend on the number of tasks and deliverables that are outsourced. If many outside parties are required to complete the project, then additional time will be needed for project communications, risk mitigation, to establish quality and service level agreements, deliver training, and to complete comprehensive knowledge transfer between internal experts and external resources, and so on.

Companies need to assess where they want to be in "x" years, and this vision should both align with the company's mission and also drive planning efforts for the technology development pipeline. If the ultimate goal is to develop revolutionary, ground-breaking technology, then the company needs to establish a project team with innovate capabilities and an organizational structure that supports such creativity. To ensure the company is bringing in revenue to support itself, the company will also need to establish teams and processes to research, develop, and launch technologies in the short and medium term.

For a technology pipeline to thrive, there should be an appropriate balance of high-risk, high-reward projects with low risk, low return on investment projects. Achieving this balance does not mean the company should develop at the very extremes; the probability of success with each project should not be 1%, and it will probably will never be 100%. Leadership should select and initiate technology development projects that are 1) supported by the core competencies of the company and the project team, and 2) align with the technology portfolio. The Technology Development Plan timeline should be summarized and included in the business plan as part of the technology plan.

4.5.6 Writing a Technology Plan

Technology strategies are often integrated with business strategies, as the two often go hand in hand. When strategizing about how to bring an innovative, new technology to market, a good place to start is to look at the potential value that the technology will have in the target market.

The path to commercial success for an innovative new technology or product is paved by consumer demand. If there is no demand for the innovative product, or a company is incapable of creating consumer demand, then the product will ultimately fail commercially. But how can you gauge the market potential of a new, innovative product or technology, if it is so novel and cutting-edge that there is little to compare it to? Careful consideration needs to be given, when assessing if an innovation is worth developing into a commercial product.

Two very important questions are fundamental in gauging the potential value of an innovative technology or product:
1. Does the technology satisfy an unmet need in the market?
2. Is the innovation a cutting-edge innovation?

When there is an unmet need in the marketplace (unmet needs are discussed in depth in Chapter 9), consumer demand generally already exists, or will exist shortly after the consuming public is introduced to the innovative product. To gauge the unmet need, consider what the customer base is and what their unmet needs are. It can be challenging to assign a value or weight to the unmet needs of the consumer.

Brainstorm about who the potential customers are who will benefit from the innovative technology or product. Be creative and cast a wide net, in order to capture the largest customer base possible. Categorize the potential customers into groups (e.g., four to six groups) based on characteristics that the customers of each group share. Using the point of view of a customer belonging to each of the categories you identified, think about the wants and needs for these categories of customers and think about how your technology or product addresses each customer group's wants and needs. Don't be afraid to consider tweaking your technology plan to better satisfy the unmet needs of the customers you have identified. If it is possible to adjust your innovative technology or product slightly to be more applicable to a larger consumer base, or to better address the wants and needs of a particular customer group, this is something that you should seriously consider doing.

It is important to articulate how you are going to develop your technology in a detailed technology development SIP, which is a detailed technical plan for how you will develop your technology, discussed more in Chapter 10. However, the technology plan portion of the business plan should include an executive summary version of the technology development SIP, which is meant to serve as a roadmap for how you are going to go from the Concept stage to the Design and Development stages, to Commercialization, through Continuous Improvement, and to the End of Life stage in the technology life cycle.

For instance, if you developed a new patient compliance monitoring application for smartphones and mobile devices that uses a unique wearable technology piece to track the amount of exercise and movement that the patient performs in a day, the app and the wearable are the two main products you have to offer. At a high level, your business model might comprise:

- A doctor prescribes a treatment that involves the patient engaging in a certain amount of physical activity daily.
- Patients download your convenient app and order your wearable technology as a way to track their compliance with their treatment regimen.
- You supply up-to-date versions of the app software for free, once consumers purchase the wearable.

From this scenario, it is clear that the app and the wearable technology are at the core of the business. These are the two main aspects of the business that require the most energy and yield the most revenue. Thus, the role of the technology is very important to the business.

Once you understand the role that your technology will play in your business, you can prepare a mission statement and vision statement for the development of your technology. A mission statement and vision statement for the company that is developing the patient compliance monitoring technology discussed above are featured as follows:

Mission Statement

Our mission is to promote and improve patient compliance of physical therapy and exercise regimens that are prescribed by physicians through the use of mobile technology.

Vision Statement

Our vision is that every patient can improve his or her compliance of a prescribed physical therapy or exercise regimen, and his or her overall health, through the use of convenient wearable movement monitoring equipment and a mobile device.

4.5.6.1 Prepare a List of Objectives

Prepare a list of objectives for the development of your technology. These objectives should be goals that your company needs to achieve in order to launch your product to market. Objectives should be measurable and observable (e.g., SMART objectives, see Chapter 8), so that you will be able to gauge your progress toward achieving them.

Sticking with the patient compliance monitoring technology example illustrated above, sample objectives might include:

- Prepare a technology blueprint for the wearable tracker device and the patient monitoring mobile application, that is, lay out the plan for developing these two aspects of the technology.
- Develop working prototype of the wearable tracker device.
- Test functionality of the wearable tracker device (meaning test the internal components for correct functionality).
- Develop and test the underlying software for the patient monitoring mobile application.
- Integrate the wearable tracker device and the patient monitoring mobile application.

4.5.6.2 Set Forth a List of Requirements, Efforts, and Milestones for Each Stage of the Technology Life Cycle

Your technology plan should lay out, at a high level, your path for moving your technology product through the various stage of the technology life cycle. For each life cycle stage, for example, concept/ideation, design and research and development, prototyping, testing, commercialization and scaling, there will be certain requirements that must be satisfied before you can move forward to the next stage of development.

A good way to go about organizing your technology plan is to start by writing down the various capabilities your technology will have. From there, you can work backward and identify what requirements will have to be satisfied, and the particular order that they will need to be satisfied, to arrive at your fully developed technology product. If your technology involves multiple parts that need to be developed separately and later integrated together, prepare a technology plan that looks at both aspects of your innovation.

Additionally, as you work through your technology plan, take note of any potential limitations or problems that could possibly slow down your progress. Note the limitation or problem and try and identify a potential solution or mitigating factor that could help address the limitation or problem.

Returning to the patient compliance monitoring technology example, the following is a sample of the technology capabilities, requirements, efforts, and milestones sections of a technology plan.

Concept Stage – Identify Technology Capabilities:

- The wearable patient compliance tracker collects data (via sensors) for various patient parameters (such as acceleration, temperature, etc.) to monitor for compliance with a given treatment regimen.

- The wearable tracker timestamps the data.
- The wearable tracker transmits collected data to a mobile device, via patient monitoring application.
- In the app, patients can review their data, set reminders, set goals (e.g., a goal to increase the repetitions of a particular exercise every other day), and so on.
- The app also can be used to transmit patient compliance data to a physician.
- The wearable tracker has a time-out function after 5 min of no sensed activity.

Design Stage

Requirement: Understand how the wearable tracker and the mobile app will "talk" to each other.

- **Effort:** Research device and mobile app communication.
  - **Milestone:** Prepare a flowchart detailing communication between the wearable tracker and the mobile app.

Requirement: Identify electronic components for the wearable tracker.

- **Effort:** Identify the electronic components that will be necessary for the intended operation of the wearable tracker.
  - **Milestone:** Prepare a list of electronic components for the wearable tracker.

Requirement: Plan circuitry for wearable tracker.

- **Effort:** Determine the circuit arrangement for the various electronic components of the wearable tracker.
  - **Milestone:** Prepare a circuit diagram for the electronic components of the wearable tracker.

Requirement: Basis for mobile app – decide whether to build from scratch or use an app builder tool.

- **Effort:** Assess mobile app development options: (1) develop mobile app in-house and from scratch, or (2) use a mobile app builder tool.
  - **Milestone:** Prepare a decision tree to analyze the cost and benefit of each path forward for mobile app development and make an informed decision regarding how to go forward with mobile app development.

Potential Limitation: A potential limitation for all early stage/research and development stage activities is the lack of technological expertise to handle any one of the requirements listed above. To address this limitation, self-study could be sufficient to overcome any knowledge gap that the developers might have. In the event that a particular aspect of development is too complex, outside professionals can be hired to achieve some milestones.

Development Stage – Prototyping

Requirement: Build POC prototype of the wearable tracker.
- **Effort:** Build circuitry for the wearable tracker and incorporate the circuitry into the housing of the wearable tracker.
  - **Milestone:** Demo the working wearable tracker prototype.

Requirement: Build final prototype of the wearable tracker.
- **Effort:** Build circuitry for the wearable tracker and incorporate the circuitry into the housing of the wearable tracker.
  - **Milestone:** Demo the working wearable tracker prototype.

Requirement: Build a prototype of the mobile app.
- **Effort:** Build a mobile app prototype.
  - **Milestone:** Demo the mobile app prototype.

Requirement: Integration of the wearable tracker and the mobile app.
- **Effort:** Implement communication protocol between the wearable tracker and the mobile app.
  - **Milestone:** Establish communication between the wearable tracker and the mobile app.

Potential Limitation: A potential limitation/problem for the prototype development stage is that particular aspects of the device may not work as intended. To address this potential limitation/problem, factor extra time into the prototype development schedule.

Development Stage – Testing

Requirement: Wearable tracker and mobile app must communicate as intended.
- **Effort:** Repeated testing and debugging.
  - **Milestone:** Achieve 50 successive, successful communication links between the wearable tracker and the mobile app.

Requirement: Wearable tracker must transmit data as intended.
- **Effort:** Repeated testing and debugging of wearable tracker.
  - **Milestone:** Achieve 50 successive, successful communication links where all of the relevant patient data is communicated from the wearable tracker and the mobile app.

Requirement: Wearable tracker must time-out or automatically shut down.
- **Effort:** Test the timeout functionality of the wearable tracker, including testing the sensitivity of the wearable tracker for sensing movement.

  - **Milestone:** Achieve 50 successive, successful automatic shutdowns of the wearable tracker.

Commercialization Stage – Scaling Up

Requirement: Production of wearable tracker must be scaled up to meet demand.

- **Effort:** Identify and implement partnerships to satisfy increased production demands.
  - **Milestone:** Fulfill 1,000 orders for wearable tracker technology.

Potential Problem: A potential problem that can arise in scaling up production is that supply of a particular component could be short. To address this problem, for each component of the wearable tracker have a primary supplier and a secondary supplier.

4.5.6.3 Prepare a Timeline for Achieving Milestones for the Various Stages of the Technology Life Cycle

Prepare a timeline to help you and your team stay focused on achieving your technology plan milestones. The classic Gantt chart style of preparing a timeline can be helpful in organizing when certain milestones need to be accomplished relative to other milestones, particularly in technology plans that involve developing two or more distinct technologies that have to work together. Figure 4.6 shows a sample Gantt chart based on the above discussed milestones.

4.5.6.4 Prepare a Budget for Your Technology Development Efforts

Your technology plan should also include a budget. How your budget should be broken down, and how detailed it is, will depend on your particular technology. For complex multifaceted innovations that require the development of several technologies in parallel, a more detailed breakdown of costs would be best, whereas simpler technologies can be represented in less. Figure 4.7 shows a sample development budget for the wearable tracker and mobile app technology example.

4.5.7 Transitioning from Research to Development and Beyond

A healthy technology pipeline involves innovative researchers working closely with team members throughout the company, and understanding how their projects meet the company's short-term needs and long-term strategy. People at all levels within the company and outside the company, for example, academic partnerships, crowdsourcing investors, venture capitalist investors, and so on, have the ability to fill the pipeline with ideas from the bottom up. It is the role of researchers and leadership to let these ideas mold themselves, without the undue influence and bureaucracy of the company.

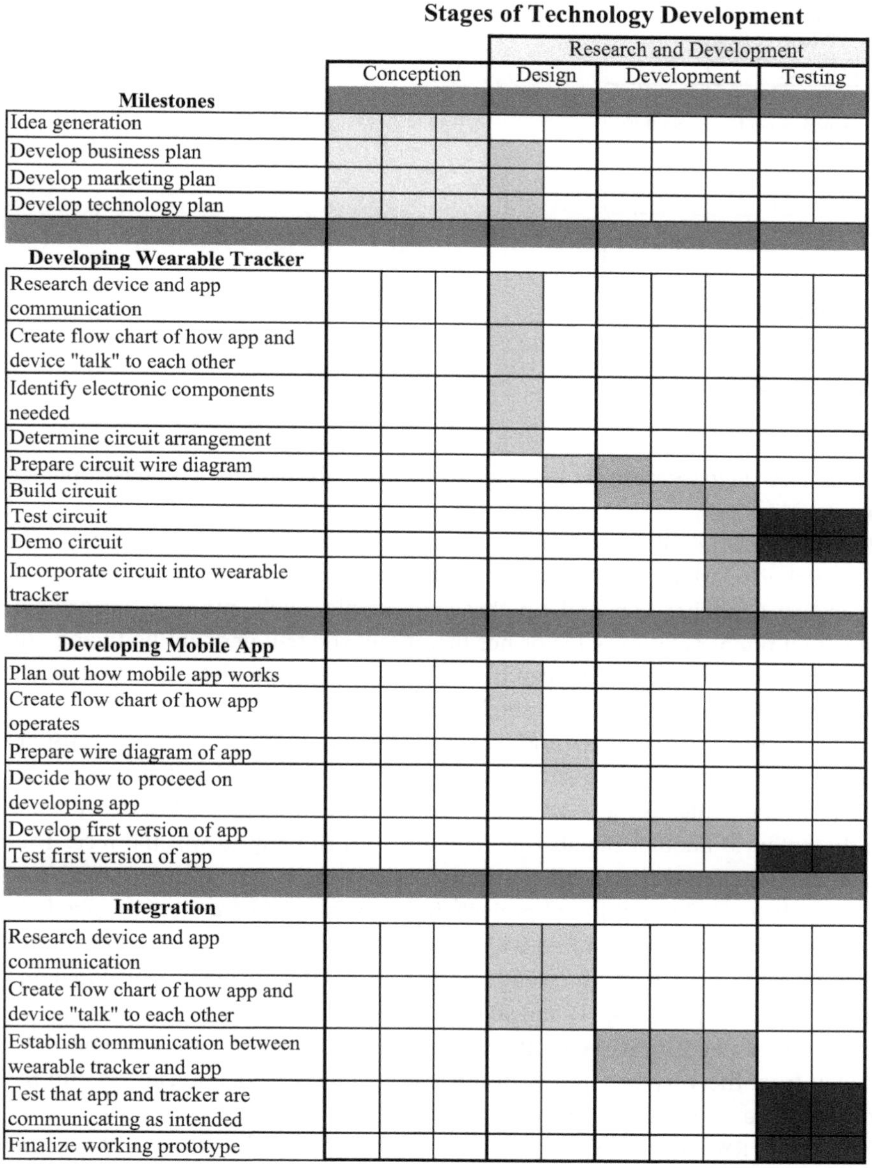

Stages of Technology Development

		Research and Development		
Milestones	Conception	Design	Development	Testing
Idea generation				
Develop business plan				
Develop marketing plan				
Develop technology plan				
Developing Wearable Tracker				
Research device and app communication				
Create flow chart of how app and device "talk" to each other				
Identify electronic components needed				
Determine circuit arrangement				
Prepare circuit wire diagram				
Build circuit				
Test circuit				
Demo circuit				
Incorporate circuit into wearable tracker				
Developing Mobile App				
Plan out how mobile app works				
Create flow chart of how app operates				
Prepare wire diagram of app				
Decide how to proceed on developing app				
Develop first version of app				
Test first version of app				
Integration				
Research device and app communication				
Create flow chart of how app and device "talk" to each other				
Establish communication between wearable tracker and app				
Test that app and tracker are communicating as intended				
Finalize working prototype				

Figure 4.6: Sample Gantt chart.

Early stage technology discovery or research projects can fail for a number of technical and nontechnical reasons, but it is disheartening to see projects fail because there is a breakdown in the handover and transition of projects from early-stage

Development Costs through Prototype Stage

Expenses	Cost
Physical hardware (electronic components, housing for tracker, etc.)	$4,000
Labor to build tracker prototype over 4 months	$7,000
Labor to build mobile app	$15,000
Labor for testing/debugging	$4,000
Total	$30,000

Figure 4.7: Sample budget.

research to late-stage research, or late-stage research to development, or development to commercialization.

Many a times, people describe this handover using the baseball analogy of a pitcher–catcher relationship. A pitcher–catcher relationship is prone to risk, however, because a pitcher can end up throwing a wild pitch, missing the catcher completely; a pitcher can miss his or her target and the ball ends up bouncing out of the catcher's glove; or a pitcher can throw a ball wide of the plate and the catcher has to get out of position in order to catch the ball.

This does not seem like a nice and easy transition of project work and responsibilities from one company project team to another. Instead, a project handover should be thought of in the context of a relay race on a track, where one member of the team hands the baton to another member as they are both running to win the race. This handover implies close contact, communication, and being in sync with one another.

The handover and transition of projects from one stage to another can also fail, if the project team is not integrated with other areas of the business. For example, Research project teams need to include experts from Management, Development, Marketing, business partners, IT, and so on, as needed.

- Management needs to agree on the value of the idea and the utility of the technology from the very beginning. because they will be allocating funds and resources to get the project finished.
- If the "R" in R&D is not engaged with the "D," especially while the POC prototype is being developed, then the design may be flawed because it is not scalable or feasible to manufacture. The benefit of Development being involved at the R&D stage is that Development can be tasked to fix the scale-up issue in parallel, while Research completes prototype design so that project timelines are maintained.
- If Research is not engaged with the product managers within the marketing team who talk to customers, then the design may not be fulfilling an unmet need with customers, or the design may not be user friendly.

- If Research is developing a software tool for engineers to use internally and Research is not engaged with IT, then the Research team may discover too late that IT does not have the database infrastructure needed to support the storage and retrieval of data that the software tool requires.

When these types of glaring communication and project management mistakes occur, management gets frustrated and may decide to delay or cancel projects. Once projects transition from research to development, Development teams should include manufacturing and commercialization experts who know what the challenges are in taking an idea from laboratory or workbench to market. Design and process decisions that are made during this stage are critical, as they affect the ability for scaling-up and COGS optimization opportunities in the future.

Once projects transition from development to commercialization, costs to sustain and support the technology post commercial launch also need to be calculated and understood. These sustainment costs cannot be overlooked in commercial financial planning efforts; companies tend to cut sustainment costs, given the opportunity to allocate more room in the budget. This end-to-end investment is needed to capture the most value from of the technology created. There is an optimal value that can be capitalized from the technology versus the amount of sustainment and operating costs the company is willing to spend. This decision will be unique for each company.

4.5.8 Planning for Special Considerations for Your Technology

Some businesses need to plan for additional, special considerations during the development of their technology. For instance, there are many industries and technologies that are highly regulated, some materials used in manufacturing must be handled and disposed of in an environmentally safe way, and some technologies might create legal liability in the company that produces the technology. Table 4.1 lists a list of common, special technology-dependent considerations.

4.6 Preparing a Business Plan

A business plan is a formal exercise for detailing how your business will work. A business plan is an ever-evolving, written document that lays out, at a high level, your plans for starting and running your business – it is a written codification of your business model that details how the business will operate, how revenues will be generated, identifies who the customer base is, and lays out what the product or service is that will be sold. Business plans are useful because they force you to

Table 4.1: Example list of technology-dependent considerations.

Specialty	Description
Medical Devices and Pharmaceuticals – Controlled Environments/Clean Rooms	Medical devices and pharmaceutical products are often highly regulated, and the manufacturing and packaging of medical devices and pharmaceuticals often must be done in sterile facilities called clean rooms. The physical construction of clean room facilities is expensive. The alternative to constructing a clean room is to rent a clean room, or to outsource the manufacturing or sterile packaging of the product to a third party. Outsourcing manufacturing, packaging, and/or sterilization can be costly in the long term.
Medical devices and Pharmaceuticals – Regulatory	Medical devices and pharmaceuticals have highly regulated, specific pathways for getting to commercialization. Consulting with regulatory strategists/consultants at an early stage will be important to both understanding what pathway you will need to take with the Food and Drug Administration (e.g., 510k, PMA, de novo, etc.) but also how much it will cost to go down that regulatory path.
Medical Devices and Pharmaceuticals – Reimbursement	Medical devices and pharmaceuticals have general pathways to get approved for reimbursement by Center for Medicare and Medicaid Services (CMS) and insurance agencies. Consulting with reimbursement strategists/consultants at an early stage will be important to both understanding what path you will need to take with the American Medical Association (AMA), specific medical societies associated with the AMA, and CMS, but also how much it will cost to go down that path.
Biological	Biological laboratories will require certification for the biosafety cabinets, and they will require workplace safety compliance and biological waste disposal protocols.
Hazardous Waste	When setting up a laboratory or workshop, there may be a need for hazardous waste removal. In the US, your facility may require an Occupational Safety and Health Administration or hazardous waste certification. Certifications cost money not only to obtain and implement, but also to execute. An example of hazardous waste removal would be a mechanic who needs to dispose of radiator fluid, transmission fluid, and motor oil.
Radioactive Waste	Radioactive waste will need to be reclaimed, safely stored, and disposed of according to state and federal regulations.
Animal Research	Animal test and research facilities will require an Institutional Animal Care and Use Committee certification to ensure proper protocols have been established for the treatment and handling of animals, including ventilation and living conditions.

Table 4.1 (continued)

Specialty	Description
Online /Cloud-Based Platforms	Data security is critically important in today's high-tech and digitally connected world, due to the need for portability and accessibility of electronic information. Depending on the requirements of your industry, certifications for data security may be required, or your company may be required to hire personnel with specific certifications.
Digital Healthcare Solutions	Companies that handle patient records are required to comply with the Health Insurance Portability and Accountability Act in the US. Health Level 7 is the standard database architecture used in US healthcare systems for interplatform communications. Certifications and/or training will need to be considered.
Software Platforms	In many cases, third-party vendors will be used for integrated services. An example of this would be a third-party vendor that would conduct identification verification of credit card purchases. There can be upfront and long-term fees associated with each.

organize your thoughts and think about important questions about how your business will function. For instance, business plans often answer questions like:
- What is the purpose of the business?
- What type of business structure (i.e., legal business entity) makes the most sense for the company?
- What is the product or service that will be sold?
- Who is the target customer base?
- How will the business establish relationships with customers?
- Which types of partners will be helpful for the business, and what resources can they provide?
- What pricing structure do you anticipate using to sell your product?
- What sources of revenue will support the business?
- How will you finance the startup costs for the business?
- Should the business structure change at a later date, for example, convert to a corporation so that stock can be issued?
- What is the exit strategy for the business?

The precise format and length of a business plan can vary, with the average business plan being about 15–20 pages in length, but business plans can certainly be longer (depending on how complicated your technology is) or shorter (investors often initially want to see only the most summarized version of your business plan). Comprehensive business plans based on complicated technology can easily be over a hundred pages in length – but it is often difficult to get prospective investors to read a lengthy business plan. If you do have a business plan that is longer than 20 pages in length, it is a good idea to prepare a full-length comprehensive business plan, and an abbreviated version of the business plan (5–7 pages), and an executive summary (no more than one page). It is often useful for entrepreneurs and startups to offer the abbreviated version of the business plan and the executive summary to those who express interest in learning more about their business. The executive summary and abbreviated version of the business plan can serve as a way to get "a foot in the door" with potentially interested parties.

As a general rule, business plans follow a specific structure, in terms of the topics that are covered. However, based on the type of business, you may have to include additional sections or may have to remove sections that are inapplicable to your business. For instance, technology companies often need to include a section on their technology plan (i.e., their plan for reaching specific technology development milestones). Conduct online research to figure out what types of information, or sections, should be included in your specific business plan, based on the type of business you plan on creating. The general structure for a business plan for a technology-based company includes the following sections.

4.6.1 Executive Summary

The executive summary is a concise overview version of the business plan that is no more than one page in length. The executive summary is intended to be engaging and intriguing – so that readers will be enticed and excited to read the full business plan. Make sure that the executive summary includes:
- The name of the company.
- Where the company is located.
- A description of the product or service sold by the company.
- The company's mission statement.
- The company's vision statement.
- Specify the intended purpose of the business plan (e.g., the business plan is meant to secure funding, solicit synergistic business partnerships, encourage merger or acquisition, etc.).

4.6.2 Company Description

The company description section provides background information on who the company is or how the company came to be, how the company operates (such as the company's legal corporate structure and ownership), and what the company's goals are, with specific discussion about the company's one year goals. This section can include a high-level discussion on the product or services offered by the company, can identify the target customer (for instance, a customer profile), and can identify the need in the market that the company is trying to fulfill. The company description can also provide an overview of the company's growth over the past year to three years.

This section should address forward-looking issues, such as the highlights of the short- and long-term goals of the company (such as planned growth rate, planned expansion, exit strategy, etc.), important financial information about the company, and high-level market analysis.

4.6.3 Marketing Plan

As discussed in Section 4.2.4, the marketing plan focuses on identifying the target market, performing competitive analysis, and anticipating the needs of the customer. The marketing plan section of the business plan needs to provide a high-level summary of the marketing plan, including a detailed description of the four Ps of marketing: the *product* or service being offered by the company, the *price* that the product will sell for (or pricing structure, if appropriate), the *place* where the product will be sold (which could include information concerning placement in a store, whether the product will be sold exclusively online, how the product will be distributed, etc.), and the *promotion* strategy that will be used (i.e., the type of advertising or promotion that will be used).

In addition to the four Ps of marketing, the marketing plan section of the business plan should give a synopsis of the market analysis (Chapter 1) and identification of the target market (Chapter 1 and Section 4.2.4.1), and an analysis of the competition (Section 4.2.2).

4.6.4 Sales Plan

As discussed in Section 4.2.4, the sales plan provides details on customers and a plan of action to make and increase sales. The sales plan section of the business plan is a summarized version of the sales plan, and can include detailed target customer profiles and a catalogue of the techniques that will be used to make sales. The sales plan section of the business plan should also provide sales forecasting information, including estimates of future sales based on assumptions that certain marketing strategies or advertising campaigns will be successful.

4.6.5 Technology Plan

Companies operating in the technology industry will most likely need to include a technology plan summary in the company business plan. The technology plan section should include an overview of the strategic technology plan objectives and goals. Specifically, the technology plan section should provide highlights of how the technology will be used, how many consumers will need the technology (e.g., the quantity of units that need to be produced, whether scaling up will be needed, etc.) and the specific techniques, tactics, or plan that will be used to achieve the strategic technology plan objectives and goals.

The technology plan section of the business plan can also include a summation of the technology development plan (which is a technically specific plan focused on the logistics and timeline associated with the development of a technology, described more in Chapter 10) and the technology development SIP (which is a detailed plan for how to execute the development of your technology from a technical perspective, discussed more in Chapter 10), including identification of the major technical, engineering, or design milestones that are involved in developing the technology into a marketable product. If there are any special considerations associated with the technology, such as concerns about the technology being at the cutting edge of innovation or specific regulatory requirements that will be imposed on the product, these concerns should be made clear in the technology plan section. Any technology-specific concern that is identified in the technology section of the business plan should be paired with a brief explanation of how the issue will be addressed or resolved.

4.6.6 Organizational Chart and Management Team

The organizational chart and management team section should lay out:
- Who owns the company (and in what percentage).
- The organizational structure of the company (for instance, if the company is broken down into various departments or divisions, it is important to include a brief description of each department or division).
- Key employees assigned to various operational and management roles in the company.

A short summary should be provided for each person on the management team that highlights his or her role in the company, responsibilities, qualifications, and level of experience. A profile such as a resume for each person that is part of the management team should be included in the business plan as an attachment.

Inclusion of an organizational chart, such as the one shown in Figure 4.8, with individuals assigned to each position, can be a useful visual aid for explaining the

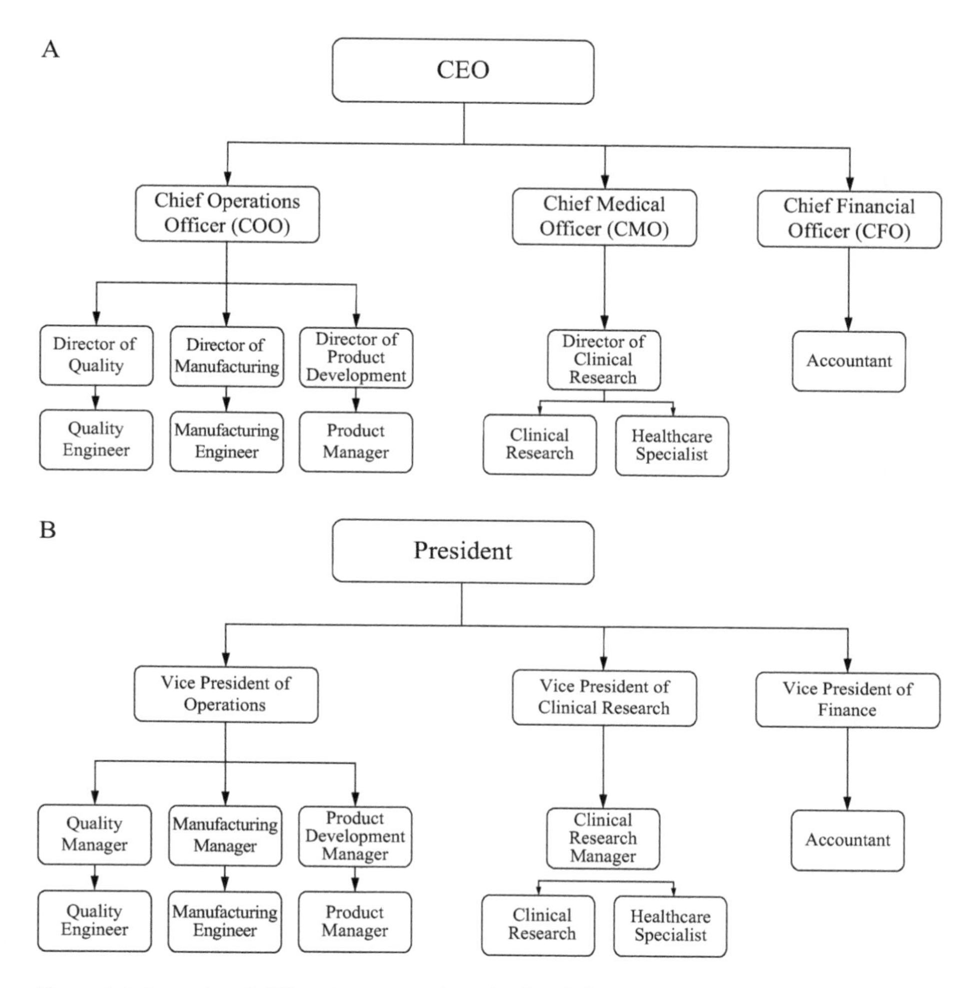

Figure 4.8: Examples of different company Organizational Charts.

company hierarchy and organizational structure. Figure 4.8 shows two different organizational structures: Organization Chart A shows a company organizational structure featuring a CEO, and Organization Chart B shows a company organizational structure featuring a president. If a position is not yet filled, include a list of desired candidate qualifications for the unfilled position.

Organizational charts are often aligned with a Roles and Responsibilities Matrix. The specific role and responsibilities for each position on the company's organizational chart are recorded in the matrix. The Roles and Responsibilities Matrix can be included in the business plan as an attachment (see example Roles and Responsibilities Matrix in Figure 4.9).

Roles and Responsibilities Matrix

R = Responsibilities
B = Back-up

	CEO/ President	CMO	CFO	COO	CTO
Business Development	R				
Clinical Affaris		R			B
Engineering				B	R
Finance & Payroll			R		
Human Resources			R		
Legal Affairs	B		B		
Manufacturing				R	B
Purchasing/Contracts	R			R	B
Quality Assurance & Regulatory Affairs				R	
Research & Development		B		B	R
Strategy	R	B		B	B

Figure 4.9: Example Roles and Responsibilities Matrix.

4.6.7 Operating Procedures

The operating procedures section is intended to lay out the logistics of how the company runs and should include a discussion about the following:

- **Location** – Where is the physical location of the company? Does the location have land and a building to house the business operations? How many square kilometers is the space? Is the building and land rented or owned? Are there satellite (i.e., off-site) offices associated with the business? How big are the satellite offices? Does the business need room to grow in the future (for example, to facilitate scaling-up of manufacturing), and how much space will be needed? How much does the location (grounds and building) cost in terms of rent/mortgage payments, utilities and upkeep?
- **Equipment** – Is there any specific technology or specialized equipment that your company relies on to conduct business? Does the company own or rent this equipment? How much does it cost to purchase/rent, maintain, and use the equipment?

- **Personnel** – How many employees are there? What individuals within the company are key to running the day-to-day operations of the company? Do any of these key personnel require certain qualifications to perform their job duties (e.g., do you need a lawyer, or a certified public accountant)? How will the company hire, train, and retain these key employees? How much does personnel cost the company?
- **Inventory and Suppliers** – Does the company hold inventory? What type of inventory system does the company use (e.g., just-in-time inventory first-in, first-out inventory, etc.)? Does the company rely upon any suppliers or vendors? How are products distributed (i.e., through what distribution channels)? What is the pricing structure for purchasing inventory?

4.6.8 Financial Models

Financial models are a series of graphs, charts, and spreadsheets designed to convey financial information to potential investors. Financial models are nothing more than a tool for telling a story, using numbers instead of words. But it is important that the story makes sense. Most data presented in financial models is an estimate at the beginning, and that is okay. The exercise of walking through the financial models, line item by line item, is not only necessary for investors, but it will predict how much money will be needed to achieve the company's valuation milestones, for example, securing IP, completing design and development, achieving regulatory approval/clearance (if applicable), conducting human trials (if applicable), conducting market testing, and market launch.

It does not matter what type of company is being formed – every company should develop financial models to help determine feasibility and to help identify costs that they were not expecting, such as the costs associated with insurance, cell phone plans, marketing, and so on. Financial models should outline 3–5 years of overhead and projected revenue. The further out one goes, the less realistic the numbers become. Again, that is okay.

Financial models are often included at the end of a business plan, or as an attachment to a business plan. Since financial models need to be updated frequently, as the business begins to take form and grow, they are usually one of the last items completed for the business plan, and they are often prepared by a certified public accountant. Furthermore, it is common practice to update financial models right before giving an investor pitch, so that prospective investors will have the most up-to-date version of your company's financial models.

Venture capital firms offer free financial model spreadsheets with interconnected formulas between data entry cells that make the task of assessing a company's financials much easier. These financial model template spreadsheets can be found online and cost nothing to download. Additionally, local economic development agencies, such as Small Business Development Centers, and some for-profit organizations may have sample financial model spreadsheets. These spreadsheets are useful for any company that is preparing a financial model to support a business plan. The details of each financial model need to be customized to the company, but the main sections of the financial model will generally be the same.

The financial models that will be included in a business plan depend on whether the company has been operational for a while or is in the startup phase. For companies that have been operational for several years, the financial models section of the business plan should include:

- **A Financial Summary/Overview.** An overview that condenses all the numbers into a yearly format that is easy to read can be helpful for conveying complex financial data at a glance. The overview is often one of the last sections to develop, because all of the numbers are collated from the rest of the financial model spreadsheets.
- **Historical Financial Documents.** Historical financial documents, such as an income statement, balance sheets, and cash flow statements from the previous 3–5 years of operation.
- **Prospective Financial Information.** Prospective financial information, including pro forma financial documents, including projected revenues, forecasted income statements, balance sheets, and cash flow statements.

On the other hand, if the company is just starting out, then there is not a lot of historical financial data from which to draw conclusions or projections. Instead, financial documents are prepared pro forma, meaning that the financial projections included in the financial documents are based on assumptions in the financial models. The financial models that should be included in a business plan for a startup are:

- **A Financial Summary/Overview.** An overview that condenses all the numbers into an yearly format that is easy to read can be helpful for conveying complex financial data in a glance. The overview is often one of the last sections to develop, because all of the numbers are collated from the rest of the financial model spreadsheets.
- **Startup Budget.** A startup budget lays out the expected costs it will take to get the company up and running. Startup budgets are usually for only one year at a time, but could stretch to two years.
- **Pro Forma Income Statement.** The pro forma income statement is a forward-looking projection of income to be earned in the future based on assumptions. The assumptions about the income projections are documented in the pro forma income statement.
- **Break-Even Analysis.** The break-even analysis explains when the startup company is likely to become profitable.

However, just as a lot of planning and ground work goes into preparing a business plan, there are many other financial documents that should be prepared as well and included as attachments to the business plan. These other financial models include:

- **Projected Revenue.** The projected revenue model takes into account customers over time, price of the product(s), estimated revenue from each customer segment over time, COGS, services provided, and profit margins.
- **Research & Development (R&D).** The R&D model outlines costs related to prototype development, testing and costs related to transitioning or transfer to manufacturing before the final COGS are established. This section can also include the cost of regulatory and/or reimbursement consultants for medical device and pharmaceutical companies, for example.
- **General and Administrative Expenses (G&A).** G&A are often overlooked, but are common expenses that most businesses encounter, such as insurance, telecommunications, IT, rent for office space, rent for laboratory space, and so on.
- **Sales.** Sales includes the resources needed for an internal or external sales force; including travel expenses, software subscriptions, such as customer relationship management software, etc.
- **Marketing.** Resources that are needed for marketing could include travel expenses, software licenses, subscriptions, cost of different modes of advertising, survey and focus group studies, participating in trade shows and conferences, and so on.
- **Staffing Expenses.** Staffing is needed throughout all business departments and functions. Expenses associated with staffing include hiring workers, how much workers are paid, costs for providing workers with benefits, and so on.

There may be company-specific additions to the financial models that are necessary, such as:

- **Depreciation and Amortization Tables.** The depreciation of business assets needs to be accounted for in financial models. For instance, when a company needs a large piece of equipment (known as a capital piece of equipment) to conduct business, the company can depreciate the value of the capital piece of equipment over time. The company might also need to borrow money. Amortization of that loan, that is, the amount of money that is repaid toward the loan plus interest, should also be accounted for in the company's financial models. An amortization schedule is a simple tool that can display the breakdown of cost and interest associated with the loan by month, quarter, or year.
- **Capitalization Table.** A capitalization table is a breakdown of all of the shares issued and to whom they are issued. If angel investors or VCs are involved, they will require this.
- **Clinical Trials.** If the company is developing a medical device or pharmaceutical product, then the cost of clinical trials needs to be accounted for in the company's financial models. The costs associated with each clinical trial are product specific

and clinical trial endpoint specific. Each trial will last over a period of time, depending on the data that needs to be obtained before the product can be brought to market. The number of trials and what data is necessary will be determined by interactions with the Food and Drug Administration and regulatory consultants.

- **Models and Metrics.** Models and Metrics are visuals that are calculated from the Overview. Visuals can be easier to understand, and many potential investors prefer to look at an easy-to-digest visual rather than a complex spreadsheet of numbers.

4.6.9 Attachments

The attachments section of the business plan is where all other relevant documentation goes. The attachments section is meant to serve as a repository for additional information that may be relevant or of interest to a reader of the business plan. For instance, additional financial models are often included in the attachments section of the business plan. Or resumes for each member of the management team can be included in the attachments section.

5 Understanding Intellectual Property as It Relates to Your Business

Abstract: In the United States, intellectual property (IP) rights can be a huge boon for a company, and particularly for an early-stage company or startup. IP rights are legal rights to an intangible asset and are protected by patents, trademarks, copyrights, and/or as a trade secret. An intangible asset is something that has value attached to it (i.e., it is an asset), even though it is not a physical thing that can be touched or physically possessed (i.e., it is intangible). IP is any product created by the human mind, whether it is an innovative new design or invention, intriguing company logo, smart tag line, or artistic graphic design, that carries commercial value for its owner. New businesses that are based on technological innovation are often teeming with IP. Company names, logos, inventions, and website addresses all are forms of IP that can have value to a business, and young businesses need to protect the valuable IP that they create – the earlier, the better. Early-stage businesses need to be concerned with securing IP rights in their intangible assets as soon as possible.

Keywords: Intellectual property (IP), patents, trademarks, copyrights, trade secrets, provisional patent application, non-provisional patent application, registration, abandonment, generic, improper licensing, misappropriation, unauthorized disclosure, reverse engineering, independent discovery, IP strategy, IP portfolio, IP audit, international IP strategy, freedom to operate (FTO), cross-licensing.

Intellectual property (IP) rights create a legal barrier to protect technology innovations. Since others are prevented from using whatever is protected by the IP rights, the owner of the IP rights enjoys a period of exclusive use as defined by the IP rights. Exclusive rights to IP allow the business to use those rights as they see fit, which often translates into a competitive advantage in the marketplace. IP rights can be:

- **Used Offensively.** Once IP rights are secured, the IP owner can actively enforce his or her IP rights by suing others for infringement.
- **Used Defensively.** IP rights can be used to prevent others from entering the IP space that is occupied by another, that is, as a barrier to entry for competitors.
- **Monetized.** IP rights can be assigned, licensed, or sold.

A few examples of how IP rights create value include:

- **Intellectual property rights form a barrier to entry for competitors.** Intellectual property rights generally prevent or prohibit others from using protected IP without permission. IP rights can be used defensively. In essence, others are

https://doi.org/10.1515/9783110521900-006

excluded from using the IP that they do not own. IP ownership can prevent competitors from making similar products or copycat products.
- **Intellectual property rights can create a unique selling proposition.** For many consumers, the fact that there are IP rights associated with a product is appealing. For instance, if a product is patented or has patent pending status, it can be a unique selling proposition in the eyes of the consumer who distinguishes the product from similar products made by competitors.
- **Intellectual property rights are monetizable through licensing agreements.** Useful IP can be licensed to competitors or anyone who is interested in using the technology in exchange for a licensing fee. Licensing is an attractive money making opportunity for IP rights holders because they retain ownership of the IP and can place restrictions on licensees, such as where the technology can be used (geographical limitations), how the technology can be used (limited scope of permitted use), and for how long the technology can be used.
- **Intellectual property rights are attractive to investors.** Potential investors are attracted to companies and startups that already have issued IP rights, since IP rights are intangible business assets that can be exercised, enforced, monetized, licensed, or sold. Investors are also interested in companies that have pending IP rights, such as pending patent applications or pending trademark registrations. In either case, the pursuit and securement of IP rights are attractive to prospective investors.
- **Intellectual property rights can enhance the valuation of a company.** If your company's exit strategy is to be bought out or acquired by a competitor, then having an IP portfolio related to your technology can increase the valuation of your company.

Entrepreneurs and innovators simply cannot develop and market an innovative new technology without securing some form of IP. Securing IP protections is an important part of the Business Development Process, as shown in Figure 5.1). IP right protection is a significant source of competitive advantage and value for a company. Generally speaking, companies obtain multiple types of related IP rights in order to build what is known as an IP portfolio. Related IP rights can have a synergistic effect when bundled together, increasing the overall value of the IP portfolio and, in turn, the company (IP portfolios are discussed more in Section 5.5).

There are several types of IP rights in the United States that are useful for companies in the business of developing a new innovative technology. These include:
- Patents.
- Trademarks.
- Copyrights.
- Trade secrets.

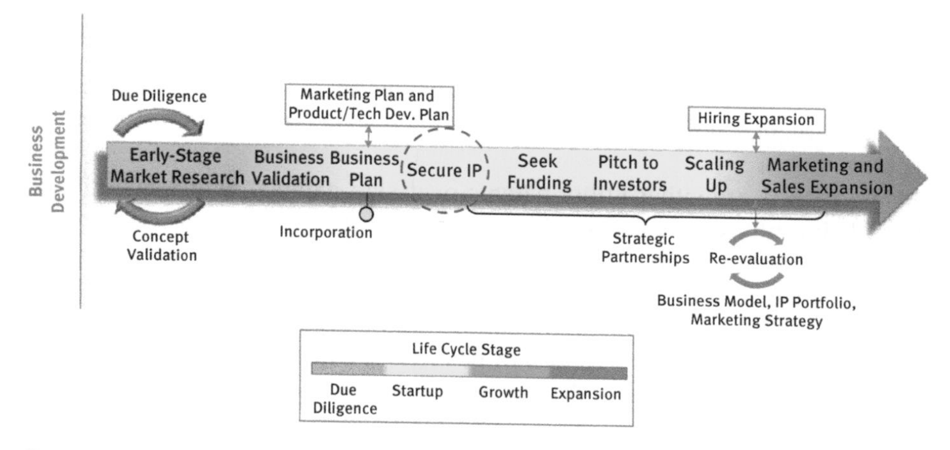

Figure 5.1: Securing IP rights and developing an IP portfolio is an important part of the Business Development Process.

When it comes to IP matters in business, it is always a good idea to speak with an experienced IP attorney. If the matter specifically has to do with patent protection, then it is important to make sure that the attorney is a registered patent attorney. Most IP lawyers, but not all, are registered patent attorneys.

5.1 Patents

What follows is a high-level discussion about US patents. Patents are a form of IP rights that can be sought to protect novel and nonobvious inventions and innovations that are based on a useful machine, process, manufacture, or composition of matter. In the United States, patent protection is sought by filing a patent application with the United States Patent and Trademark Office (USPTO). The protected part of an issued patent is the patent claims, which legally define the metes and bounds of the invention.

Although not legally required, it is highly recommended that you hire a registered patent attorney or patent agent to help secure patent protection. Patent attorneys (i.e., lawyers who have a qualifying technical or engineering background) and patent agents (i.e., highly trained scientific or engineering experts with qualifying educational credentials) must take a special registration examination in order to practice patent prosecution before the USPTO. Drafting broad, yet well-defined, patent claims is more of an art than a science. The patent prosecution process is highly complex, as there are specific aspects of patent rights that are incredibly nuanced. The guidance of a patent professional can help avoid costly mistakes during the patent application process.

Once patent protection is secured, patent right owners have the right to exclude others from making, using, selling, offering to sell, or importing the patented technology for the term of the patent. For this period of time, no other person or entity can make, use, or sell the technology in the United States without permission from the patent owner.

Most patents are either utility patents or design patents, with a majority of patents being utility patents. A utility patent is a patent that protects a novel and nonobvious invention or innovation that is based on a useful machine, process, manufacture, or composition of matter. Patent protection on an issued utility patent lasts for 20 years from the date that the utility patent application was filed (the effective filing date). A design patent is a patent that is sought specifically to protect the ornamental design of an invention, which is how the invention looks. Design patent protection lasts for 15 years from when the design patent is granted.

Patent owners can authorize others to use their patented technology through a licensing agreement. A licensing agreement is a formal contract that details the rights and responsibilities of both the licensee (the party that gets permission to use the patented invention) and the licensor (the party that agrees to allow the licensee to use their patented invention in exchange for valuable consideration, such as a licensing fee or use of other IP rights owned by the licensee).

Once the patent protection on the technology expires (i.e., the 20-year patent life terminates), the technology is considered to be part of the public domain, and anyone can copy or use the technology at that point without permission from the patent owner.

Case Study

During the 1990's, the pharmaceutical industry made a regular practice of securing patent protection for blockbuster drug formulations, that is, extremely popular and highly prescribed drugs. Any big-name drug you can think of was covered by patent protection, at one point. In late 2011, Pfizer's patent protection for Lipitor (chemical name atorvastatin), a statin drug used for controlling cholesterol levels, expired. During the period of time where Lipitor was marketed to the public, Pfizer made more than US $120 billion.

Patent protection enabled Pfizer to secure a large return on its research and development investment in the statin drug market. Once the Lipitor patent expired (after a patent term of approximately 20 years), the chemical formulation of the drug (atorvastatin) became public knowledge. After becoming public knowledge, any generic manufacturer could produce the chemical atorvastatin without incurring costly research and development expenses and market it directly to wholesalers in the pharmaceutical industry. Pfizer was able to continue to charge a premium compared to generic versions of atorvastatin, but Pfizer knew it was no longer financially feasible to charge the same price for Lipitor, when so many generic drug competitors were available on the market. The price of Lipitor dropped, but it could not remain competitive with the generic drugs that were better able to move into the statin market, after the Lipitor patent expired. Consequently, Pfizer lost 19% of its profits in the first several months following patent expiration.

Patent protection is useful because it can limit competition in the market and can provide a source of value. Patent rights are often monetized by patent owners, for example, through licensing agreements or used as assets to collateralize loans. Additionally, prospective investors, such as angel investors and venture capitalists, are often attracted to businesses that have secured patent and other IP rights on their technology, because a comprehensive IP portfolio demonstrates the company's commitment to the development and commercialization of the technology.

5.1.1 Provisional and Non-Provisional Patent Applications

The United States employs what is referred to as a first-to-file patent system, meaning that patent rights for an invention are awarded to the first inventor to file a patent application with the USPTO. With that being the case, it is critical for inventors to get their patent applications in as soon as possible to secure the earliest possible filing date. You want to beat out others by getting your application on file sooner rather than later, thus preventing others from securing patent protection on your invention!

To that end, patent applications exist in two varieties: provisional patent applications and non-provisional applications. Applicants can opt to start the patent application process with a provisional application (a lower-cost, low-burden initial filing option), or can skip straight to starting the process by filing a non-provisional application (a full application that is ready for review by the USPTO).

A provisional patent application is a preliminary version of a patent application that grants the applicant an effective filing date, upon filing with the USPTO. The filing requirements for submitting a provisional application are less strict than the filing requirements associated with a non-provisional patent application, and it is more affordable to file a provisional application than it is to file a non-provisional one. Once a provisional application is filed, an applicant has up to one year to convert the provisional patent application into a non-provisional application. So no matter which filing path you choose for your company, you will ultimately have to submit a non-provisional patent application, if you want to try and obtain a patent from the federal government.

Many cash-strapped startups and small companies can extract a lot of value from filing a provisional patent application to start. Provisional applications are not reviewed by the USPTO, so filing a provisional application effectively "holds your spot in line" and buys some time to (1) complete the preparation of a non-provisional application, and (2) to muster up the funds that will be required to pay for the patent application process, which is not cheap.

A word of caution about using provisional patent applications: While a provisional application has less strict filing requirements than a non-provisional application (for instance, no patent claims are required in a provisional patent application filing), it is important that the provisional application be as complete and thorough in terms of what it discloses as possible.

Having a provisional patent application filed with the USPTO can provide startups and small companies with a number of benefits. For instance, once a patent application is filed with the USPTO, applicants can use the designation "Patent Pending" on their products and product packaging. The use of "Patent Pending" labeling on products has three important effects:

1. Consumers tend to more readily purchase products that are labeled with a "Patent Pending" designation.
2. The "Patent Pending" designation puts potential infringers on notice that your company takes IP protection seriously.
3. Investors like to see that startups are taking IP protection seriously. If you believe in your product or innovation so strongly that you are willing to front the cost of securing IP protections, investors will not fail to notice.

Figure 5.2 shows a chart depicting the typical utility patent application life cycle in the United States.

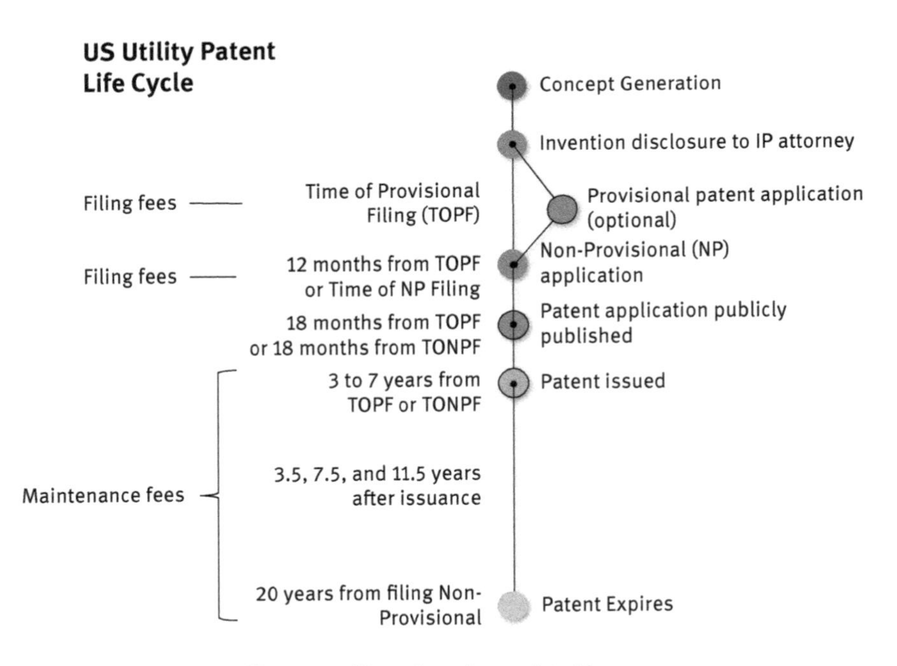

Figure 5.2: Typical US utility patent life cycle and associated fees.

5.1.2 Example Questions to Ask Your Patent Attorney

Doubtless you still have many questions about inventions, claims, disclosures, prior art, patents, and so on. It is highly recommended that you consult with an experienced patent attorney about your invention and your particular situation. Following are a few good questions that you should ask your patent attorney, which can help guide your conversation:

1. What is the best patent application filing pathway based on my business goals?
2. What is a claim?
3. What are the types of utility patent claims?
4. How long does it typically take for a patent to issue?
5. What information in my patent application is made public and when?
6. What protections does a provisional patent application provide?
7. Can you talk to investors about the invention? Can you give presentations on the technology?
8. What information related to the invention can you disclose?
9. What is a utility patent?
10. What is a method patent?
11. Why would you want a method patent?
12. Is software patentable?
13. What are some patent protection options for software IP?
14. Are there filing fees and maintenance fees, and how often does a company need to pay them?
15. Are there any other costs associated with owning a patent?
16. What is a patent portfolio?

5.2 Trademarks and Service Marks

Trademarks and service marks are a form of IP that are designed to identify or distinguish a source of goods or services. Consumers come to associate trademarks with the maker or manufacturer of a product over time, because the trademark often indicates a level of quality, brand and/or a corporate identity to the consumer. For example, a word, symbol, phrase, design, jingle, logo, and sometimes even a signature smell or color can be secured as a trademark for a particular company or product. Trademark protection prevents others from using the protected mark, as well as any mark that is considered to be confusingly similar to the protected trademark.

It is very common for a business to trademark the name of their company, and then to use that trademarked name as a domain name for the company's website. Companies also often obtain trademark protection on specific product designs, product packaging, logos, and unique product names.

Trademark rights can generate value for the trademark owner. Permission for others to use the protected trademark can be granted in the form of a licensing agreement. Additionally, investors are often attracted to companies that have secured trademark protection for their company name, logo, and product names and designs.

Anyone can simply begin using a trademark (for use with a good, such as a product) or service (for use with the provision of a service) so long as they sell a good or service. Use of a trademark or service mark creates what are called "common law" rights to the mark. This protection allows for trademark holders to designate their protected mark with a ™ symbol (for a product) or a ℠ mark (for a service). Trademark rights based on common law are fairly weak defensive rights, from a legal perspective.

To improve the strength of trademark and service mark rights, trademarks and service marks can be registered with the federal government by filing a trademark application with the USPTO. Once registered, trademark owners can designate their registered trademark or service mark with the symbol ®. Trademark protection for registered marks is national and can last indefinitely, so long as the trademark owner takes care to ensure that the trademark protection is not lost. Trademark registrations must be renewed every 10 years, and the mark must continue to be used in commerce. Trademark protection can be lost through:

- **Abandonment.** Typically, if a trademark owner discontinues use of the trademark for a period of three consecutive years, or never actually uses a registered trademark in commerce, then the trademark rights will be considered abandoned and no longer enforceable.
- **Generic use of the trademark**. When consumers start to use a trademark name of a product in place of the general name for the product, trademark protection can be lost through what is called "genericide."
  - For instance, if you have a medical device company that makes bioabsorbable bone screws, and your company has obtained trademark protection for a specific brand of bone screws called "Translatio," the brand name "Translatio" should always be used as an adjective for your brand of bone screws, that is, Translatio bone screws. In writing, the term should also be distinguished from the rest of the text, for example, TRANSLATIO bone screws, or *Translatio* bone screws. When customers casually refer to bone screws as "Translatios," or use the word "Translatio" to refer to bone screws generally, you should be worried about genericide of your trademark.
- **Improper licensing**. Trademark protection can be lost due to improper monitoring of licensing activities. If a company chooses to license its trademarks to third parties, it is imperative that the company/licensor maintain a level of control over how the licensed trademark is used by the licensee. Failure to retain adequate control over the user of trademarks under a licensing agreement, or failure to enforce the terms of a licensing agreement, results in what is called a "naked license." Since the licensor has failed to exercise adequate control over the use of the trademarks, trademark protection may be lost.

5.3 Copyrights

Any creative work that is fixed in a tangible medium of expression can be protected by copyright. To say this another way, any original, creative expression of an idea that is recorded in some physical way (electronically recorded counts too!) is a form or IP. If you can write it, draw it, paint it, record it, videotape it, or save it to a hard drive, you can most likely protect the IP via copyright. A copyright protects the creative work from unauthorized duplication, that is, unauthorized copying.

A copyright is created in the work the moment it is recorded. For example, by simply writing this sentence, a copyright is created. No one can copy a copyright protected work without permission (with a few exceptions). If someone does copy a copyright protected work without permission, then the creator, author, or owner has grounds to enforce the copyright in a lawsuit. Damages for unauthorized copying are limited to only actual economic losses due to the unauthorized copying of the work.

However, if a copyright protected work is registered with the United States federal government at the U.S. Copyright Office, then additional damages can be sought under copyright law.

Companies often register copyrights on:

- Business plans.
- Marketing materials.
- Investor pitch decks.
- Graphic content related to branding and marketing.
- Commercials.
- Jingles.
- Commercial brochures and pamphlets.
- Advertisement copy.
- Print advertisements.
- Web page layout and content.
- Software code.
- Logo designs.
- Stylized product packaging or labeling.

5.4 Trade Secrets

A trade secret is any confidential business information or data that gives a company a competitive advantage in the commercial marketplace, based on the fact that the information is kept secret. Trade secret protection is conferred to confidential business information by not disclosing the information to the general public. Companies need to take active steps to prevent the accidental, inadvertent, or deliberate disclosure of the trade secret information.

Many new companies and startups that have limited financial resources use trade secret protection to their advantage because, for the most part, it is free. Maintaining a trade secret takes minimal effort, and through the diligent use of non-disclosure agreements (NDAs) and keeping trade secret information on a strictly need-to-know basis, misappropriation (i.e., the inadvertent, accidental, or deliberate disclosure of the trade secret information) can easily be avoided.

The trouble with trade secret protection is that once the secret is out, the protection terminates. Furthermore, while trade secret protection might be an effective "starter" form of IP, the trade secret route of IP protection is not always the best option for protecting valuable IP. For instance, keeping a new invention a secret by treating it as a trade secret is a good idea to keep competitors from copying your invention before you have a chance to file a patent application, but patent protection must be sought in order to get the benefits afforded by patent protection.

Case Study

One of the most famous trade secrets in American business is the secret formula for the soft drink Coca-Cola, which has been kept a trade secret for more than 130 years. Over the years, The Coca-Cola Company took painstaking care to ensure that the secret recipe for their signature brown, bubbly soda was protected. Initially the formula was only shared by word of mouth to a handful of individuals on a need-to-know basis. Later, once the formula was written down on paper, the written recipe was kept inside a vault. For the 125th anniversary of The Coca-Cola Company, the recipe was removed from the vault and brought "home" to the World of Coca-Cola in Atlanta, Georgia. To this day, Coca-Cola is vigilant in its efforts to keep the secret recipe a trade secret.

The trouble with trade secret protection is that once the secret is out, protection is lost. Trade secret protection can be lost in a number of different ways:

- **Misappropriation.** Misappropriation of a trade secret occurs when someone who is authorized to have access to trade secret information commits a deliberate bad act and discloses or uses the information without authorization to do so. Misappropriation usually takes the form of theft of a trade secret, corporate espionage, or hacking. When a trade secret is misappropriated or stolen, the trade secret owner can seek legal recourse with a lawsuit.
- **Unauthorized disclosure.** Unauthorized disclosure is when someone who is authorized to have access to trade secret information makes an unauthorized disclosure. The unauthorized disclosure can be deliberate, inadvertent, or accidental. Trade secret owners can seek legal recourse through a lawsuit for trade secret loss due to unauthorized disclosure.
- **Independent discovery.** Independent discovery of a trade secret occurs when another party lawfully discovers the trade secret information through their own independent means. The other party could have been looking to

solve the same problem as the trade secret owner and just happened to come up with the same solution as the trade secret. When trade secret information is found out through independent discovery, the trade secret owner has no legal recourse.

- **Reverse engineering.** Reverse engineering occurs when someone figures out the trade secret information through purposeful, lawful efforts. Loss of trade secret protection in this way often happens because someone legally obtains a product containing trade secret information, and then works backward to reverse engineer the secret to its success. There is no legal recourse when a trade secret is fairly obtained through reverse engineering.

Case Study

WD-40 is an incredibly handy multipurpose product. The formulation of WD-40 is a trade secret. Here are some hypothetical examples of how the secret formula of WD-40 could be lost:

- **Misappropriation.** If an employee who has knowledge of the secret formula for WD-40 steals the formula and sells it to a competitor, it would be misappropriation of the WD-40 formula trade secret.
- **Unauthorized disclosure.** If an employee who has knowledge of the secret formula for WD-40 talks about the formula at a big party, it would be unauthorized disclosure of the WD-40 formula trade secret.
- **Independent discovery.** If WD-40's biggest competitor independently arrives at the correct formulation through experimentation, it would be independent discovery of the WD-40 formula trade secret.
- **Reverse engineering.** If WD-40's biggest competitor uses a new and highly advanced chemical separation technique to identify the exact composition of WD-40, it would be reverse engineering of the WD-40 formula trade secret.

5.5 Developing an Intellectual Property Strategy and IP Portfolio

Developing an IP strategy that is in alignment with the company's business and marketing strategies can be hugely beneficial to a company with an innovative new technology. In most cases, IP legal counsel will be involved. The particular circumstances surrounding your technology, how best to bring it to market, and identifying how your IP rights will provide the most benefit can help you formulate an IP strategy that is tailored to your company's needs. A company's IP strategy is part of its larger business strategy, and it is important for getting and maintaining a competitive edge in the market.

Securing, maintaining, and exercising IP rights grant the company access to sustainable profits. In the case of developing IP around an invention, often an IP portfolio can be established. An "IP portfolio" refers to a collection of related or unrelated patents, trademarks, and copyrights held by a single individual or company. A visual representation of an IP portfolio for the TRANSLATIO bioabsorbable bone screws is featured in Figure 5.3.

5.5.1 Start by Deciding on Which IP Rights You Need

When it comes to developing an IP strategy, start with a plan of what type of IP rights your company will need to obtain in order to protect the technology, company, and brand. This plan should identify all of the company's relevant intangible assets (e.g., potential IP rights), and then evaluate which form or forms of IP protection would provide the most value. In some cases, a cost–benefit analysis can be conducted, when two IP protection options seem like a good fit (e.g., deciding whether patent protection or trade secret protection would be the best approach). Sometimes it may be necessary and smart to double up on IP rights (e.g., obtaining both patent and copyright protection on software).

Table 5.1 shows a sample list of intangible assets and a plan for securing IP rights for each asset (see Table 5.1).

5.5.2 Developing a Timeline

Next, it is important to determine a timeline for obtaining the IP rights that your company needs. Certain IP rights have specific time frames in which protection must be sought, or the opportunity to secure IP rights could be forfeit. There are also certain legal requirements for filing or registering for various IP rights, which must be satisfied before you can apply or register your IP. An IP lawyer can help your company figure out the timelines you should follow for procuring your IP rights. A sample timeline is shown in Table 5.2.

Use whatever timescale makes sense for your business and IP plan. Remember that your IP timeline is a guide, and is not set in stone. Don't feel constrained or pressured to stick strictly to your planned timeline.

5.5.3 Developing an International IP Strategy

Finally, if your marketing and sales plans involve moving into international markets, then it is prudent to consider securing IP rights in the countries where you plan on conducting business. IP laws vary from country-to-country, so your international IP strategy should reflect the laws of the specific country or countries that are important to your business. Consider working with an IP lawyer to develop an international IP strategy that is tailored to your circumstances.

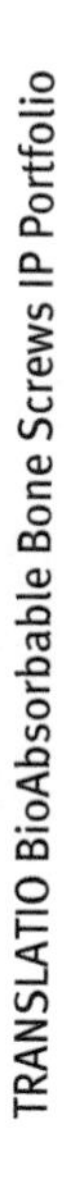

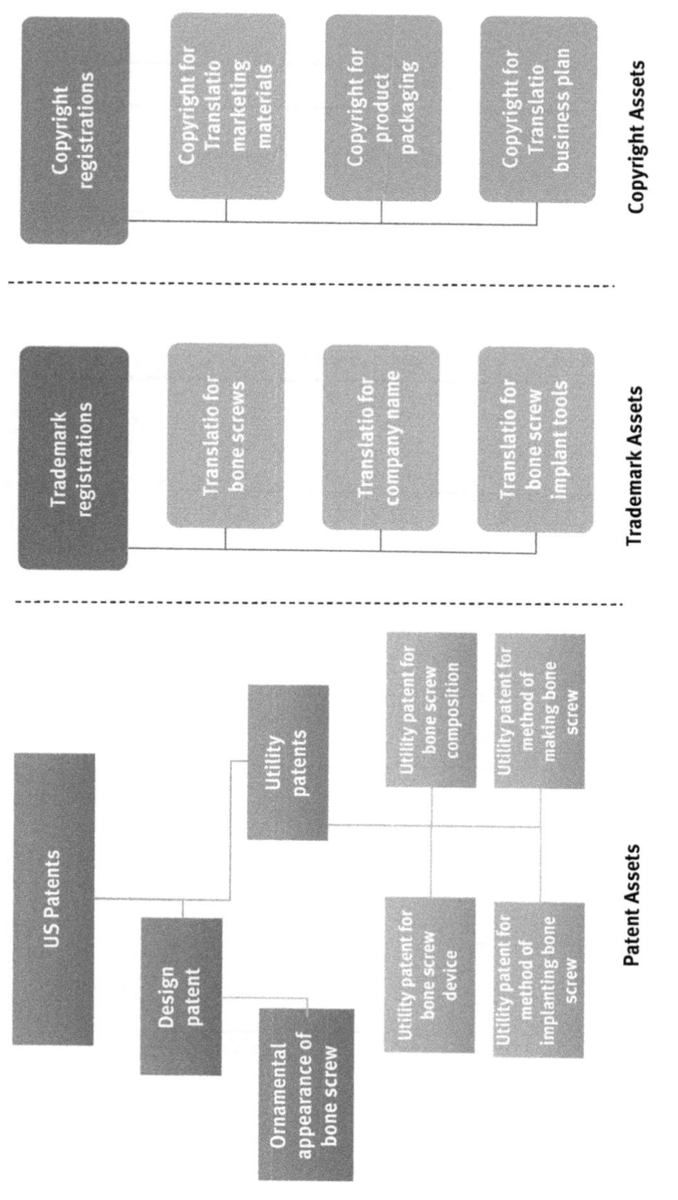

Figure 5.3: Example diagram of related intellectual property rights in an IP portfolio.

Table 5.1: Sample list of intangible assets.

Business Asset	IP Protection
Company Name	Trademark
Company Logo	Trademark and/or Copyright
Product Name	Trademark
Company Domain Name/Website Address	Trademark Protection/Register Domain Name
Technology/Product	Patent or Trade Secret
Method of Using the Technology/Product	Patent or Trade Secret
Business Plan	Copyright or Trade Secret
User Manual	Copyright
Software Used in Technology/Product	Copyright and/or Trade Secret and/or Patent

Table 5.2: Sample timeline for obtaining IP rights.

	Q1 2020	Q2 2020	Q3 2020
Patent Protection			
Technology/Product		X	
Method of Using the Technology/Product		X	
Software Used in Technology/Product	X		
Trademark Protection			
Company Name	X		
Company Domain Name		X	
Company Logo		X	
Product Name			X
Copyright Protection			
Business Plan		X	
Website Layout			X
User Manual			X
Software Code Used in Technology/Product			X

One of the most useful forms of IP protection abroad is patent protection. Each country has its own legal requirements to obtain a patent, but, it can be a very expensive international IP strategy to go to each country individually to file patent applications. Fortunately, there is an international patent treaty, known as the Patent Cooperation Treaty (PCT), that allows those seeking patent protection in multiple countries to file a single application in a format that is accepted by the patent offices of all participating PCT countries. More than 150 countries are participants in the PCT.

Filing a PCT application gives the applicant 30 months to file formal applications in any PCT participant country's respective patent office where the applicant would like to seek patent protection. For example, a company can file a PCT application with the USPTO and within 30 months of filing the applicant can use the PCT application to file a patent application at the USPTO, the European Patent Office, the Japanese Patent Office, and the Australian Patent Office.

5.5.4 Perform an IP Audit to Determine if You are Missing Any Key IP Rights

Once you have secured some IP rights, your company will need to conduct regular IP audits (annually or more frequently) as part of keeping your IP strategy on target. An IP audit effectively involves taking an inventory of all of the company's IP rights, evaluating the scope of each IP asset, and identifying any potential holes or gaps in the company's IP portfolio.

Remember that you don't have to create all of your own IP rights from scratch. You could also obtain IP rights by purchasing them from others (and having ownership of those rights assigned to your or your company). You could also fill gaps in your IP portfolio by licensing the IP rights you need from others. Licensing will be discussed more in Section 5.7.

5.5.5 Monitor Your Intellectual Property Rights, and Your Competitors'

In addition to procuring IP rights, it is prudent to continuously monitor your own IP rights and to also carefully monitor the IP rights status of your closest competitors on a continuous basis. Certain IP rights that you might hold may require payment of occasional maintenance fees (for instance, US utility patents require regular maintenance fees be paid to the USPTO), or may require that you police their use in order to maintain protected status.

Additionally, keeping an eye on what your competitors are patenting can help you and your business:

1. Keep up to date on how your competitors‘ technologies are progressing.
2. Spot when your competitors begin to infringe on your IP rights.
3. Avoid potentially infringing on the IP rights that belong to your competitors.

Don't make the mistake of undervaluing the benefit that IP can have for your technology.

5.6 The Importance of Freedom to Operate

In the realm of IP law, there is a term of great importance to startups and small companies – it is "freedom to operate (FTO)." FTO means that the company is free to make a product or use a process that it wants to use without risk of committing IP infringement. The term "FTO" can be used in two contexts. The first is the form of a legal opinion known as a FTO opinion. The second is in the form of achieving FTO through the use of a license agreement.

5.6.1 FTO Legal Opinions

A good legal professional or two can be very handy, as you begin your innovation adventure. You can always hire an IP lawyer to provide you with a FTO opinion as part of your due diligence efforts. A FTO opinion is a legal opinion provided by an IP lawyer concerning whether your proposed invention/innovation would infringe on any existing IP rights held by others.

An FTO opinion is based on a search that is conducted by a trained IP attorney, who surveys the current state of the art for existing IP rights which your planned IP rights may infringe upon. While FTO opinions are usually discussed as relating to patent rights, they can also be used to evaluate other forms of IP rights, such as trademarks. In the case of an FTO opinion related to patent rights, it is important to make sure that the FTO opinion is produced by a registered patent attorney.

An FTO opinion based on patent rights looks at existing patents and published patent applications that are directed to similar inventions as the invention you plan on claiming in your own patent applications. The search will involve reviewing both foreign and domestic patent databases and documenting any relevant patents or published patent applications in a report. From the identified patented documents, the patent lawyer can conduct an infringement analysis to determine whether your proposed invention, innovation, or product infringes any currently issued US patents. The analysis might also include identifying any published patent applications that you may be at risk of infringing in the event that the patent application is ultimately issued as a valid patent.

An FTO opinion is a way to evaluate the IP landscape that your innovative product will ultimately enter into, to make sure that there is room for you and your innovation to participate without infringing on the IP rights of others. The opinion will evaluate the market you plan to enter for IP barriers to entry and will assess the business risks associated with bringing your product to market. From the results of the FTO opinion, you can decide whether pursuing IP protections, such as patents, would be a good investment of time and resources for your company.

The only way to get an FTO legal opinion is from an attorney. And if the FTO specifically involves assessing patent rights, which is most likely the case, it is prudent to obtain the opinion from an attorney who has the added credential of being a registered patent attorney.

Following are few useful questions which you can ask your IP attorney about FTO opinions:

1. Does everyone need to do an FTO?
2. How does one do a preliminary FTO?
3. What does it mean to infringe on someone else's IP rights and how is that defined?
4. If you get an FTO legal opinion and it is wrong, then who is liable?

5.7 Licensing Agreements

FTO can also be achieved through the use of IP licensing agreements. To reiterate, in an IP licensing agreement, two parties agree to an arrangement where one party is permitted to use the IP rights of the other in exchange for some form of compensation, such as a licensing fee, or other benefit, such as the right to use the other's IP rights in return. Licensing agreements can be one-way licensing agreements (meaning that only one party needs the IP rights of the other), or they can be two-way licensing agreements (referred to as a cross-licensing agreement, where the parties each need the freedom to use the IP rights owned by the other).

Licensing agreements can be useful in overcoming IP obstacles that might prevent your business from moving in the direction that it needs to go in order to grow and be successful. For instance, if you obtain an FTO legal opinion that identifies a patent which poses a risk of being infringed by your planned business activities, you could mitigate your risk by approaching the owner of that patent and asking about licensing opportunities. Or, you could offer to purchase the patent outright, if buying the patent would be beneficial for your company's IP portfolio and funds permit.

A cross-licensing agreement is when two parties need or want to use the IP rights owned by one another, and they agree to permit each other the right to use their IP rights freely. Cross-licensing agreements are a common practice, especially in the technology sector.

Cross-licensing agreements can be limited in terms of what rights are being licensed (the parties might only agree to license to one another certain patents that they each respectively own), duration of the agreement (the cross-licensing agreement is only valid for two years), and scope (one company might only be permitted to produce a certain number of authorized copies of copyright protected software code owned by the other company in the cross-licensing agreement).

Virtually every major industry has cross-licensing. Google, for instance, is involved in several cross-licensing agreements; Google has a broad cross-licensing agreement in place with Samsung, as well as cross-licensing agreements with SAP and Tencent. Apple and Microsoft have been engaging in cross-licensing of patents for decades. Microsoft and Xiaomi are also in a cross-licensing agreement, and so are Ericsson and Huawei. Going outside of the technology sector, there are examples like Hyundai and Audi, which are both parties to a patent cross-licensing deal, and Garmin and Navico, which entered into a cross-licensing agreement to put an end to some long-running litigation.

Licensing in order to achieve FTO is a popular technique for avoiding infringement, because it is considerably more affordable to license IP rights from others than to be sued for infringement in court. The duration of the licensing process depends on how long the parties take to negotiate a licensing deal. The costs associated with a license are also dependent on what the parties agree to.

6 Funding Methods

Abstract: Finding the necessary funds, or capital, to help your company develop its product and to grow is a long, complicated, and ongoing process that is necessary for most companies at some point in their life cycle. You will need to have a good idea of what the funds are designated for (e.g., prototype development and growth/expansion), how much you will need, what sources of funding are most appropriate for your company (e.g., debt funding, equity funding, a blend of the two, and funding from IP licensing revenues), and what type of funding arrangements are acceptable to you (e.g., loans and convertible notes). This chapter explores the various types of funding (dilutive funding sources, such as equity financing, debt–equity hybrid funding and other forms of securities-backed financing; and non-dilutive funding like bootstrapping/self-funding, loans, lines of credit, crowdfunding, venture debt, grants, and awards) and investigates the advantages and disadvantages of each type. This chapter also discusses strategies for securing funding, including tips for making pitch presentations to potential investors and monetizing intellectual property (IP) assets (such as licensing IP assets and royalty-backed funding) as a means for producing revenue to fund the growth of your company.

Keywords: Exit strategy, lifestyle, acquisition, initial public offering (IPO), pre-seed funding, bootstrapping, self-funding, seed funding, early-stage funding, growth funding, crowdfunding, equity, investors, shares, stock, venture capital, angel investors, series financing, convertible notes, debt, bridge, venture, royalty-backed, grants, competitionawards, awards, pitching to prospective investors.

One aspect of the Business Development Process that nearly all companies encounter is the need to secure funds, in order to finance the future growth and development of the business. As shown in Figure 6.1, the stages of Seeking Funding and Pitching to Investors are often key steps in the Business Development Process, as funds are often needed for any number of reasons, such as to conduct additional research and development, finance prototype development, facilitate scaling-up of manufacturing capabilities, or to further the expansion of the company (for instance, to a new facility, geographical location, etc.).

6.1 Start with the End in Mind: Know Your Exit Strategy

To formulate a funding strategy for your business, it helps to start with the end in mind, that is, what is the exit strategy? The "end game" or finish line for your company will dictate the funding options that are preferable. The three most popular exit strategies are lifestyle, acquisition, or initial public offering (IPO). When

https://doi.org/10.1515/9783110521900-007

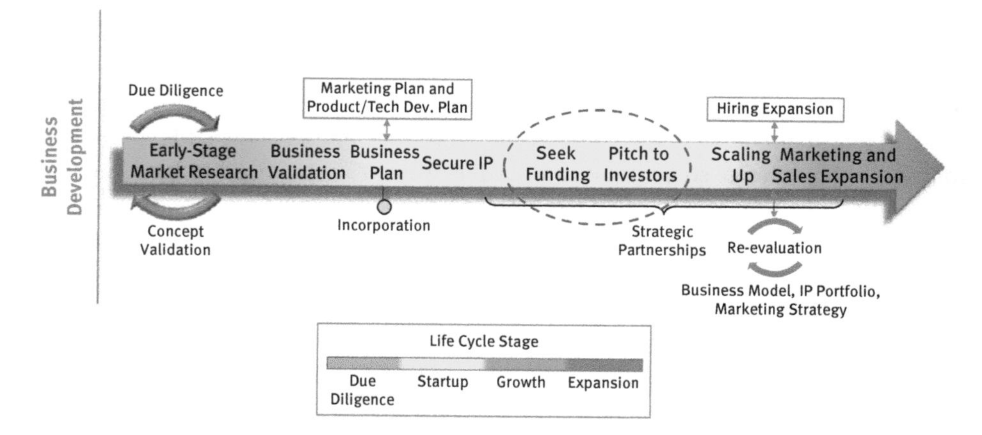

Figure 6.1: Seeking funding and pitching to prospective investors are key steps of the Business Development Process.

acquisition or IPO occurs, the investors get "cashed out." This means that each investor is looking for an ideal monetary return over an acceptable time period. Each industry and type of investor will have different criteria. For example, a venture capitalist (VC) may be responsible to their fund to show a ten time return in five to seven years, which can be accomplished via acquisition or IPO.

Lifestyle – The choice **not** to exit is still a strategy. A lifestyle business is one where you do not intend to sell, for example, a mom-and-pop gas station, a gourmet candy store, a custom furniture manufacturer, and an art boutique. Optimal funding for this exit strategy would be personal savings, bank loans (debt financing), and/or funding from friends and family (i.e., personal loans).

Acquisition – Acquisition is when a company is purchased by another company. Developing a company with the intent to be acquired can utilize different approaches, such as:
1. Development of a patent portfolio owned by the company.
2. Development of a product with the intention of growth.
3. Development of a company with the intention of product line expansion.

The development of a company with the intent of acquisition must focus on being a long-term, growing company. This includes multiple stages of growth, where a funding round will precede a period of growth. Different stages of growth can determine what type of investors will be interested and/or involved.

IPO – An *initial public offering* (IPO) is when an investment bank makes the securities (i.e., shares of the company) publicly available on the stock market, and the company exits from being a privately held venture. An IPO is conducted once a company is stable (most likely has conducted many funding series rounds) and after careful analysis and due diligence. If successful, this particular exit strategy will maximize the returns of early investors and raise a significant amount of capital for the business's next expansion.

The exit strategy, whether it is to (1) stay private and find a niche for operating the business in the long term, (2) increase the valuation of the business to prepare for acquisition in the near to mid-term, or (3) expand your portfolio, go public, and grow across multiple markets, possibly even by merger and acquisition, will influence many aspects of the business.

The exit strategy will influence the business's:

1. **Business Model:** If a company plans to offer its products and services within a small market and does not intend to be acquired or go public, then the business model will not need to be as complex. For instance, it may take a smaller number of marketing channels to engage customers, and it may take a smaller number of company partners to accomplish company activities. This type of company should aim to simplify its cost structure and maximize revenue from its identified revenue streams.
2. **Go-to-Market Strategy (Commercialization Strategy):** For instance, a company that is positioning itself to go to market in specific markets needs to realize that there is the possibility of being acquired by a strategic partner before another round of fundraising is required, if certain key valuation milestones are reached. Acquisition may be more attractive for the Board of Directors and for majority shareholders, because another round of fundraising would dilute their shares in the company and dilute their investment.
3. **Technology Development Plan:** A company that is developing a new product or technology needs to be mindful that the various stages of the technology life cycle reflect technology development milestones. A company's exit strategy can influence technology development plans. For example, a company plans to offer a product in a range of different sizes and, just before Verification and Validation (V&V) of the design is about to commence, leadership realizes that a lot of time and money can be saved if V&V is completed with most but not all of the size offerings. The sizes that leadership wants to focus on represent a majority of the sizes that will be used by customers, which would also satisfy a majority of the customer requirements and workflows. Leadership is confident that an earlier completion of design and development and transition to manufacturing will lead to an earlier exit.
4. **Risk Acceptance:** If the investors and Board of Directors of a company want to get to market as quickly as possible, in order to be acquired post-product launch, then the development team will prioritize finalizing development activities and will refrain from making additional change to the design prior to market launch – even if there are some defects or anomalies, which the company is willing to accept. If the Board of Directors receive an offer to

merge with a strategic partner in earlier rounds of funding prior to market launch, then the company needs to decide if the return on investment (ROI) is lucrative enough and if there are opportunities to own a share of the combined organization after the merger.

5. **Organizational Strategy:** Is your business going to grow and actively hire full-time employees or remain lean with as few full-time hires as possible and rely on temporary staff or consultants? Is the business going to outsource development and manufacturing and distribution of products or develop and grow these capabilities internally? Is the business going to hire consultants for R&D, Quality, or Regulatory needs? To remain lean, it is typical for companies to outsource shared services, such as IT, payroll and accounting, and legal.
6. **Legal Structure:** The legal structure of a business plays a role in a company's exit strategy as well. For instance, a limited liability company (LLC) may be the right structure for a private business, but prospective investors will often require a corporation structure.[15]

6.1.1 Stages of Funding

Companies need money in order to finance growth and development. Funding for startups and fledgling companies proceeds in stages, based on where the company is in its life cycle. The stages of funding include pre-seed funding, seed funding, early-stage funding, growth-stage funding, bridge (mezzanine) funding, and public funding.

- **Pre-seed Funding.** Pre-seed funding is the earliest stage of funding available to a startup. Pre-seed funds are usually provided by the founder of the company, that is, the pre-seed funding stage is a self-funded stage. Self-funding is referred to as bootstrapping, which will be discussed in more detail in Section 6.2.2.
- **Seed Funding.** Seed funding, which is also sometimes referred to as startup funding, is the next stage of funding. At the seed funding stage, other investors may begin to get involved in the process of funding the startup. Loans can come from close friends and family, personal debt vehicles (such as credit cards and small personal loans), and the internet (through crowdfunding).
- **Early-stage Funding.** Early-stage funding is when professional investors, such as angel investors and VCs, begin to get involved in the funding of the startup.

15 Once a startup gets to a level where multiple investors are getting involved and there are a lot of equity stakes in the company, investors will push the startup to convert to a corporate structure, such as a C corporation, because these corporate structures provide certain legal protections for shareholders. For instance, with a C corporation structure, priority is given to shareholders who invested early. If the company fails, the early shareholders are prioritized, in terms of getting paid out.

Second, larger rounds of crowdfunding also occur at this stage. Debt vehicles, such as larger bank loans, can also be sources of early-stage funding.
- **Growth/Expansion-stage Funding.** Additional series of equity financing and additional loans are sources of growth-stage funding. In the growth stage, companies need additional influxes of capital to grow and expand the business. Two important types of growth/expansion-stage funding include:
  - **Bridge or Mezzanine Funding.** When fairly established companies need to finance big moves in the continued growth of their company, such as a merger or acquisition, they can seek out bridge or stop-gap funding, or mezzanine financing, which is a hybrid blend of debt and equity financing.
  - **Public Funding.** Once a company is off on a solid trajectory, the company can seek additional funding from the public in the form of an IPO and stock or bond issuances.

Stages of funding correspond to the stages of the life cycle of a company and are as follows: the due diligence or seed and development stage, the startup stage, the growth stage, the expansion stage, and the maturity and possible exit stage. Refer to Figure 6.2.

Life Cycle Stages of a Business

Due Diligence	Startup	Growth	Expansion	Maturity/Exit
Gestation period of a business. Due diligence, initial market research, and validation techniques are used to decide if business should go forward.	Company is officially formed and launched. Very exciting, but also most risky stage of life cycle. The search for funding starts here.	Business grows into an established company. Stability improves, revenues rise and the company scales up to meet demand.	Company makes expansion moves, such as expanding the company's workforce, product lines, moving into new regions, and/or markets.	Company either continues business as usual (as a lifestyle business) or makes an exit.

Corresponding Stages of Funding

Pre-seed funding. Seed funding.	Early-stage funding.	Growth/expansion funding.	Expansion funding. Public funding. Bridge funding. Mezzanine funding.	Bridge funding. Mezzanine funding. Public funding.

Figure 6.2: Business life cycle stages and corresponding funding stages.

Pre-seed and seed funding are generally obtained in the due diligence or seed and development stage of a company's life cycle. Early-stage funding is usually obtained during the startup stage of a company's life cycle. Growth/expansion funding is usually sought during the growth and expansion stages of a company's life cycle. Finally, bridge and mezzanine funding are sought in the expansion stage, as well as the maturity and exit stage of a company's life cycle. The stages of funding, and the various funding formats, will be explained in greater detail in this chapter.

6.2 Funding

6.2.1 Risk and De-Risking

Financial "risk" refers to the degree of uncertainty for, or chance of loss in, an investment. The more uncertainty there is upon investment means there should be a higher chance for reward for the investor(s), once it is time to be compensated. Higher risk often translates to higher reward. When investors look at your company, they will do their best to assess what are the risks surrounding the company and technology at a specific point in time and then calculate how much it is worth for the investor to get involved. Investors will assess the following areas of risk during their due diligence:

1. People – Is the management team the right team, and do they practice good leadership skills? Is the team staffed with the appropriate experts?
2. Market – Has the company identified a genuine market need? Is the addressable market accurate?
3. Competition – Who are the closest competitors and how does the technology compare to the gold standard? Is there a gold standard?
4. Technology – Is it technologically feasible to develop the product or process?
5. Operations – Can the company strategically and tactically implement the innovation?
6. Financial – Is the technology financially viable? Is the company charging customers the right amount to procure the technology (not too much, not too little)? Can the company realistically sell the product or service for more than it costs to make it?
7. Legal – What is the status and quality of the patents, patent portfolio, freedom to operate opinions, and licensing agreements that are associated with the technology?
8. Regulatory – What are the foreseeable regulatory hurdles and how has the company accounted for them?
9. Systemic – Is there something inherently wrong with the company's ethics, approach, team, or structure?

In many cases, startups will notice that they may have trouble getting "first money in," commonly also referred to as an initial investor. This is where non-dilutive funding from sources such as competitive awards and grants (Section 6.2.3.4) can help bring in initial funding. But more importantly, the company can show potential investors that a respected authority, such as the National Institute of Health (NIH), National Science Foundation (NSF), and Department of Defense (DoD), has reviewed an organized and structured funding proposal, and a peer-reviewed panel of experts have found the intellectual merit of the proposal worthy of being funded.

One should never mislead investors about the capabilities of your concept, prototype, minimum viable product (MVP), or final product. An investor should understand that a company is "selling" itself, meaning that potential investors should conduct their own due diligence on the viability of the business model and product.

Startups often try to protect themselves by only accepting investment from accredited investors. At a very high level, an accredited investor is a person, business, or organization that has demonstrated to the US Securities and Exchange Commission (SEC) that they have the monetary means to invest. Nonaccredited investors are limited in their investment choices, and the SEC enforces this rule to protect people who are unfamiliar with the risks associated with investing. This rule prevents nonaccredited investors from getting involved in risky investments. Each country has its own criteria and certification process for investors. In the United States, accredited investors are defined in Rule 501 of Regulation D codified in 17 C. F.R. § 230.501 (Title 17 at Section 230.501 *et seq.*, of the Code of Federal Regulations).[16] Each country provides its own regulations and exemptions surrounding investment and corresponding risk disclosures.

6.2.2 Pre-seed Funding/Bootstrapping

At the beginning, your innovative idea and business model are untested, so funding is tough to come by. Most innovators and entrepreneurs resort to initially funding their idea by bootstrapping, that is, internal funding and stretching these funds as far as possible. If you are independently wealthy and can bootstrap the development of your technology from the ideation stage through development, that is wonderful for you! While you are taking on a lot of potential risk, your rewards could be vast if you are successful. However, not all engineers, scientists, and doctors who turn into entrepreneurs are able to bootstrap their technology and their company's development alone. Bootstrapping is a necessary means for attracting external funding. Once the proof-of-concept or early prototype is developed, then the entrepreneur is in a better position to secure outside funding. A majority of companies, especially young technology startup companies, often have to rely on securing funding from outside sources, in order to finance the growth of their company.

In the earliest days of your company, your potential funding sources will be fairly limited. You can always produce some of your funding through personal financing, but new entrepreneurs often have limited personal funds available to invest into their company. Between living your life, paying bills, and trying to bring your new technology to market (which is likely to consume a significant portion of

16 Rule 501 defines what an accredited investor is, but Rule 506(b) discusses exemptions to the SEC's registration requirements for accredited investors.

your free time), it is often impractical for entrepreneurs and innovators to pour much of their own money into developing their company.

Typically, founders of a startup company are willing to put up some initial capital to get their technology off the ground. For instance, they might self-fund an initial prototype of the technology. It is not uncommon for founders to invest several thousands of dollars of their own funds into their companies initially, usually in conjunction with executing an operating agreement that provides a clause concerning how their investment will be repaid to them over time by the company. This initial infusion of cash is pre-seed funding.

You should only attempt to self-finance your company if that is what you want to do and you are comfortable doing so. If you are not in a position to self-finance, don't worry – there are other potential sources of early-stage financing available that you can attempt to secure.

6.2.3 Types of Funding: Equity, Debt, and Non-dilutive Funding

There are several types of funding which correspond with the company's growth life cycle stage (for instance, pre-seed/bootstrapping funding corresponds to the due diligence stage; refer back to Figure 6.2). Each type of funding has different advantages and disadvantages. The most frequently encountered types of funding include equity, debt, and non-dilutive funding.

6.2.3.1 Equity

Equity funding is often the preferred means of securing funding for young companies in need of capital. In exchange for upfront capital, the investor gets a share of ownership of the company. The share in the company is referred to as an equity stake in the company, or simply equity.

An equity funding arrangement is often convenient for the company because it is a way to obtain funding easily. However, giving away equity is effectively giving away a piece of your company, thus diluting your ownership share. Giving away too much equity to investors from the very beginning can leave founders with very little ownership in their own company in the future, because equity is given away with each round of funding.[17] Think carefully about how much equity in your company you are willing to part with in exchange for investment capital, and consider carefully the long-term implications of equity deals you make in the early days of your company's growth. If you give away too much equity in your company, you may lose control over the company you created.

17 As a general rule, the founder(s) of a company should still collectively own over 50% of the company, after completing a Series A round of fundraising.

It is important to note that equity financing is sophisticated and nearly always requires registration with the SEC in the United States. There are exceptions to the registration requirements, but these exemptions have very specific criteria. You should always work with an experienced securities lawyer to make sure that any equity financing agreement is in compliance with SEC and state regulations. Additionally, individual states and countries have their own regulatory agencies and registration requirements.

Equity Funding at the Seed Stage and Series Funding

At the seed funding stage, equity funds can come from a variety of sources, including the internet, that is, crowdfunding, and equity investors, such as VCs and angel investors. Equity funding usually involves the investor providing a loan to the company that is in need of funding, in exchange for a share of equity in the company and repayment of the loan amount at a certain interest rate within a given time frame. In many seed funding situations, the loan might take the form of a convertible debt note, which means that the debt is treated like a loan (and requires repayment like a loan) until (or unless) the company successfully conducts an additional funding round, such as a Series A round of funding. At that point, the original investor's convertible debt note converts into shares in the company, and the debt is released.

Crowdfunding

First, let's look at seed stage fundraising through equity-based crowdfunding. Crowdfunding can be a very successful way of procuring the investment capital that your company needs by empowering many small-time investors to pool together money to create one large investment total for your company to use. Crowdfunding is immensely popular in the technology space, because it offers low risk for the investor and can serve as an affirmation (validation) for you and your company that there is interest in your product in the market.

The crowdfunding process works like this: You post a sales pitch, which includes a description of your company and technology, on a crowdfunding website like KickStarter or GoFundMe. On your view homepage, you can include product demonstration videos, testimonials, and so on. Potential investors review your page and decide if they are interested in pledging an investment amount to your project. Crowdfunding platforms often require that a minimum amount of capital be secured by crowdfunding efforts in a limited window of time. If the minimum funding goal is not met, the investors get their money back and you get nothing. Conversely, if your pitch is well received, you could get substantially more investment than your minimum funding goal.

Usually, the companies seeking investment capital through crowdfunding platforms offer to give investors a reward, such as a sample product or beta version of the technology, based on each investor's respective investment contribution. Companies receiving crowdfunding can also offer securities in exchange for the investor's respective investment contribution (but this is often a complicated process that requires federal SEC filings).

Venture Capital

Equity funding can also come from outside, professional investors who do not have a preexisting relationship with members of the startup. VCs and angel investors often provide equity funding to startups at the seed and early-stages.

Venture capital is equity funding that comes from VCs. The investment totals for venture capital can be significant – this can range from hundreds of thousands to multiple millions of US dollars. VCs often make equity deals where the company gets the capital it needs to grow in exchange for an equity stake in the company. For example, a venture capital firm may take a 25% equity stake with an expected ROI of ten times in five to seven years. Under these conditions, the venture capital firm will have the right to exert control of the company, if the ROI conditions are not on track. Additionally, the venture capital firm will require an appointment to the company's Board of Directors and may require a vote in all major business decisions. If the company does not meet the expected milestones, the venture capital firm may exercise its right to replace the management team.

Angel Investors

Angel investors are wealthy individuals who have an interest in investing in companies that have a working prototype or a demonstrated successful business model. Their investments are often made from the angel's own personal funds. The amount of investment made by angels varies, and whether the angel wants to play a significant role in the operation of the company will depend on the individual angel investor.

Since angels often invest in industries and types of technologies that the angel is familiar with through prior work experience, hobbies, and so on, angels often are a valuable resource to the companies that they invest in, because they are professionally well connected and very knowledgeable.

After bootstrapping and seed fundraising, a company has started to make some progress in developing its product and securing intellectual property (IP), and it becomes easier for a formal business valuation to be conducted which leads to a Series financing round. Once this happens, potential investors now have a basis to evaluate the company and judge the risk associated with investing in the company; thus, the company is being "de-risked." This opens the company up to more opportunities to raise additional funds from equity funding sources like VCs and angel investors. Additional stock issuances become a critical way for companies to secure additional funding for growth and expansion.

Series Financing

Series financing refers to rounds of equity funding that occur in sequential order. The first round is referred to as Series A, the second is Series B, the third is Series C, and so on. Each investment round begins with a valuation of the company (discussed in more detail below), and investors can make an offer to the company to be part of the particular series of funding. In exchange for the funding, the company issues equity in the form of shares of the company. The shares are referred to as Series "X" shares ("X" corresponding to the Series letter that they are associated with, for example, Series A shares are issued for providing Series A funding).

As a general rule, companies use Series A to raise the funds needed to finish developing the technology, and Series B to grow the company (for instance, into a new geographical region, or to bolster the company's workforce). Series C is often sought when the company needs funds to scale-up production, but can also be sought for other big ventures, like purchasing a smaller company or securing licenses for technology needed for the company to become more efficient or competitive. But this is not always the case. Some technology companies use Series A, B, and C funding for product development and additional product testing prior to commercial launch.

Ownership and Valuation

The incorporation of your company will set the stage for how your company is able to grow and expand. When incorporating, the corporate structure will determine how ownership is distributed in your company. Ownership is the percentage of the company that someone owns. For companies that issue stock, the percentage of equity is determined by the number of shares someone owns relative to the total number of shares that have been issued. Many people are concerned about percent ownership, because it can determine who ultimately has control of a company.

For companies that have issued stock, also known as shares,[18] the initial number of shares issued is an arbitrary number. The value of those shares is based off how much the company is worth. The meticulous process of determining the value of the business is called a business valuation, or "valuation." A valuation can also represent the final value of the company determined from the business valuation process used to estimate the economic value of a business.

At the beginning of a startup this process is relatively simple because the value of each company share will be the value of the total amount of capital (for instance, cash in the bank, equipment, etc.) the founders put into the company divided by the number of shares issued. Many startups will issue anywhere from

18 Different types of shares can be issued (such as preferred or common). The type of share will dictate voting rights and who gets compensated in which order, if a company is sold.

10,000 to 10,000,000 shares during incorporation. As investment occurs and the company begins to take shape, the valuation becomes much more complex, due to the different types of capital raised, equipment purchased, prior investments, and debts that need to be accounted for. During the seed fundraising stage, an accurate valuation is difficult to calculate, because the company is pre-revenue. To help startup companies raise funds, convertible notes have been developed so that investors can be rewarded for putting money in early while, at the same time, waiting until a true valuation can be conducted (often conducted just prior to the Series A round). During the valuation of a business, a third-party assessment is conducted to determine the current worth of the business by evaluating every aspect.

Capitalization Tables and Dilution

As mentioned before, the number of shares issued by the company relative to the number of shares purchased per investor will determine the percent ownership of the company of each investor. This is documented in a capitalization table. The capitalization table and all related contracts are usually developed and maintained by corporate attorneys, so that there is consistency between funding rounds and so there is a central authority that keeps all investment documents organized. As more funding rounds occur, the more companies will need this organization.

The concept of dilution is a good example for why companies need this level of organization. Dilution is when additional shares are issued, in effect, diluting down the percent ownership of the individual, individuals, entity, or entities that owned the company. This dilution occurs with every funding round. Losing or reducing one's percent ownership sounds bad, but percentage is not the important number; rather, the price per share is what is important. With each funding round, the valuation of the company is increased (hopefully), thus increasing the value of each share. Issuances of additional shares are used to bring in more capital to grow the company so that the value of the company continues to increase, thus increasing the value of each investor's shares.

Convertible Notes

Convertible notes are often used as a vehicle for raising money for the company. Typically, convertible notes are issued during a funding round. Convertible notes are a type of short-term debt that convert into ownership equity. The advantage of convertible notes is that the company and investors do not need to determine the value of the company prior to series fundraising rounds. The value will usually be set during the Series A round, when the company is more stable and the valuation will be more accurate. With convertible notes, investors loan money to a startup, and instead of a return in the form of principal plus interest investors receive shares of stock in the company equal to the value of the loan, the interest, and in most cases a discount rate on the value of the shares set by the valuation.[19]

Though there is a lot involved with convertible notes, the most significant points are:

1. **The discount rate.** The "discount" rate represents the discount received when a valuation is eventually conducted.
2. **Valuation cap.** The valuation cap is an agreed-upon dollar amount between the company and the investor stating that if the business valuation exceeds the cap, then all calculations are conducted from the agreed upon cap amount. The valuation cap ensures the convertible noteholder is rewarded appropriately in case the company does extremely well very quickly; otherwise, their investment can become severely diluted.
3. **Interest rate.** The interest rate is typical of any loan. Since the investor is lending money to the company, convertible notes can accrue interest as well. The main difference between a traditional loan and a convertible note is that instead of being paid back in cash, this interest is applied to the principal invested, thus increasing the number of shares issued when the note is "converted" to shares.
4. **Maturity date.** The maturity date is the date at which time the company needs to repay the note.

The following examples are intended to demonstrate the calculation process for different parts of a convertible note. Interest is not given in the examples to simplify the numbers, but simply, if an investor loans $10,000 at 8% per annum, at the one year mark $10,800 is used to purchase shares, rather than $10,000. Three scenarios are given to help explain the intricacies of convertible notes:

1. Valuation Cap, No Discount Rate (Cap, No Discount)
2. No Valuation Cap, Discount Rate (No Cap, Discount)
3. Valuation Cap and Discount Rate (Cap, Discount)

1 Valuation Cap, No Discount Rate (Cap, No Discount)

An investor invests $10,000 in your startup's seed round using a convertible note with a $5M cap. If during the Series A the pre-money valuation[20] is determined to be $10M, resulting in the price per share being set at $5.00 (i.e., $10M at $5 per share = 2,000,000 original shares), since the cap is $5M, then

19 If the company fails, then the convertible note is strictly a loan.

20 "Pre-money valuation" refers to how much a company is worth before an investor puts cash in. As investors add cash to a company, the value of the company being invested in will be different before and after the investment. The pre-money valuation during a seed round (before a formal valuation is conducted) will be the estimate of the amount of money you will have raised by the end of the seed round (e.g., a company needs to raise $1M for their seed round, so the pre-money valuation will be stated to be $1M).

the price per share would be $2.50 per share (or $5.00 × $5M cap $10M pre-money valuation) for the seed round investor. In this scenario, the cap would apply therefore converting the $10,000 note at $2.50 per share, giving the investor 4,000 shares of Series A Preferred Stock (i.e., $10,000 at $2.50 per share), where the investor coming in at the Series A round would only get 2,000 shares, if he or she invested $10,000.

2 No Valuation Cap, Discount Rate (No Cap, 20% Discount)

An investor invests $10,000 in your startup's seed round using a convertible note with a 20% discount. In this scenario, the pre-money valuation determined during the Series A round dictates price per share. For this scenario, assume the pre-money valuation set the price at $5 per share. The 20% discount means that the investor only pays $4 per share. The $10,000 convertible note translates into 2,500 Series A Preferred Stock, whereas the Series A round investor would only receive 2,000 Series A Preferred Stock for a $10,000 investment.

3 Valuation Cap and Discount Rate (Cap, 20% Discount)

In the case of a cap and a discount, either the cap or the discount is triggered. In other words, the better calculation on behalf of the investor prevails, the cap or the discount. An investor invests $10,000 in your startup's seed round using a convertible note with a $5M cap and a 20% discount.

3a. **Discount trigger.** If during the Series A, the pre-money valuation is determined to be $5M, and the price per share is set at $5.00, then the discount will make the initial seed round investor pay $4.00 per share (or $5.00 –20%), whereas the Series A round investor will be paying the full $5. Respectively, the seed round investor would get 2,500 shares, where the Series A investor would get 2,000 shares for the same amount. Since the cap was equivalent to the pre-money valuation, the discount is applied.

3b. **Cap trigger.** Alternatively, if during the Series A the pre-money valuation is determined to be $10M, since the cap is $5M, then the price per share would be $2.50 per share (or $5.00 × $5M cap $10M pre-money valuation) for the seed round investor. In this scenario, the cap would apply, therefore converting the note at $2.50 per share and giving the investor 4,000 shares of Series A Preferred Stock (i.e., $10,000 at $2.50 per share), whereas the Series A investor would only receive 2,000 shares.

As can be seen from the above-mentioned examples, the value of the cap and discount is to offer alternative ways to incentivize the investor, regardless of the pre-money valuation determined during the due diligence conducted for the Series A round. Regardless of the outcome, the earlier the investor puts in, then the reward should be greater because of the "de-risking" that has occurred due to the early investment. The "convertible" part of the note is that the money put in by the investors is converted into shares of the company.

6.2.3.2 Debt

Debt funding is another option for financing your company's growth, but it is often one of the least attractive funding options that exists. A lot of entrepreneurs are reluctant to take on pure debt, by taking out loans from a bank or putting expenses on credit cards, in order to fund their company's growth. These pure debt-financing options are undesirable because they pose a risk of loss to the entrepreneur – if the company fails, the entrepreneur is still burdened with the debt load. The debt could be dischargeable through bankruptcy, but bankruptcy is detrimental to your credit rating and can be quite a hassle that takes years to fully resolve in bankruptcy court.

One way that many young technology companies utilize debt-financing in a way that is slightly more palatable is by using IP assets for securing a loan. Not all financiers are willing to accept this type of transaction, but there are many banks that are willing to offer financing that is backed by valuable IP assets. This is a good funding option, if you have or can obtain a valuation of the established IP assets that belong to your company. For instance, if your company holds a few patents that are being licensed, the patents clearly hold a monetary value for valuation purposes. The value from these patent assets can be pooled and used to secure a loan.

Alternatively, some banks will allow IP rights to be used as collateral for a loan. In exchange for a loan amount, a valuable IP asset is assigned to the lender, until such a time that the debt balance is repaid by the company. When the debt balance is repaid, often with interest, the lender will assign the IP rights back to the company. If the company fails to repay the loan, the lender reserves the right to sell the IP asset to recoup its losses on the company's unpaid loan. Collateralizing IP assets means that you could ultimately lose the assets if the debt is not repaid, but many entrepreneurs are more comfortable with this debt-financing approach than with other debt-funding options.

It is understandable that you might want to avoid pure-debt financing, but sometimes it is your only viable funding option. Think carefully about whether you really want to use debt vehicles to finance your startup. Perhaps you should make sure to exhaust all other avenues of funding that are available to you, before opting for debt financing.

6.2.3.3 Growth Stage Debt Funding Types: Bridge, Venture, and Royalty-Backed

At the growth stage, companies often need larger investments, in order to finance the big moves the company is making. At this point in the life cycle of the business, the owners (and shareholders) may not be interested in obtaining more financing through equity, since such arrangements can lead to the dilution of each shareholder's stake in the company. Instead, the company may turn to debt funding sources to finance growth. Additionally, since the company is fairly established and is likely producing consistent revenue streams in the growth stage, the company may be better equipped to handle debt financing than when it was in the startup stage and early-stage of its life cycle.

There are several debt-funding options at the growth stage of a company's life cycle, namely bridge financing, venture debt, and royalty-backed funding.

Bridge financing, or a bridge loan, is a type of short-term funding solution meant to "bridge-the-gap" between an immediate funding need and securing long-term funding, which may take time to put into place. Bridge financing can either be debt-based, in the form of a loan, or as an equity investment. Debt-based bridge financing could be as simple as a company taking out a short-term, high-interest loan while waiting for an approved, long-term funding strategy to be fully implemented. For example, if a commercial real estate investor is setting up an investment fund (which can take up to a year) and sees that a property becomes available that would be part of the commercial development site, the investor can purchase the property using a bridge loan, knowing the property will be refinanced or purchased using the funds, once the appropriate paperwork is complete. While bridge loans are often immensely useful, they are still risky. Interest rates can be so high that it can cause downstream financial struggles if the repayment term becomes longer than expected.

Another bridge-financing approach is to sell equity to a venture capital firm. Equity-based bridge financing avoids debt, but usually at a relatively high cost in terms of equity. In this case, a company would likely seek out an established venture capital firm for a bridge-financing round to provide the company with capital, until it can raise a larger round of funding in the future.

Venture debt is debt financing provided to venture-backed companies by non-traditional lenders, in order to minimize equity dilution for investors. Venture debt can complement venture capital fundraising. Venture debt lenders combine their loans with the right to purchase equity, known as a warrant, to account for the high-risk nature of the investment.

One final form of funding that is used at the growth stage is a debt-like type funding known as royalty-backed funding. In a royalty-backed funding arrangement, a company leverages a percentage of its future revenues in the form of regular royalty payments, in order to obtain funding capital. By way of example, let's say a company has IP assets that it licenses for royalties to a third party (a third-party licensee pays for the right to use the company's IP). The company needs a short-term loan immediately and makes a royalty-backed loan arrangement with a bank. The bank gives the company an up-front sum of money in the form of a loan, that is to be repaid over time with interest through the royalty payments that the company receives from the third-party licensee.

6.2.3.4 Non-dilutive Funding Sources

Non-dilutive funding is funding that does not require the distribution of ownership equity or incur debt. Non-dilutive funding can come in the form of grants, bank loans (debt financing), cash awards from competitions, and so on. This funding can sometimes fund a seed round or a lifestyle company, but is usually not enough to fully fund a company that intends to grow quickly.

Grants

Grants are a good source of funding at any stage, because grants are non-dilutive. Grants come from a number of sources, including private firms and foundations, state and local governments, the federal government, and academic institutions. While grants are often a desirable source of funding because they do not require repayment, the application process often takes a lot of work, eligibility requirements are often strict, and in most cases, will not cover operational business expenses, especially with for-profit companies. The odds of being awarded a grant are generally slim, and if your business is awarded a grant, there are often restrictions on how the grant funds can be spent.

If applying for a grant seems like a source of funding you want to pursue for your business, carefully evaluate the grants you are considering. The application process is often lengthy and work intensive, so make sure you gauge whether the up-front effort necessary to do a thorough job completing the application is worth the likelihood of being awarded the grant. Weigh the effort versus the potential reward carefully.

Now, that is not to say that applying for grants is not worth the time and effort it takes to try and get them. The process of submitting an application becomes easier with each subsequent submission. To determine if your company is a good fit for the grant, evaluate whether your company, business goals, and strategic implementation plan (discussed in Chapter 7) meet the criteria for the grant. If your company, and your cause, seem like a great candidate for the grant you are seeking, it may be worthwhile to complete the grant application and see what happens. Some grants are for designated causes, such as veteran, minority, or women-owned businesses, which can make it more likely to get funding, if you qualify for one of these categories.

In the United States, grants are generally awarded by federal, state, and/or local agencies, as well as foundations. Federal and state grants often have a long grant cycle, meaning that it can be 9–12 months before funding is awarded. Local and foundation grants are usually smaller amounts of money, but funds can be dispersed more quickly. The most common programs for federal funding are the Small Business Innovation Research (SBIR) and Small Business Technology Transfer (STTR) programs. These are overarching programs that are affiliated with each federal agency and are sources for early-stage capital for technology commercialization. For example, you could apply for an SBIR at the National Institute for Health (NIH) in its orthopedics division or you could submit an SBIR application to the NSF in its Division of Biomechanics. Applications will depend on the policy interests of the division within the agency. It is best to speak to a program officer at the agency of interest to you, in order to determine if you are a good fit for a particular grant.

Another potential source of non-dilutive funding is business competitions. Competitions can be hosted by academic institutions, accelerators, incubators, or foundations. They are often directed to fund startups, sometimes in specific fields. When getting involved in a competition, be sure to read the terms and conditions. Some competitions may ask for equity, if you decide to accept their award.[21] Lastly, you may need to ensure that your IP is protected before applying.

In summary, the chart shown in Table 6.1 below displays the types of funding most commonly encountered by companies and whether those funding types can be dilutive or non-dilutive.

Table 6.1: Funding examples.

Dilutive	Non-Dilutive
Equity Financing	Bootstrapping
VC Investments	Loans
Angel Investments	Lines of Credit
Series Financing	Reward-Based Crowdfunding
Securities-Backed Crowdfunding	Bridge Financing
Hybrid/Blended Equity-Debt Vehicles	Venture Debt
Convertible Notes	Grants
Bridge Financing	Competition Awards

6.3 Lessons Learned from Pitching to Prospective Investors

Obtaining funds from investors is a process that can be richly rewarding if funding is successfully obtained. The process starts by getting the attention of prospective investors. Prospective investors, such as angel networks and VC firms, often seek proposals from companies interested in securing funding. If the proposal is intriguing, the company will be invited to present a fundraising pitch to the prospective investors.

Asking people for money is a humbling experience. Knowing that the continued progress of your idea or the growth of your company is dependent on strangers putting their trust in you is daunting and stressful. Many companies work very hard to prepare a stellar sales pitch for prospective investors. It can be tough to put yourself and your company out there in this way. Even though innovation, hard work, and dedication all factor into a successful fundraising round, luck can't be ignored,

21 In most cases, competitions will only ask for equity if the company accepts the award. This means that a company can still compete and win without accepting the terms and conditions, or the prize.

which is sadly out of everyone's control. The process of seeking funding from VCs and angel investors is also unfamiliar to most young technology companies, and so the process can be challenging, stressful, emotional, and sometimes disappointing, when funding is not obtained.

It is very unlikely that you will get funded on your first try. You probably won't get funding on your second or third pitch either. Keep your expectations realistic and know that many prospective investors will not be interested. Instead of losing hope, or doubting yourself, focus on the positive and strive to become better. Every pitch that you and your company do is a learning opportunity and a chance to improve your pitch presentation. Remember that persistence and patience are traits that are a blessing in the world of company fundraising.

The fundraising pitch presentation is tailored to your business and the story you want to tell to your prospective investors. There is no "right way" to prepare your pitch. What is key is that your pitch deck engages and excites your audience of prospective investors. You are unlikely to walk away from your first pitch presentation to prospective investors with funding. Instead, you want to walk away from your first pitch presentation to prospective investors with a second meeting. Some general guidance can be helpful in preparing a fundraising pitch deck.

Preparing Your Pitch Deck

Business pitches are usually between 10 and 20 slides, beginning with a slide that explains your company's vision and value proposition. Make clear from the outset that what your company is doing is worthwhile and creates value. Next, identify what problem your innovative technology is used to address, what target market(s) could benefit from your innovative technology, and where there are potential market opportunities.

Remember, your technology is a solution, and you need to explain your technology to your prospective investors. This can be challenging, especially if your technology is complicated. Storytelling could be a useful technique for explaining your technology and business, as could visual aids. You want to be able to explain your technology in simple terms that anyone can understand.

Your pitch presentation should address what competitors exist in the market and how your company plans to deal with the competition. Identify what alternative solutions are available in the market and take note of any deficiencies or weaknesses of those solutions. Your presentation should make clear where your company and technology fits in the market and should highlight any advantages that your company offers over your competitors.

Investors also want to know what your company plans to do going forward. A roadmap can be helpful in identifying to prospective investors the key milestones your company is striving to achieve. Explaining your company's marketing and sales strategy to your prospective investors will also shed light on how your company plans to conduct itself in the future.

Investors are interested in how your company will make money and how your company will operate. A slide or two of your pitch presentation should be dedicated to explaining your business model and how revenues will be generated. Include an explanation about how certain aspects of your business are or will be funded, the pricing structure used by your business, and an explanation of why your pricing makes sense in your given industry. If you have current sales, you can use the sales as evidence of validation of your technology and business model. Most investors want to see important financial documents, such as sales forecasts, profits and loss statements, and cash flow forecasts into the future. About three to four years' worth of these financial projections is suitable in most situations.

Along similar lines, investors want to know how much investment capital you are asking for, and what you plan to do with the funding, if you get it. You need to be able to justify the amount of funding you need to achieve your business milestones; otherwise, investors will not be likely to invest in your company.

Finally, your pitch presentation should provide the prospective investor with an opportunity to learn more about you and your team. Investors want to know who is running your business operations, and if these people can be trusted. Introduce your team and highlight their qualifications, in your fundraising pitch presentation. If you are missing key players, don't be shy about that fact – the prospective investor might know someone who would be a good fit, and the available position on your team might serve as an incentive for an investor to take the plunge with your company.

An example pitch deck outline is featured below. Remember, this deck is representative of your elevator pitch (i.e., a pitch version of your business idea and product that you could give to someone in as little as 30 s – or the length of an elevator ride) and is a teaser for the business plan.

Cover Slide:

[Insert Company Logo]

Title: Company Name

Insert Company Slogan or Tagline

Team Slide:

Who is on the team?

- What people with opposite/complementary skill sets
- Build a business board with strong and relevant network of connections
- Technical advisory board vouches for our solution
- Why are we the right team to hit the next value creating milestone?
- Have I quit my day job? How much money do we need to raise to pursue your startup full time?

Problem Statement Slide:
What is the problem?
- What is the need?
- Who are our customers?
- What are their pain points?
- How large is the market (in terms of $, incidence, prevalence, etc.).

Solution Executive Summary Slide:
What is the solution?
- Technology description – Do we have media or graphics to show?
- What are the positive attributes of the technology?
- What are the technology's indications for customer use/claims?
- What is unique about our solution?
- Is there third-party evidence that validates our solution, such as grant awards, publications, etc.?
- Is there evidence that customers will buy our product/service?

Value Proposition Slide:
What is the value proposition?
- Why should anyone invest time/money/energy in our idea/company?

Intellectual Property Slide:
What is the IP status?
- Who owns the company's IP? Who holds licenses?
- If we do not have a license, what is our path to acquiring a license?

Competitive Analysis Slide:
Who are the competitors?
- There are NEVER no competitors – Who are the current and emerging competitors?
- What is our competitive advantage?
- How will competitive advantage and differentiation be sustained?

Regulatory Strategy Slide (if applicable):
What is the regulatory path?
- How long will it take?
- How much money will it take?

Revenue/Reimbursement Strategy Slide (if applicable):
What does reimbursement look like?
- Who is the user/payer/customer?
- If they are not already paying for it, prove that customers will pay for it.
- How much will we as a company get paid?
- Sales estimate: peak sales with justification for pricing/reimbursement rates, Cost of Goods Sold.

Implementation/Execution Strategy Slide:
What is the execution timeline?
- What is the current state of development (e.g., beta prototype and customer trials)?
- What are the key milestones?
- What are the timeframes to achieve significant revenue and/or cash flow positive?

Fundraising Strategy Slide:
How much money will it take?
- What is our financial status?
- How much investment are we seeking?
- What are we going to achieve with the funds? What is the budget for each milestone on the previous slide?
- In what form (seed, Series A, etc.)? What value will we create?
- Exit strategy – How will we make money for our investors?

Your fundraising pitch deck should be a highly polished presentation (e.g., using a consistent, visible, and crisp font and using icons and graphics that are not too busy or confusing), lasting about 30 min on average (some are longer). But just as important as the slide deck is the pitch presentation itself. Below are a few pointers to help you prepare for your fundraising pitch presentation.

Before the Pitch Presentation

Getting Prepared
- **First impressions mean everything and professionalism of your pitch presentation is key.** Your pitch deck needs to be in tip-top shape, polished, and professional in appearance. If your team is not good at graphic design, consider hiring a creative design firm to help you with your presentation materials. Polished materials go a long way and first impressions are important.
- **Rehearse, rehearse, rehearse.** Practice the pitch deck until the "story" is second nature. You want to sound casual and comfortable in what you are saying during your pitch to investors. Try rehearsing in front of a full-length mirror and try recording yourself during rehearsal. By watching yourself in a mirror, you can see what the audience will see as they watch you during your presentation. By recording yourself, you can hear how you will sound to your audience. You might be surprised at how many times you hesitate or say "um" when you speak!
- **Know the room you are presenting to.** Do some reconnaissance on the prospective investors you are pitching to in advance of your meeting. Research your audience and know the key dealmakers in the room.

- **Have backups!** Nothing is worse than setting up a pitch meeting, taking up the time of your potential investors, and then losing your presentation materials just before you have to present. Save your presentation materials in several different formats, for example, local hard drives, thumb drives, and on the cloud, to make sure that you will be presentation ready for your pitch meeting.
- **Have an understudy ready to go.** Make sure more than one member of your team can give the pitch to prospective investors. You can never be too prepared – people can fall ill, emergencies occur, travel can be delayed, and so on.
- **Avoid last minute changes to presentation materials.** Do not make last minute changes to your presentation materials. Do not wait until the last minute to print your materials. Do not allow last minute content changes to impact your printing decisions. With regards to presentation materials, there will always be room for improvement, and over time many changes will be made to your content.

What to Wear

- **Look good, feel confident.** On the day of the pitch presentation, pick an outfit that you feel comfortable and confident in. Get a haircut or anything else you need to make you feel like you look good – it will do volumes in terms of your confidence levels during your presentation. A suit or dress suit is usually a good bet on what to wear, and refrain from wearing anything that is too tight or too flashy.
- **Be comfortable.** Wear comfortable shoes and make sure that the outfit you choose for the pitch presentation is comfortable as well. You might be slotted for an hour pitch presentation, and you will likely have to be on your feet the whole time. Furthermore, if the prospective investor is very interested in what you have to say, your one hour presentation might unexpectedly turn into several hours.

Essentials for Doing Your Best

- **Rest up.** Get your sleep the night before your pitch presentation to avoid dark circles under your eyes in the morning.
- **Lay off the coffee.** Do not drink lots of coffee just ahead of the presentation. The caffeine will make you jittery and you may have to leave during the presentation for a restroom break.
- **Do not skip meals.** Eat before your pitch presentation, but do not eat anything that will upset your stomach. Also, eat before getting dressed – you wouldn't want to spill anything on your presentation outfit.
- **Arrive early.** If you are on time to the meeting, you are already late. Get to your meeting location early, so that you have some time to get familiar with your surroundings and so that you can have a chance to settle in.

- **Bring a small water with you to the presentation.** In case your mouth goes dry, have a small amount of water ready.

During the Pitch Presentation

- **Make eye contact, and ask if there are any questions so far.** As you present, make sure to practice good eye contact with those in the audience. You also want to take care that the audience is following along as you present. Make sure to periodically ask if the audience has any questions during the presentation.
- **Investors want to know about you.** While investors are interested in knowing more about the opportunities your company can offer them, investors are primarily interested in you. Investors need to trust you, if they are expected to invest in your company. Let them know more about you to reinforce to them that you are a solid investment.
- **Expand on your slides.** Investors want to hear your story, not a recap of what appears exactly on the slides of your pitch deck. The prospective investors are perfectly capable of reading the content of your pitch slides, so use the presentation as an opportunity to go beyond the content of the slides.
- **Pace your speaking.** When people get nervous, they have a tendency to speak more quickly than they normally would. This is particularly true during fundraising pitches because presenters are usually uncomfortable with making the pitch, asking for money, and putting themselves out there. Measure your speech when you present your pitch to investors. Use pauses and inflections in your voice to draw in the audience.
- **Do not sell to investors, *engage* with them.** Many presenters mistake their fundraising pitch for a sales pitch on why investors should fund their company. You don't want to sell your prospective investors, you want to inspire them on a personal level to get involved in your company's journey.
- **Tell a story.** Storytelling can be a powerful presentation technique when executed properly. Through the use of a compelling narrative, and visual aids where possible and appropriate, you can help your prospective investors to stay engaged and can promote better understanding about your company. A picture or video has the ability to say more about an idea or concept than you ever could.
- **Answer questions honestly.** There is no doubt that the investors will have questions, and in all truthfulness, you aren't always going to have an answer. And that is okay! When asked questions, do your best to answer it, and do not lie. If you do not know the answer to a question, acknowledge the speaker and the question, ask for clarification if necessary, and commit to follow-up later.

After the Pitch Presentation

- **Follow-up is just as important as the pitch.** After your pitch presentation, make sure to follow-up with the prospective investors. You might consider sending a thank you note, or the answer to a question that they had during your presentation. Do not delay in following up. If the investors asked to see an executive summary or business plan, make sure to send a copy of the requested document immediately.
- **Thank those who made the pitch presentation possible.** It's more likely than not that someone set up or facilitated your meeting with the prospective investors for you. In such a case, send a personalized token of thanks to this person for their help.
- **Ask for a second meeting.** If it seemed like the pitch was a good meeting, it might be prudent to ask for a follow-up meeting or discussion at the earliest convenience to sustain the momentum created during the pitch presentation. If the prospective investors ask questions or provide feedback during the pitch presentation, be prepared to address the issue at the next conversation.
- **Ask for feedback.** At the conclusion of the pitch, if there are no questions, or it seems like the investors might not be very interested in funding your company, don't let a golden opportunity for feedback slip by. Ask for feedback about your presentation from the prospective investors. Even if the pitch was a total disaster, which does happen from time to time, you aren't going to make the situation any worse for yourself by asking for constructive feedback.

Moving on from the Pitch Presentation

If the pitch presentation generates a follow-up meeting and things seem to be moving in a positive direction, prospective investors will ask for due diligence packets for the company. Prospective investors need to do due diligence on your company, just like you did due diligence when deciding to pursue commercialization of your innovative product or technology, so that they can determine whether to invest in your company.

In their own due diligence research, prospective investors are checking to see that everything is in order and that you have been truthful in the claims you have made to them. Prospective investors often request due diligence packets that include:

- **Copies of pitch materials.** Investors often want copies of the pitch presentation, along with an executive summary for the business. Additionally, investors want to see a detailed business plan, complete with thorough financial documents, as part of their due diligence research. Other plans may also be requested, such as a technology development plan, marketing and sales plan for the company, a commercialization plan, IP procurement plan, and so on.

- **Copies of organizational and management documents.** Organizational documents, such as annual reports, state filings, authorizations to operate, and so on are used to prove that the company has a legal corporate structure, and management documents such as copies of employee listings, employment agreements, operating agreements, and listings of board members, are used to demonstrate that the company is situated to operate in a clear way. Investors might also ask for copies of any permits or licenses that are required for the company to conduct business.
- **More information about board members.** Investors often request background information about board members, including listings of their qualifications, biographical information about the board members, and curricula vitae for all board members. Contact information for each board member is also typically requested.
- **Evidence of intellectual property assets.** Investors often request documents related to the company's IP portfolio. The investors take an inventory of the IP assets that are owned by the company and sometimes seek a valuation of the portfolio as part of their due diligence.
- **Any legally binding documentation.** Investors often want to review any and all contracts that the company is a party to, in order to get a better understanding of the company's obligations to third parties. This includes material contracts of all types, including founders' agreements, operational agreements, employment agreements, supplier or vendor agreements, leases, licensing agreements, royalty agreements, any prior equity or financing agreements, loans, and so on.
- **Asset listings.** Assets take many forms, and investors want to know what assets the company holds before investing. Lists of physical assets, IP assets, and other intangible assets, like customer lists and trade secret assets, may be requested by the investors.

6.4 Understanding Investor Expectations

If investors give your business money, then they will expect a return of some sort. Each type of investor is looking for something in particular, depending on what stage your company is in. Before taking investment funds, conduct the research necessary to understand what is typical to include in the agreements between businesses and investors. For example, a very early-stage company that only has a concept may need to offer a convertible note with a 30% discount and 8% interest to an angel investor, whereas the same company at a slightly later stage, for instance, at the stage where the company has developed a prototype, may only need to offer a convertible note with a 20% discount and 8% interest to an angel investor.

Most startups are unfamiliar with the motivations, responsibilities, and structure behind how VCs do business. Typically, a company set to receive the funds does not understand the responsibilities that the VC has to his or her fund, and/or does not fully understand the terms and conditions of the venture capital. Venture capital money is expensive, but it has its place, especially when a company is planning a large expansion in preparation for an IPO. VCs will usually not invest in early-stage companies because (a) the investment is still too risky, (b) the VC will want too much of the company at too early of a stage in the company's life cycle, and (c) the investment will most likely be too small to satisfy the VC investment fund.

Additionally, once a business receives investment funds, then investors have a degree of "say" in that business. After each series round, an investment representative will usually receive a seat on the board of directors to ensure that round's investors are treated fairly and to maximize their returns. Before a business accepts investment, the business should conduct due diligence on the prospective investor to ensure that they are a good fit for the team and that their business practices are ethical.

6.5 Common Early Startup Challenges

Startup challenges can vary depending on the stage, or maturity, of the startup. In general, early-stage technology and innovation startups are early in the concept generation and research life cycle phases, and are also early in the fundraising process, whereas late-stage technology and innovation startups are later in the development life cycle and have met fundraising goals. Startups begin pre-revenue and therefore rely solely on fundraising and investments. The requirement for early-stage startups to operate and develop their concepts and business models while surviving in the absence of a steady cash flow is forefront in the mind of the startup's leadership.

Early-stage startups need to come across as very polished to secure funding and partners. Everything that is produced for external viewing and that can be observed by others (e.g., executive summaries, business plan, business cards, videos, website design, how the team prepares for meetings, and follows-up with prospective investors and partners) says something about the startup. Professionalism, competence, and confidence is key – and a unified team with great leadership outwardly exhibits these qualities.

Startups need to be fiscally conservative. For instance, the startup team needs to be savvy about minimizing board meeting fees, traveling expenses, technical conference fees, the cost of marketing materials, business development fees (such as dining and entertaining expenditures), and business operation costs (such as the cost of office space and telecommunication services). Remember, you are not spending your own money, you are spending a finite amount of the startup's investors' money. Therefore, the spending rate or "burn rate" needs to be in control, so that your startup "runway" of cash is as long as possible in duration for the team to achieve the required milestones to increase the valuation of the company and so that investors have continued interest in keeping the startup afloat. A formal financial model (Chapter 4) is an excellent way to display these expenses in context.

Startup operations also need to maintain good credit by keeping track of purchase orders and paying invoices on time (to avoid paying unnecessary interest charges and late payment fees). A development project can quickly get derailed, if a vendor or supplier refuses to provide a service or ship needed materials, if the startup has a habit of being late with payments. Ensure your financial credibility and keep progress moving as a startup by staying on top of accounts payable!

Financial Credibility Example

New vendors will most likely ask a startup to fill out a credit application and request the name of several vendors that do business with the startup. If these references do not come back with good reviews of their working relationship with the startup, then vendors will be forced to ask for all the money on a purchase order up-front and/or net terms on invoices will be reduced (Net 15 days to pay the balance versus Net 30 days to pay the balance). This makes it difficult for accounting personnel to keep up with outstanding invoices, especially if accounting is relying on the postal service, which does not always deliver on a timely basis, to send checks.

Every employee in a startup needs to be independent, flexible, and quick to learn. In a startup, for instance, as an engineer you could be asked to review test protocols and data, assess suppliers and vendors, negotiate contracts, prepare quality and regulatory documents, hire interns, review marketing materials (such as press releases and executive summaries), or to attend trade shows. In a startup, every employee is a valuable asset – the company is only as strong as its weakest link. Therefore, it is important to encourage one another and establish a "team" mentality. Early-stage startups need to be focused on establishing a solid team and advisory panel with experience that is commensurate with the technical and non-technical challenges that lie ahead.

Once funding is in place, businesses need to put their strategies into action. Strategic and tactical implementation of the business plan is discussed in Chapter 7.

7 Launching Your Innovation to Market – Strategy then Implementation

Abstract: Once funding is in place, then it is time to implement the business plan into executable actions. Launching a product, process, or technology requires finding the necessary resources your company needs to be successful. Resources can be funds, people, materials, vendors, time, and energy. This chapter discusses the shift from planning to implementation and considers the resources needed to execute the interdependent strategies that are involved in launching your innovation to market. The high-level plans that are outlined in your business plan require preparation, team communication, and adaptation to become a detailed, multifaceted Strategic Implementation Plan (SIP). The SIP integrates a company's objectives, timeline, financial resources, and deliverables into an actionable plan. This chapter focuses on the implementation of various business-related aspects of launching to market, including personnel hiring, vendor management, marketing and sales, and postmarket launch customer surveillance and analytics.

Keywords: Implementation, Strategic Implementation Plan, resources, personnel hiring, vendor management, vendor search, assessing vendors, vendor qualification, vendor approval, approved vendor list, vendor risk management, sole-source vendor, single-source vendor, request for proposal (RFP), statement of work (SOW), scope creep, marketing and sales, customer relationships, communication channels, postmarket launch customer surveillance and analytics.

7.1 Introduction

Ideally you have reached your fundraising target, so you can start implementing the strategy that you have outlined while preparing your business plan (Chapter 4). Now it is time to shift from due diligence (Chapter 1), analysis (Chapter 2), and asking for money (Chapter 6) to refining the business strategies for execution through a Strategic Implementation Plan (SIP).

The planning associated with a SIP occurs simultaneously while developing the Marketing Plan, Sales Plan, and Technology Plan (which further involves the simultaneous preparation of a Technology Development Plan and corresponding Technology Development Strategic Implementation Plan)[22] for

22 A Technology Development Plan is a technically specific plan focused on the logistics and timeline associated with the development of a technology, and is described more in Chapter 10. A Technology Development Strategic Implementation Plan is a detailed plan for how to execute the development of your technology from a technical perspective, and is discussed in detail in Chapter 10.

https://doi.org/10.1515/9783110521900-008

the Business Plan. In other words, each "plan" underlying the business plan has a corresponding SIP for its execution, and all of the interdependent SIPs are consolidated into a single SIP. Importantly, the SIP identifies the actionable steps necessary and assigns resources (such as funds, people, materials, vendors, time, and energy) to the various tasks that must be completed in order to execute the SIP. Execution of the SIP usually occurs once funding is in place and concludes with the successful launch of *the* product or technology to market. Figure 7.1 shows where Strategic Implementation fits in the Business Development Process.

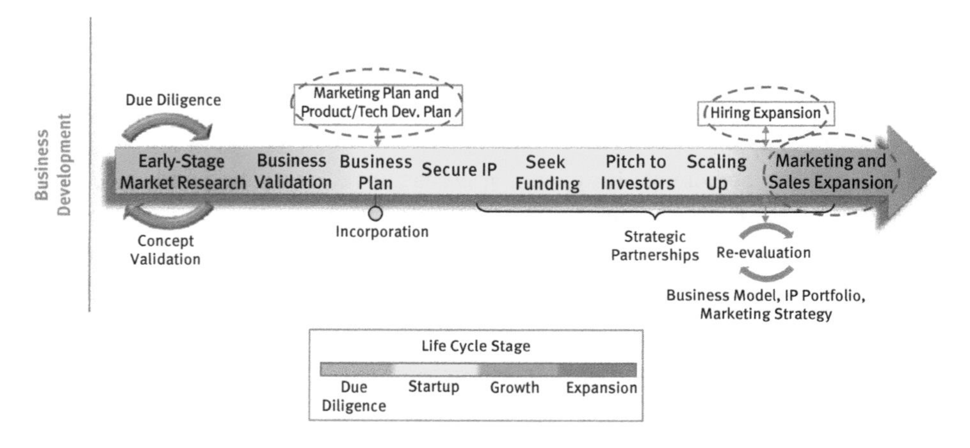

Figure 7.1: Strategic Implementation planning activities throughout Business Development.

Strategic Implementation is the point during the Business Development timeline where the high-level discussions in the business plan start blossoming into the multiple, in-depth implementation strategies that overlap and interconnect (as represented by Figure 7.2). These strategies are captured in an integrated SIP; the execution of which results in the launch of a product or technology to market.

Just like the business plan, the implementation strategies will be ever-changing, and therefore, the SIP will be an ever-changing document. It will be necessary to continually revise the SIP that you put in place for the business plan, and for investor pitches, to incorporate your learning along the way.

This chapter focuses on the implementation of various business-related aspects of Strategic Implementation, including personnel hiring, vendor management, marketing and sales, and postmarket launch customer surveillance and analytics.

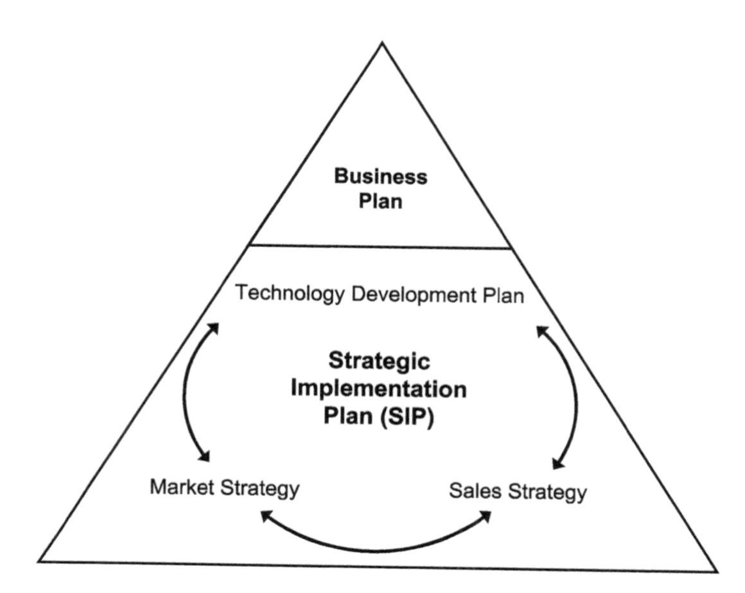

Figure 7.2: Relation of Business Plan to Strategic Implementation Plan.

7.1.1 Strategy versus Implementation

Strategy is the process that dictates the company's direction. Implementation is the execution of processes that ensure that the company will move forward in the chosen strategic direction. An implementation strategy is the planning process for how to execute the company's direction, while taking into consideration the multiple facets of the company, available funds, and time. Common hurdles for implementation include:

- There may not be enough money to execute due to scope creep (Section 7.3.4).
- Fundraising targets are slow to achieve.
- It may take longer to hire qualified personnel than expected.
- Slow hiring will affect productivity of multiple departments.
- Employee onboarding and training is a lengthy process.
- Vendor onboarding, contracting, and/or licensing may take longer than expected.
- The initial vendor(s) may not give the quality of product promised.
- Business and/or project priorities shift.

It is highly likely that something (or many things) will divert your implementation strategy from its intended timeline. Diversions can range from discovering that a component of your technology is not as scalable as once thought, to a vendor not producing a quality final product as contracted, to a key partner organization taking months longer than anticipated to get an agreement into place, to difficulties recruiting personnel that are a good fit for the company culture. Putting the plan into action is exciting but complicated. Staying on task is key, and this is where communication and planning will determine everything from the attitude of your workplace environment to the productivity of your company.

7.1.2 Strategic Implementation Plan (SIP)

An implementation strategy is the process of defining how to bring the business plan to life. Your SIP takes the components of your business plan and organizes them into actionable items. The SIP starts by defining your timeline and objectives. It consists of the actionable items needed to ensure that the individual, granular components of the business plan (i.e., technology plan, marketing plan, sales plan, workforce hiring and development plan, etc.) can be executed concurrently. To execute the objectives outlined in the SIP, you must define how you will execute each component. This means accounting for funding, resources, hiring, training, contracts, licensing, development times, marketing techniques, analytics, personnel, and much, much more. Without a comprehensive implementation strategy, it can be difficult to identify how you will achieve each of your stated goals and objectives. As mentioned before, tools such as Gantt charts (Figure 7.3) are helpful for correlating objectives to the individual actionable tasks that overlap in each of your plans and the resources (e.g., people) needed to accomplish them. In the example Gantt chart, you can see overarching tasks, correlated subtasks, milestones, and resources. The completion of tasks can be dependent on one another or on resources.

7.2 Hiring Strategy Planning and Implementation

People (such as personnel, employees, workers) are a company's most valuable resource. Implementing a hiring strategy requires methodical planning, persistence, and patience. Your business will not immediately run like clockwork with teams of people knowing exactly what to do and how to work together, but eventually this is achievable. It is important to take the time early in the hiring process to think about which skills are needed to complete the work ahead to reach critical valuation milestones for your company.

7.2.1 Identifying Types of Resources

Before the hiring process can begin, day-to-day work, tasks, and deliverables need to be identified and mapped out, to identify the resources needed to complete the work. Business leaders can utilize multiple tools as inputs into the hiring plan, such as investor pitch decks, project roadmaps, and valuation milestone timelines,

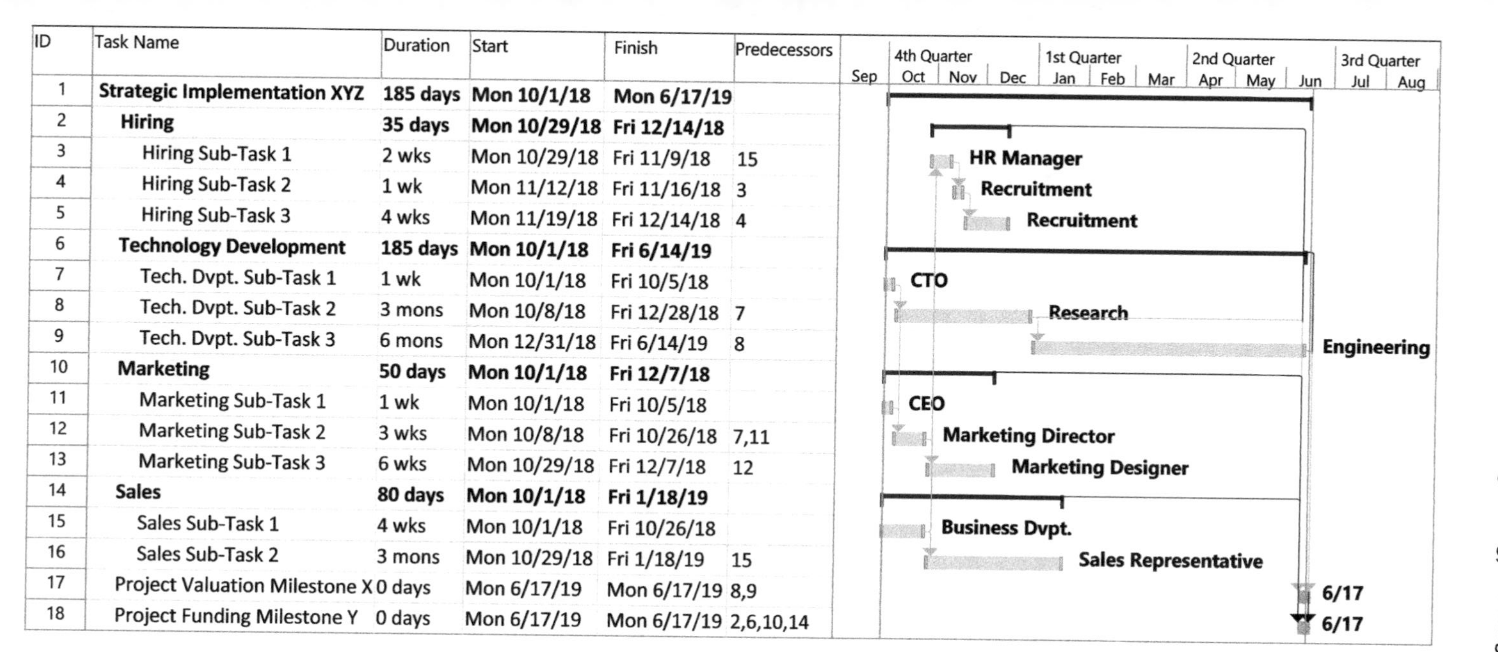

ID	Task Name	Duration	Start	Finish	Predecessors
1	**Strategic Implementation XYZ**	**185 days**	**Mon 10/1/18**	**Mon 6/17/19**	
2	**Hiring**	**35 days**	**Mon 10/29/18**	**Fri 12/14/18**	
3	Hiring Sub-Task 1	2 wks	Mon 10/29/18	Fri 11/9/18	15
4	Hiring Sub-Task 2	1 wk	Mon 11/12/18	Fri 11/16/18	3
5	Hiring Sub-Task 3	4 wks	Mon 11/19/18	Fri 12/14/18	4
6	**Technology Development**	**185 days**	**Mon 10/1/18**	**Fri 6/14/19**	
7	Tech. Dvpt. Sub-Task 1	1 wk	Mon 10/1/18	Fri 10/5/18	
8	Tech. Dvpt. Sub-Task 2	3 mons	Mon 10/8/18	Fri 12/28/18	7
9	Tech. Dvpt. Sub-Task 3	6 mons	Mon 12/31/18	Fri 6/14/19	8
10	**Marketing**	**50 days**	**Mon 10/1/18**	**Fri 12/7/18**	
11	Marketing Sub-Task 1	1 wk	Mon 10/1/18	Fri 10/5/18	
12	Marketing Sub-Task 2	3 wks	Mon 10/8/18	Fri 10/26/18	7,11
13	Marketing Sub-Task 3	6 wks	Mon 10/29/18	Fri 12/7/18	12
14	**Sales**	**80 days**	**Mon 10/1/18**	**Fri 1/18/19**	
15	Sales Sub-Task 1	4 wks	Mon 10/1/18	Fri 10/26/18	
16	Sales Sub-Task 2	3 mons	Mon 10/29/18	Fri 1/18/19	15
17	Project Valuation Milestone X	0 days	Mon 6/17/19	Mon 6/17/19	8,9
18	Project Funding Milestone Y	0 days	Mon 6/17/19	Mon 6/17/19	2,6,10,14

Figure 7.3: Example Gantt chart.

executive summaries, the business plan, and financial models to understand the bigger picture and determine what work needs to be completed to achieve that bigger picture. Once the work packages (or groups of deliverables) are defined and put in sequence at a high level, it will be easier for business leaders to see what types of resources will be needed to complete those work packages.

For example, a business leader can study the project roadmap to see which types of resources are required by specific points in time in order to complete the work packages so that valuation milestones can be achieved with the least amount of resources, keeping investors satisfied and the business afloat. In Figure 7.4, it is evident that technical staff (e.g., scientists, engineers, and designers) are needed immediately to begin research and development activities and to prepare invention disclosures. Legal experts will also be needed to prepare the appropriate patent applications. Regulatory experts will be needed, once the project moves into late stage design development, to prepare for a regulatory submission (if applicable). Marketing and sales staff will be needed later in the game, when business development activities are required in preparation for commercial launch. Business leaders can use this information to create resource hiring plans.

In a different example, business leaders can study the financial models (in conjunction with the Gantt chart) and compare them to available funds and the monthly expenditure goals, or monthly cash burn rate, to determine how much money can be best utilized. Resource costs are more than just salary – permanent hire resources typically require benefits, insurance, employment taxes to be paid by the business on behalf of staff, and training. There are also general and administrative costs that are associated with hiring permanent hires, including

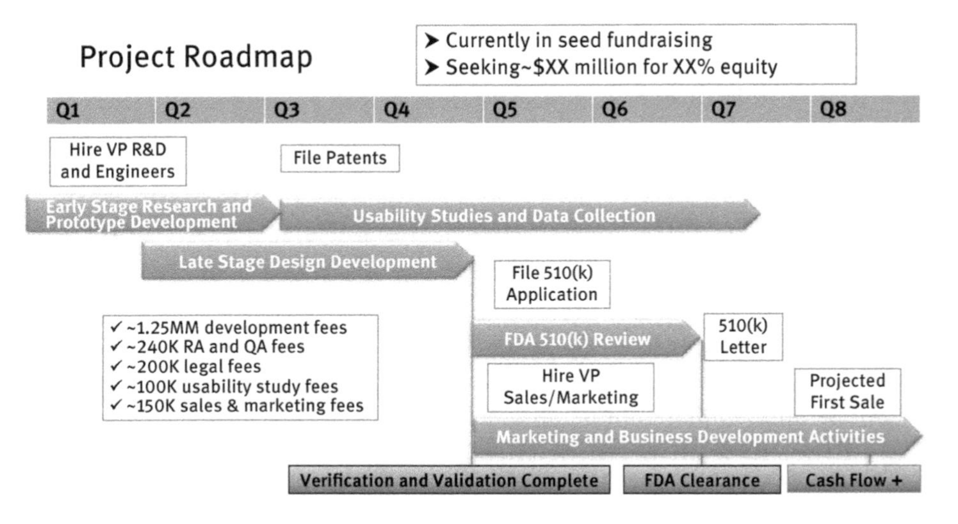

Figure 7.4: Using Project Roadmaps to create resource hiring plans.

computers, software packages and licenses, phones, internet, office space, and the business needs to carry insurance for those hires. By incorporating these costs into the financial model and comparing against project plans and company balance sheets, business leaders can define a resource plan that will meet project and financial objectives.

7.2.2 Hiring Strategy Planning

A hiring strategy includes other aspects besides the types of resources that are required to complete the work packages. A hiring strategy needs to account for:

- Organizational Structure
- Employee Status
- Outsourcing Needs
- Employee Location
- Employee Work Experience

Organizational Structure

Regarding organizational structure, business leaders need to decide how many layers of management should exist between the C-level executives down to the most junior member of the company. As a leader, do you want there to be many layers in this structure: C-level executives, presidents, vice presidents, directors, managers, supervisors, and so on, or do you want the company to be relatively flat where most employees are at the same rank and level?

Warning: The more layers there are in your organizational structure, the more bureaucracy and process inefficiencies there will be to slow down work progress.

Instead of a classic organizational structure, would you like to create a matrix organization and hire a program manager to oversee multiple projects and hire a team of project managers to oversee individual projects? In the matrix organizational structure, resources report to a leader within their own department, for example, Marketing, Sales, Quality, Research, Engineering, and so on, and have accountability to perform work for a project manager as well.

Employee Status

Regarding employee status, business leaders need to decide if work can be completed through a balance of permanent hire resources and temporary resources. Temporary resources can be helpful in completing a defined work package during a defined period of time, without charging the company for benefits and insurance. Temporary resources can also be observed to see if they would make valuable,

permanent additions to the company. Examples of work that can be performed by temporary resources include:
- Preparation for a financial audit.
- Designing a computational modeling simulation to test the prototype designs in various environments with exposure to various stimuli.
- Revising quality policies and procedures to comply with new regulations.
- Performing a failure investigation as to why the software algorithm is not working or why the manufacturing subassembly process is producing inconsistent results.
- Project management support, such as monitoring project schedules, integrating multiple schedules into a program plan, and so on.

Outsourcing Needs

Regarding outsourcing, not every function or department needs to be operated within the company, that is, in-house. Business leaders should form teams that focus on the core competencies and mission of the business, so if there are external companies that can perform a function better, more efficiently, and for less money, then it would be wise for business leaders to outsource that function to save on monthly expenditures. For example, human resources (HR) can be conducted internally or can be outsourced to an HR firm that has all the legal and training documents readily available.

In a different but similar example, most businesses outsource payroll, because there are payroll providers that are more familiar with local and state level tax laws and tax reports. Many businesses, at least in the startup stage, do not carry their own legal staff, as this would be an expensive team of permanent hires.

As a technology business, you can even outsource various aspects of design and manufacturing, if you cannot hire a core group of engineers with the exact technical expertise that is needed for the project, or if you do not have the facilities to operate a manufacturing line.

Employee Location

Regarding employee location, business leaders need to decide if they are willing to hire resources that do not need to be or cannot be colocated with the rest of the team. Sometimes it is required for employees on the same team to be colocated together, for example, a test engineer needs to be colocated with the laboratory and other engineers to perform engineering tests. But there are situations where it is not required for employees to be colocated together, for example, there are aspects of financial management, project management, quality, and regulatory management that do not need to be performed at the same business location or job site. In other situations, employees may wish not to move from their current residence or they may need to take care of loved ones at home and therefore they cannot work onsite;

however, they are willing to work for the business remotely. Business leaders need to determine whether the work requires employees to be colocated, and whether the employee is too valuable not to hire. If leaders determine the hire is too valuable to overlook, then leaders need to ensure the staff member has the right information technology and telecommunications capabilities at home to perform the work required. That might be a cost that can be negotiated between the new hire and the business.

Employee Work Experience

Lastly, as part of the hiring strategy, think about the level of experience required to perform the work. As a business leader, would you prefer to hire experienced working professionals or recent graduates? Are you looking to groom talent or need resources to hit the ground running as main contributors and leaders?

If the business is looking to hire recent graduates, which schools and universities are you targeting? Businesses should develop a graduate recruitment strategy based on familiarity with the schools and their employment assistance departments. These departments will indicate to businesses when is an appropriate time to participate in career fairs, for example, during the autumn or spring. A business may prefer to hire recent graduates from a university that has a strong co-operative education program, perhaps because that graduate may have worked at the business already as an intern or co-op student. A business may prefer to hire graduates from top-ranked university programs, or a business may prefer to hire graduates from schools that predominantly represent females or minorities.

If the business is looking to hire experienced professionals, how can business leaders reach the professionals with the desired credentials and work history? Many business leaders utilize their professional and personal networks to ask for introductions to aspiring job candidates. If the right individual cannot be found through current networks, then business leaders can advertise with local and regional interest groups and organizations, for example, Life Sciences Pennsylvania, Society of Women Engineers, American Society of Mechanical Engineers, American Society for Quality, Project Management Institute, and so on. Business leaders can also utilize credible job listing websites to post job ads. If the business needs to hire executives, then leaders should consider seeking outside executive recruitment firms that specialize in executive level recruitment to execute the hiring strategy, because they have access to the top candidates.

7.2.3 Hiring Strategy Implementation

Implementing a hiring strategy is carried out in multiple steps, typically that occur concurrently. These steps include:

- Writing job descriptions
- Preparing training materials.
- Selecting candidates for interviews.
- Preparing for interviews.
- Conducting interviews.
- Establishing new hires for success.

Writing Job Descriptions

Think of a job description as a blank canvas for describing everything a candidate needs to know to determine if he or she is a good fit for the role. Job descriptions should include a summary of what the role is and what will be accomplished through the role, the day-to-day tasks and responsibilities, objectives, education requirements, credentials and qualification requirements, desired prior job experience, and general skills.

Example C-Level Job Description

Chief Executive Officer (CEO) Job Description

Summary

The CEO is a leader and manager who guides the company to consistently meet its financial objectives.

Essential Duties and Responsibilities

- Develop and execute commercial and business strategy
- Chair executive meetings with Board of Directors, investors and company cofounders
- Develop a strong, creative business model As a leader, coordinate activities throughout all company operations
- Evaluate the effects of external business and market forces on the company and integrate appropriate mitigations into the company's short- and long-term strategic plans
- Establish formal responsibilities and hold management staff accountable for their performance

Qualifications

- BS degree in Finance, Business, Science, or Engineering
- MBA preferred
- 20+ years of strategic leadership experience, preferably in the XYZ industry

Experience and Skills

- Experience in leading teams
- Strong track record in fundraising and sales
- Demonstrated ability to build and maintain relationships
- Highly analytical and strategic thinker
- Solid business judgment and maturity
- Demonstrated communication and organizational skills
- A leader with emotional intelligence

- Effective and decisive under pressure
- Ability to prioritize and meet deadlines
- Effective risk taker

Example Senior Level Job Description
Director of Marketing –Nursing and Rehabilitation Healthcare System
Seattle, Washington (United States)
$110,000–$115,000 Annual Salary
Job Type: Full-Time
Manages and coordinates a cohesive, creative marketing strategy and supporting activities across multiple nursing and rehabilitation facilities within the healthcare system. Responsible for 10% year on year growth by identifying new patient populations and obtaining new patient admissions. Must be exceptional at developing strategic, external partnerships and relationships across the city of Seattle.

Responsiblities
Implements mission and vision of the healthcare system by planning and executing strategic marketing activities that are designed to increase visibility of the healthcare system across the Seattle region and increase patient admissions in nursing and rehabilitation centers.
Leads the Marketing team that is composed of Managers and Analysts.
Works closely with Business Development to maximize the sales funnel process: leads, referrals, clients.
Develops educational, communication, and promotional materials that can be utilized by Sales
Conducts market and competitive analysis.
Maintains Marketing budget and presents monthly updates to Executive Leadership team.

Education Requirements
Bachelor's degree required. Master's degree in Business Administration preferred. A minimum of 10 years of experience in healthcare and healthcare marketing experience preferred. Healthcare sales experience required.

Preparing Training Materials

While business leaders prepare job descriptions, they should interface closely with HR professionals to determine what types of training materials should be offered to resources, once hired. Training courses can be delivered live by instructors and business leadership in a classroom setting with teams of new hires, or resources can complete self-training through online, e-learning modules, and by studying policies and training manuals. Training can include on-the-job activities, for example, facility tours, learning the different manufacturing processes and operating those processes on the manufacturing line, or by attending a simulation or engineering test to learn about the technical aspects of the products that the business designs. New hires can also be expected to take tests, depending on their role.

Selecting Candidates for Interviews

While business leaders and recruitment professionals review candidate applications and resumes, it is important for them to remove personal bias from the selection process. A candidate should not be overlooked because he/she took time away from the workforce to serve in the military, raise a family, travel internationally, learn a new skill, and so on. Gender and age should also not prevent a candidate from being interviewed. Diversity enables creativity and different ways of looking at problems and finding solutions.

Preparing for Interviews

In preparation for the interview process, the experience should feel very seamless to the job candidate. Businesses that take a long time to e-mail or call back candidates to schedule interviews and follow up with candidates following interviews tend to lose these candidates to more efficient, expeditious businesses. Also, a candidate will be alarmed if he/she calls the business and the hiring manager is not aware of the candidate's application status. If there are multiple interviews as part of the hiring process, businesses should schedule these interviews to follow closely together. At all times, businesses need to come across to job candidates as polished, professional, and organized. Business leaders do not wait until the interview to read the candidate's resume – study it closely and prepare questions in advance!

Conducting Interviews

Once prospective candidates are brought in for their interview, it is beneficial for the business to ask multiple employees to interview the candidates and to get different perspectives and answers to different questions. For example, a member of a business's technical staff may only be interested in a candidate's ability to explain his or her role and individual contributions on prior research and design projects, whereas a member of a business's management team may be studying a candidate more closely for interpersonal skills and an ability to think quickly in responses to industry-related scenario questions.

Businesses are legally not allowed to ask personal questions during interviews, so business leaders should work with their recruitment professionals to come up with a list of questions that would be appropriate for interviews. Interview questions are broad and may include the following:

- Tell me about a time you needed to lead a technical project; what were the outcomes and lessons learned.
- What industry-relevant trend excites you the most?
- Which types of people do you tend to work well with, and which types of people have you struggled to work with in the past?
- What was your most valuable experience in school?

- Why are you a good fit for this company?
- What are your 1-, 3-, and 5-year professional goals?
- How can you help us to achieve the next valuation milestone?
- What questions do you have, or what more do you want to know about this role?
- What do you consider to be your biggest weakness at work?

Establishing New Hires for Success

Once candidates are selected and accept your job offer, ensure their first few weeks and months on the job are successful. It can take a lot of time and money to hire a candidate, so do not trust that they will be happy right away and work well with the other employees. Establish a trial period upon hire (typically 3–6 months), so the hire can learn the role and then write his or her own goals and objectives for the role. Ensure that managers provide frequent feedback. It is also good practice for businesses to assign a mentor and an office buddy t3 Vo provide coaching and support for new hires.

7.3 Vendor Management

A vendor is an overarching term for a person or an organization that your company will rely on for a service (from a contractors or consultants) or part (a screw, electrical resistor or conductor, etc.), component (a subassembly or chemical compound), or product (a final product). By way of an example that is focused on a vendor who sells and services camera technologies, the vendor could:
- supply parts for the camera (such as screws, electronic components, or lenses),
- supply components for the camera (such as the camera housing, flash subassembly, or digital display assembly), and/or
- supply a camera product (such as specific models or brands of cameras).

To clarify, suppliers are vendors, contract manufacturers are vendors, and service providers are vendors. Every company will have vendors that they work with to complete tasks the company is unable to do itself or does not have the resources to accomplish internally. Vendors can range from web developers to manufacturers.

The division or department that utilizes the vendor often conducts the identification and search for specific vendors, whereas a company's Quality Department or Quality Manager oversees vendor onboarding and continual management, to ensure that each vendor is compliant with your quality requirements. Lastly, the authorization of each vendor should go through the company specific chain-of-command to ensure that the necessary players within the company (e.g., Project Manager, Director of Product Development, Quality Manager or Director of Quality,

Comptroller or VP of Finance, and President or CEO) know what is being accomplished and where funding is going.

To give an analogy that is outside the realm of technology development, your roofer, plumber, electrician, and even the local hardware store that you frequent are all personal vendors. It is one thing if you need any of these resources when an emergency arises (for instance, a pipe bursts or there is a leak in the roof), but if you were going to build a house from scratch, how do you organize the various vendors (such as the architect, construction crew, lumber supplier, plumber, electrician, roofer, drywall, etc.) that you will need to get the job done? What if you plan on scaling up to build a series of houses in a neighborhood? How do you make sure each vendor is doing quality work?

If you were building just one single house, you would start by figuring out what you need to make it functional and to look the way you want. You would only need the roofer, plumber, electrician, and so on for the single project of building one house. When building a series of houses, however, you will need to take into consideration the long-term costs, quality, and reliability of each of the vendors and try to have replacement vendors that you have already assessed, in case one vendor/resource is unavailable (e.g., your primary plumber is not available for work on houses number three and six). You are going to want to make sure each vendor is consistently good at the job (Quality is discussed in Chapter 11), that he or she shows up when needed, and that his or her prices are reasonable and do not fluctuate. Now, how do you do this when you work with hundreds, possibly thousands of vendors?

The answer to this question is by making good use of vendor management. Vendor management consists of (a) initially qualifying each vendor, (b) continually monitoring each vendor's performance, and (c) performing regular audits of key vendors to verify that the vendors are meeting your company's requirements.

7.3.1 Assessing Vendors

A vendor qualification process is a process for qualifying, or "vetting," a vendor (Figure 7.5). The process for qualifying each type of vendor is the same, although the qualification requirements can differ, depending on the type of vendor. With regard to vendor management, each department within a company is usually responsible for qualifying their own vendors in accordance with company policy. In regulated companies, such as companies that must comply with ISO standards developed by the International Organization of Standards, the vendor management policy will be located in the Quality Manual and overseen by the Quality department. In most cases, the legwork to qualify a vendor is conducted by the corresponding department, then approved by the various levels of the company's chain

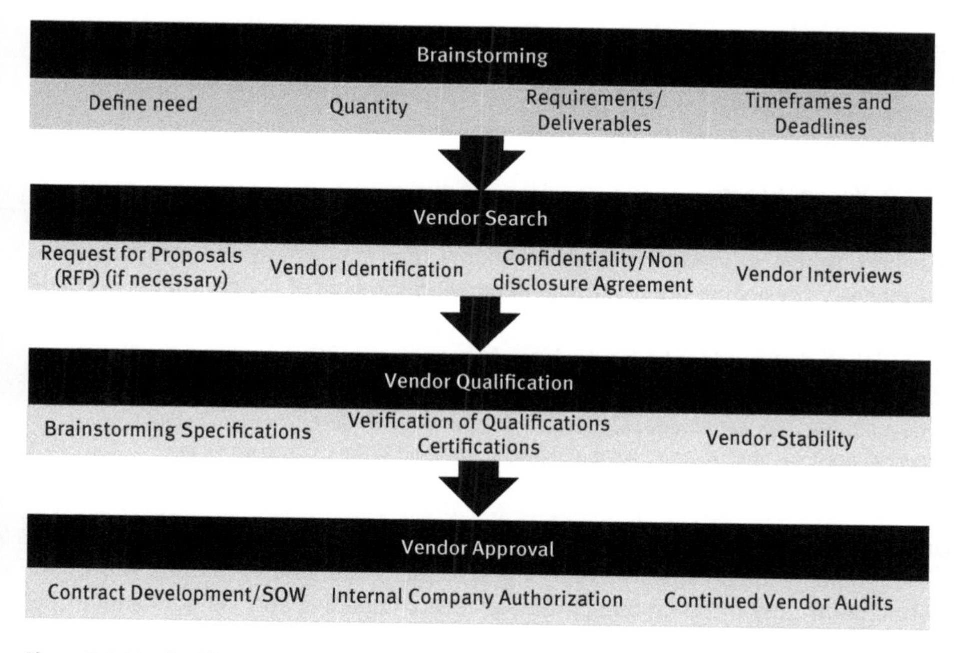

Figure 7.5: Vendor Management Process.

of command. Though the vetting process is company specific, the overarching process for qualifying a vendor (shown in Figure 7.5) consists of:

1. Brainstorming
2. Vendor Search
3. Vendor Qualification
4. Vendor Approval

7.3.1.1 Brainstorming

A brainstorming work session should be conducted to identify (a) what part, component, or service is needed; (b) the quantity of the part, component, or services needed; (c) the requirements or deliverables (including risks) associated with the part, component, or service; and lastly, (d) estimated delivery time frames and required deadlines. This internal discussion will allow a company to filter through vendors efficiently. Drawings, acceptance criteria, and any special provisions to the vendor contract can be determined based on the outputs of the brainstorming work session. In the case of contract manufacturers and service providers, the information gathered from the brainstorming work session will likely be used to develop a Request for Proposal (RFP). An RFP is a high-level summary of the outcomes, requirements, and final deliverable(s) that vendors are asked to detail in their response to the RFP (assessing service providers and contract manufacturers is discussed more in Sections 7.3.1.3 and 7.3.1.4).

7.3.1.2 Vendor Search

Next, companies should conduct a vendor search, based on the brainstorming work session outputs, to identify potential vendors that may satisfy the company's requirements. The search approach will depend on the type of vendor that is needed. The internet, social media, journals, reviews, papers, and/or referrals will provide an initial pool of vendors from which to choose.

Note: A full RFP does not always need to be developed for manufacturers and suppliers, since the requirements are more straightforward and will be specified in the contract agreement.

Once identified, companies contact each of the identified vendor's sales department to determine if the prospective vendor is interested and able to perform the work. At this point, any information shared between companies and vendors should be nonconfidential and at a very high level. If needed, this is the point to execute any nondisclosure agreements and/or confidentiality agreements between the company and vendor. Afterward, the RFP can be distributed to the vendor. Once vendor interest is ascertained, a comprehensive list of potential vendors should be assembled by the company for a due diligence committee to qualify and approve.

7.3.1.3 Vendor Qualification

As a company relying on another company to serve as a vendor, there will be many obstacles and processes out of your control, even though you are depending on the vendor to provide your company with a specific part or component. The process of qualifying and onboarding a new vendor can take months. Risk management of vendors is a necessary quality management activity for every company. If you are looking to be an ISO-certified company (for instance ISO 9001 certified), vendor risk management will be mandated. Mitigating the risks associated with the quality and reliability of vendors is necessary for maintaining and promoting seamless operations within your company. When working with vendors that provide parts, components, or finished products, vendor qualifications should be based on:

1. Need – Can the vendors develop the sourced item or perform the service needed?
2. Quality (Chapter 11) – Can the vendor repeatedly meet the specifications the company needs and that are outlined in the contract between company and vendor?
3. Price – Is the vendor's proposed pricing acceptable?
4. Reliability – Is the vendor able to fulfill orders when needed?

Just as your investors will want to conduct due diligence on your skill sets, capabilities, and fiduciary responsibilities, you should similarly conduct due diligence on key vendors. During the qualification process, or "discovery," it is reasonable to conduct background checks on vendors. Example discovery questions for service providers and contract manufacturers could be:

- What certifications does the vendor have, when did they get the certification, and when do the certifications expire?
- What is the vendor's stability (e.g., reviewing the vendor's credit history via their DUNS number,[23] customer references received by the vendor, leadership experience, etc.)?
- Does the vendor have competitors and who are they?
- Why is this vendor better than its competition?
- Does this vendor have online reviews? If so, when you filter through the unhappy client reviews, do you see trends of viable complaints?
- Can the vendor demonstrate sufficient proof of insurance?

In addition to initial vendor qualification, vendor management also includes the process of continually assessing vendors. Even though a company may not need to be compliant with ISO standards, the habit of monitoring and reassessing vendors is a good practice. As a company grows, the company will interact with more and more vendors. It is important to continually monitor your vendor's:

- Financial stability – Will the vendor continue to remain in business to fulfill your next order?
- Reliability – Does the vendor fulfill your orders in a timely manner?
- Quality – Does the vendor produce parts, components, or services that meet your company's requirements?
- Certifications – Is the vendor maintaining pertinent certifications (e.g., ISO 9001 or ISO 13485), insurance, and so on?

It is also important to regularly reassess vendors, to ensure that they remain cost effective for your business. Quarterly to annual reviews of each vendor's contracts, quality agreements, services rendered, waste, and overall cost should be performed.

---

23 The Data Universal Numbering System, or DUNS, is a unique numeric identifier developed and regulated by Dun and Bradstreet (D&B). The DUNS number is used to establish a D&B business credit history. This identifier is used by potential lenders and potential business partners to assess financial stability. In simple words, the DUNS number is a free identifier assigned by D&B that provides the equivalent to a personal credit score for corporations.

7.3.1.4 ISO-Regulated Industries

The International Organization of Standards, referred to as "ISO," is an international organization that develops and standardizes documentation related to requirements, specifications, and guidelines for specific industries. In this book we refer to ISO 9001, the standard that specifies requirements for quality management systems (QMS) throughout the world, and ISO 13485, the standard that specifies requirements for QMS for medical devices throughout the world. There are thousands of ISO standards, and each industry has its own applicable ISO standards.

In ISO-regulated industries, during the onboarding process, a Supplier Questionnaire is issued to the vendor to confirm that ISO requirements are being met, ranging from proof that the vendor has insurance to whether the vendor has its own Quality program (Chapter 11). The content of the Supplier Questionnaire is outlined by the ISO, but generally asks the following questions:

- Does the vendor have an ISO certification (is the vendor ISO 9001 certified)?
- Does the vendor have a Quality manual?
- Does the vendor use statistical process control methodologies?
- Will the vendor allow you to conduct an on-site process validation?
- Will the vendor allow you to conduct follow-up on-site audits?

7.3.1.5 Vendor Approval

The vendor approval process is company specific and depends on how titles and responsibilities are allocated within an organization. Certain job positions within a company could be tasked with performing vendor approval processes. These responsibilities are often assigned within the company's Roles and Responsibilities Matrix, which aligns with the company's organizational chart (see Figures 4.8 and 4.9 in Chapter 4). Once vendor due diligence, inspections, acquisition of necessarily documents, and, finally, contract negotiations have been completed, then your organization's chain of command will want to review and authorize the vendor.

7.3.2 Vendor Risk Management

Once a vendor is qualified and approved, then they can be placed on the company's approved vendor list, which is a centralized list of approved vendors that is continually updated with all pertinent vendor information and significant dates (e.g., certification due dates, dates of prior audits, dates of most recent assessment). Approval for continuous inclusion on an approved vendor list can require periodic re-evaluation at a frequency that is determined by the risk imposed by the vendor, that is, high-risk vendors get evaluated quarterly, and low-risk vendors get evaluated annually. During the reevaluation, scheduled vendor site visits, requests to inspect and review the vendor's most up-to-date certifications, and review of recent performance metrics are common practice.

In an ideal world, your company would have multiple vendors that can provide each and every element needed for your final product – in case something goes wrong. In many instances, there may only be one vendor that produces a part or component. A "sole-source" vendor is a vendor that supplies a key component to your company that no one else can produce or offer. A "single-source" vendor is a single vendor that your company has selected to contract with, although there are other available vendors. In many cases, single-source vendors are conducting services that, if transferred to a different vendor, require the new vendor to start from the beginning or might take an extremely long time to reproduce.

Sole-source vendors or single-source vendors are sometimes necessary, but if possible, it is best to avoid these vendor situations. As with all vendors, you will need to manage the quality of the product or service being provided, but in the case of the sole-source or single-source vendor, there is no one else to turn to, if the vendor becomes unable to meet your requirements. Your company could suffer significant losses in terms of productivity and profits, if your sole-source or single-source vendor cannot fulfill its obligations to your company.

When working with vendors, one of the greatest risks is inconsistent quality. Quality problems can occur for a number of reasons. Often, change in quality occurs due to a change in the vendor's sourcing or manufacturing process. Being able to identify when a change occurs that is impacting quality at the vendor level is imperative.

One way to hedge the risk associated with vendor quality issues is to specifically "contract the matter" by preparing a quality contract or quality agreement. When developing a quality contract or quality agreement with a vendor, it is reasonable to request:

1. Purchase orders to always include part numbers, revisions, documentation requirements, packaging requirements, shipping requirements, and the requested delivery date(s).
2. Engineering drawings or schematics to be included with every purchase order, even when there is no change to the drawing/schematic.
3. An order confirmation, including all the details of the purchase order, such as a copy of the purchase order, specifications, and drawings.
4. A Change Notification, including a requirement that the vendor notify the company of all process and specification changes prior to order confirmation. Additionally, if a change is made or is being considered, a contract requirement can be made that any changes require approval notice by the customer/company.

It is also reasonable to include incentives or penalties in the contract, to encourage the vendor to satisfy your needs, and/or to discourage the vendor from failing to meet your needs. For example, if the vendor commits to delivery of a part in 4–6 weeks, as per the vendor agreement, but fails to meet the delivery timeline, then the vendor should be penalized. Contractual clauses can be included, so that companies can sue for damages in extreme circumstances. In reality, however, it is difficult to enforce penalties, unless there is a catastrophic failure that will most likely end, or severely impact, your product line or company's ability to meet customer orders. Additionally, companies should want to maintain a good working relationship with their vendors, because high-performing vendors end up becoming company partners over the long-term relationship. By enforcing penalties, companies jeopardize that partnership.

7.3.3 Assessing Service Providers and Contract Manufacturers

As discussed before during vendor brainstorming, the leadership sets the vision for the technical team, then the technical team develops the RFP. The RFP is reviewed by leadership to address budgetary and contract requirements, and then the RFP is finalized. When developing an RFP for service providers and contract manufacturers, the technical team should plan to conduct vendor interviews as part of the vendor qualification process. Vendors utilize the RFP to outline a detailed scope of work (SOW) (also called "statement of work"). The proposal that the vendor submits in response to your company's RFP will eventually be adjusted and integrated into the contract agreement as the final SOW. The SOW should contain:
1. Overview
2. Deliverables
3. Milestones
4. Timeline
5. Budget
6. Use of Funds
7. Reports and Reporting Mechanisms

The *overview* is an abstract discussing the overall project. The *deliverables* are the parts or subcomponents of the project that need to be accomplished. Examples of deliverables could be to produce a prototype chassis, software wireframes, or capabilities of hardware/software interactions. *Milestones* are key time points for the project. Milestones are usually specific accomplishments or actions, such as submission to the FDA, obtaining a certification, or obtaining a specific dollar amount in sales (the latter is applicable for sales and marketing vendors). The *timeline* is the overall time frame to complete the SOW, lengths of time for each component or subcomponent, and milestones in relation to all of the deliverables and milestones that need to be accomplished. The *budget* outlines estimated costs and how fees for service will be addressed, if or when the services vary from the

SOW. *Use of funds* is the granular identification of how the funding will be spent. *Reports* are part of the deliverables that provide the opportunity for the vendor to summarize its work, have its work reviewed, and complete any technical and/or contractual changes that need to be addressed. *Reporting mechanisms* are the identification of when or how often reports will be delivered and reviewed with customers and, most importantly, who is authorized to make technical and contractual changes.

Case Study

A telemedicine startup company was looking to develop a software platform. The company held brainstorming sessions to determine the criteria, specifications, and requirements of the novel platform. The brainstorming sessions determined that the startup company is looking for a single-source vendor. The criteria, specifications, and requirements of the platform generated during the brainstorming session were then translated into an RFP. Upon conducting an exhaustive web search and receiving multiple referrals, the startup company found well over 20 vendor candidates.

The company required that an NDA be put into place with each of the vendors, and only vendors that were willing to sign the NDA received the RFP. The startup company conducted multiple discovery rounds, where the best vendor candidates were identified through interviews and evaluations based on the vendor's responses to the RFP. The initial discovery phase filtered the vendor candidate pool down to four qualified vendors. The team next updated the RFP for a final round to reflect the concepts learned during the discovery phase. The remaining four candidates were given the updated RFP and were asked to give in-depth budgets, answers to additional questions, and to update their responses to the updated RFP. A final decision was made after a few days of deliberation. The selected vendor then provided an exhaustive SOW including the overview, deliverables, timeline, milestones, budget, use of funds, and reporting mechanisms. The SOW was used as a legally binding exhibit in the contract prepared by the startup company's lawyer.

7.3.4 Scope Creep

When developing a SOW, scope creep (or "everything but the kitchen sink" syndrome) should be taken into consideration. Scope creep is when changes to a project start adding significantly to the vendor's SOW or shifting away from the original SOW. Scope creep can occur when the SOW is not well defined, the project deliverables change, or changes are not recorded and authorized regularly. When scope creep occurs, the vendor needs to manage the company's expectations that there will be changes to the budget and/or timeline. It is up to the vendor to guide the company (their client) to either get back on track with what was proposed in the original SOW, or update the contract and be clear about additional costs. A problem most SOW has is a lack of specificity, usually because flexibility is needed, especially at the beginning of a project. Actions that vendors (and companies) can take to manage scope creep are to develop an in-depth SOW, with a reporting mechanism to formally review and update the SOW on a monthly or quarterly basis.

7.3.5 The Importance of Vendor Management

Vendor management is a tool set that protects a company from activities outside the company's control. Factors can range from regional natural disasters (for instance, fires, flooding, blizzards, hurricane, etc.), a vendor's loss of a key employee, equipment failure or wear, poor training, and so on. The ability to track, monitor, and/or follow "paper trails" can save lives, prevent lawsuits, and reduce operational downtime.

Case Study

A case study to show the need for vendor management and open communication between different company departments/teams involves the manufacture of a malaria stain diagnostic kit. The purpose of the stain kit is to detect the presence of malaria in a biological sample. A saliva swab is taken from the patient. The culture is put on a microscope slide and stain is added to the culture. After waiting a few minutes, a decolorizer is added. Under a microscope, if the stain is still present in the cells, then the patient tests positive for the presence of malaria and is sent for additional, more extensive testing.

A United States-based manufacturer of the final stain kit received a comment from one of its customers abroad. The customer reported to the sales representative for the stain manufacturer that they "liked the new color, but it did not seem as efficacious as previous batches."

A corrective and preventive action, or "CAPA" (discussed in Chapter 11), was submitted by the company's sales representative to Quality, as required by the company's QMS. The CAPA notification informed the President, Vice President, Quality Manager, and Operations Manager in the company that something was not right with the new, final stain kit. An internal investigation was initiated. The investigation began with a review of the stain kit's bill of materials, the shipping and receiving documents, the manufacturing process standard operating procedure (SOP), and corresponding manufacturing batch record used for development of the stains. The bill of materials showed that there were 10 chemical components that composed the final stain kit product. The shipping documentation was needed to determine which product batch and lot of the outgoing malaria stain kit was under scrutiny. The Receiving Department provided documentation on the receipt, inspection, and certifications of each of the 10 components that went into the malaria stain kit. The standard operating procedures and batch records were pulled from the QMS.

The result of the investigation concluded that the primary approved vendor for one of the components of the stain was shipping inconsistent lots of that specific component to the US-based manufacturer. This was determined from a combination of observations. First, when comparing the certificates of conformance for multiple batches, the manufacturing dates were different, but the analytical values did not change from lot to lot over two years. Once this was noticed, analysis of multiple lots of the component stain in question was conducted. It was observed that even within the same lot, the color of the stain was different.

The US manufacturer quickly took action. The first step was to ensure that bad product did not leave the US manufacturer again. The next step was to inform the vendor that it was being removed from the approved vendor list and why. Lastly, the process for validating vendors was implemented for assessing alternative stains and their vendors. It took the US manufacturer three days to determine the cause of the problem and three months to find and approve replacement vendors.

Though an extreme case, the above example illustrates the reason why it is important (1) to continually assess vendors, and (2) to have a QMS (see Chapter 11 for more discussion on QMSs). The consequences of a "bad" component supplied by a vendor can be immense. For instance, in the above example, a bad component of the stain produced by a vendor caused problems for the final malaria stain kit manufacturer, and if the problem with the quality of the stain component had not been caught and addressed by the manufacturer, the resulting distribution of malaria stain kits having diminished efficacy could have contributed to a malaria epidemic.

7.4 Marketing and Sales Strategy and Implementation

Marketing and Sales teams play distinct, interconnected roles within a company. Great marketing and sales resources will help shape a company's brand, customer engagement and customer content, public relations, and customer service strategies. Both Marketing and Sales focus on data and customers, but Sales is typically more focused on customer engagement and business development, whereas Marketing is focused on analyzing market and competitive data to find new ways of communicating (e.g., animation, illustration, infographics, interactive digital content) the company's product or service value proposition(s).

Marketing teams work primarily with internal stakeholders. Marketing uses data-driven analytics to create content that can be utilized by Sales to generate prospective customer interest in the company and promote customer engagement. Marketing conveys complex information about a company's product or service in simple ways, using visual content as much as possible.

Sales teams work primarily with external stakeholders. Sales utilizes the content that is created by Marketing to establish and grow customer accounts in the current market segment, to develop business in new market segments, and ensure that there is a steady, increasing stream of revenue into the company. Sales usually is given only a short amount of time (such as the amount of time it takes to deliver an "elevator pitch") to get prospective customers interested, so more and more marketing materials need to be both entertaining and informative, that is, "movie trailer" quality.

Both Marketing and Sales teams facilitate the implementation of the company's marketing and sales plans. In technology innovation programs, and especially in matrix organizations (see Chapter 8), Marketing and Sales teams play an indispensable role in the company, because both teams also convey market and customer feedback to the various departments in the company, such as Research and Development, Manufacturing, Quality, and Regulatory.

Role(s) of Marketing

Marketing provides key customer insights to Research and Development. Product Managers within the marketing organization gather customer feedback (such as feedback about product aesthetics and effectiveness of marketing content during trade shows, conferences, and focus groups) and gather customer experience from product sales data. Product Managers provide this information to Research and Development, to inform strategic planning decisions for next generation product and service innovation.

In early stage companies, Marketing designs advertising and promotional content for the company website, product information brochures, and perhaps labeling and press releases. Before unveiling new content to the public, Marketing needs to review this content with Regulatory to ensure that content is not misleading and that all claims are compliant with federal and state product labeling requirements.

Marketing should also review this content with Quality (or the Editorial staff) to ensure that formatting, spelling, grammar, page numbering, and so on is correct. Additionally, the Marketing team will give feedback on the initial customer response concerning promotions, potentially affecting current and future product aesthetics.

Role(s) of Sales

The Sales team can shape the next-generation product or service line by providing feedback to Research and Development. To do this, Sales identifies and engages with existing and prospective customers and acquires feedback about product performance, product usability, the effectiveness of the service, and so on. It is the responsibility of Sales representatives to listen effectively, carefully, and accurately document the customers' ideas and suggestions, and to ask follow-up, clarifying questions. However, the Sales team needs to be careful not to agree with customers or guarantee customers that their ideas will be implemented in the next product or service launch. This provision of a guarantee to customers without getting approval from other departments within the company may create friction between Sales, Marketing, Research and Development, Quality, and Regulatory.

The Sales teams need to manage customer expectations and balance realistic production goals. Sales should work closely with Product Managers and the Manufacturing team to ensure that customer demand forecasts are accurate, so that Manufacturing can ramp up or ramp down production as needed. Sales teams are not meant to push the limits of Manufacturing. It can be a very unpleasant situation when Sales promises customers that exceptionally large orders can be fulfilled without appreciating what Manufacturing will have to do to be able to fill the order. It is not unheard of for Manufacturing to experience periods of downtime, such as when maintenance and calibration is required or when incoming components are delayed due to issues with suppliers.

7.4.1 Marketing and Sales Strategy

Marketing and sales strategies (see Chapter 4 for more discussion on marketing and sales strategies) need to begin first with a company:

1. Identifying desired customers and market segments.
2. Defining the type of relationship the company wishes to have with those customers.
3. Recognizing which channels need to be utilized to engage those customers.
4. Creating content for those channels.
5. Hiring personnel to carry out the activities defined in the strategy.

7.4.1.1 Customer and Market Segment Identification

Market segment identification and personnel hiring are probably the most important aspects in your marketing and sales strategy to get right, because customer and market targets will drive the rest of the strategy and the allocation of a company's budget.

7.4.1.2 Establishing Customer Relationships

With regard to customer relationships, some organizations may not need to emphasize the quality of relationships with customers, in order to earn the customer's brand loyalty. These companies may have a monopoly in the market, and they do not need to worry about earning repeat customer business. For example, in the United States each state has only one state agency where drivers can acquire driver's licenses – typically called the Department of Motor Vehicles. Customers can have exasperating experiences filling out paperwork, waiting in lines, failing driving exams, and there is no alternative. Conversely, automobile manufacturers, such as Mini Cooper, are great at managing customer relationships, because they provide an exceptional in-store experience for prospective customers, and their existing customers enjoy a hassle-free maintenance experience. Other automobile manufacturers are so well known for their reliability, for example, Subaru and Honda, that their customers are notorious for being loyal to the brand. These companies spend lots of time and resources on programs to request customer feedback, reward customer referrals, and establish online customer communities.

7.4.1.3 Marketing and Sales Communication Channels

Marketing and sales channels are lines of communication that Marketing and Sales use to engage with the customers. These communication channels come in many forms, and the form that the communication channel takes depends on the type of communication (two- or one-way communication) companies wish to have with customers. Communication channels can be digital (through the company websites, social media platforms, videos, e-mail), nondigital (by way of television, radio, newspaper and magazine print ads, billboards), and/or in-person (such as in-person interactions occurring at brick-and-mortar store fronts, conferences, focus groups, trade shows/expos).

The marketing strategy and customer segments that are identified will greatly determine which communication channels are prioritized. For instance, if the target market is an older population, marketing information should be communicated to the target market through channels that this group is more comfortable with, such as print advertising in targeted magazines and newspapers. Conversely, if the target market is a younger, technology-savvy population, a marketing campaign that is launched solely on social media platforms may be a better strategy for reaching and communicating with the target market.

Marketing and Sales communication channels with customers and internal stakeholders are represented in Figure 7.6.

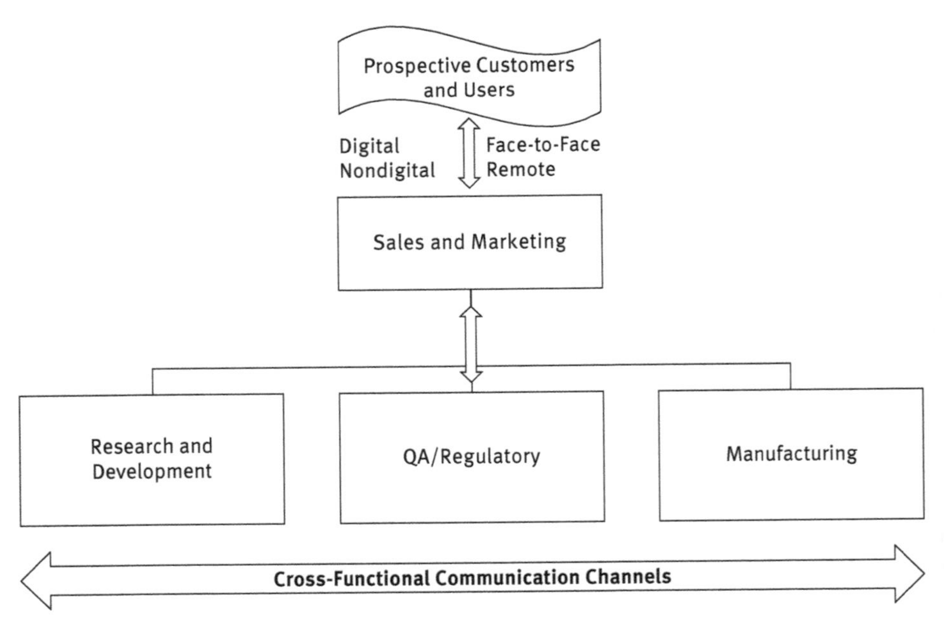

Figure 7.6: Marketing and Sales communication channels.

7.4.1.4 Marketing and Sales Content

Customer content that is created by Marketing and Sales needs to be crisp (use non-conflicting color patterns and fonts), memorable (think catchy slogans and jingles that are played on television and radio), and easy to understand, using as few words as possible. Customers have limited attention spans and are constantly bombarded with other advertising messaging, and so customer content needs to communicate its main point quickly and succinctly. Customer content that is utilized by Sales representatives needs to be portable and durable (for instance, use heavy cardstock pamphlets and brochures or develop small prototypes) because Sales is responsible for conducting product and service demonstrations to prospective customers, and Sales personnel may be required to travel to meet individual and team sales targets.

7.4.1.5 Hiring Sales and Marketing Personnel

Remember that people are a company's most valuable resource! Without people, marketing and sales strategies cannot be implemented. Regarding the hiring of Marketing teams, companies should look for individuals who are creative, great at oral and written communication, have a knack for design, are strong problem-solvers and strategic thinkers, can conduct research, and are adept at using advertising and promotional marketing and data analytics technologies. As mentioned above, Marketing interfaces with multiple teams within the company, as well as customers, so Marketing personnel need to be personable, good listeners, and have the ability to work well under stress, in order to meet many stakeholder expectations.

Regarding the hiring of Sales teams, companies should look for individuals that engage well with others, are persuasive, patient and persistent, have a track record of sales experience within the desired sector, and who are familiar and passionate about the products and services that the company is striving to sell. A prior track record in the desired sector is important, because companies need sales personnel who are familiar with the regulatory do's and don'ts (if any) of the sector and who can quickly reach out to his or her network within the sector to establish new customer accounts and promote company products and services. Sales personnel with prior sector experience will also know what types of customer training or education will be required and can convey those requirements back to Marketing.

The hiring of internal Marketing and Sales resources is contingent on available capital and the need for expansion in the market. However, the use of outside firms for both Sales and Marketing is common for startup companies. Contractual agreements to trade either equity, commission, royalties, or a combination thereof are common practice.

7.4.2 Marketing and Sales Implementation

The following case studies provide examples for how Marketing and Sales teams can implement a company's marketing and sales plans.

LactLuster Beachhead Market Example (introduced in Chapter 4)
Recall the LactLuster milk additive startup that was discussed in Chapter 4. Based on LactLuster's Marketing and Sales SIP, LactLuster Executive Leadership determined early on that they only had enough resources to pilot their product in a single beachhead market. Before expanding to new markets, LactLuster wanted to verify that their product was effective, that their marketing content was engaging, and that their sales pitch was convincing in a single beachhead market.

LactLuster did not have the financial resources to hire a Business Development Director and the necessary associated Sales representatives, and so they partnered with a local sales and marketing consulting firm to provide this expertise. The CEO of LactLuster decided to assume the Business Development role in preparation for the beachhead market launch. The sales and marketing consulting firm advised LactLuster's CEO to choose a beachhead market that was in geographic proximity to their headquarters to save on travel costs and to encourage face-to-face interaction, and to select a beachhead market that was teeming with dairy farmers to maximize the number of sales calls and collect a large amount of feedback from prospective customers in a shorter amount of time. In doing so, LactLuster's external Sales team received feedback on the performance of the product, suggestions for further product development, and confirmation that the milk additive free samples, an idea generated by the LactLuster Marketing team, were provided in a format that was easy for dairy farmers to use to test the product.

Based on the outcomes of the beachhead market launch, LactLuster external Sales team provided feedback to LactLuster's Marketing team that providing free samples to prospective customers was a great sales technique, and one that should be continued throughout market expansion. The LactLuster CEO presented the successful results of this pilot from the beachhead market to prospective investors, who then agreed that the company was ready to enter into additional dairy markets in nearby states through regional expansion. Investors pitched in additional funds, which allowed LactLuster to hire the additional Marketing and Sales resources that were needed to implement their market expansion strategy and develop additional website, digital media, advertising and promotional content, to reach a wider audience of prospective customers.

Pharmaceutical Example A pharmaceutical company is developing an immunotherapy drug they believe will be highly effective at treating certain types of cancer. The immunotherapy drug works by targeting a particular protein pathway that prevents immune cells from identifying and killing cancer cells. While Marketing and Sales teams prepare Marketing and Sales Strategy SIPs, the pharmaceutical company researches the various market segments and reviews prior clinical research, both research conducted by their own company and by competitors, to estimate and assess how successful their treatment will be with prospective patient segments. The pharmaceutical company knows that there is potential to promote the new drug with multiple cancer patient segments, but first they will need regulatory approval to treat one type of cancer. The company needs to prove that the treatment is safe and effective for that one patient segment, before applying for and receiving regulatory approval and promoting the drug to other patient segments.

The pharmaceutical Marketing and Sales teams then perform a strengths, weaknesses, opportunities, and threats analysis to evaluate their competitors' existing Sales, Marketing, and Supply Chain assets and resources versus their own to determine whether it would make sense to develop, launch, and promote this drug internally or license it to a competitor. Since the pharmaceutical company has a successful track record and portfolio of cervical cancer diagnostic products, the company already has a customer network of gynecologic oncologists, an internal sales force that is familiar with the gynecological oncology sector, and existing marketing content and contracts with external third parties (such as digital images and animations, specialty packaging, electronic manuals, conference kiosks, and presentation materials) that can be utilized to educate existing customers. Therefore, the pharmaceutical company's Executive Leadership team chooses cervical cancer as its beachhead market for its immunotherapy drug. Once the beachhead market was determined, then the Technology Development, Marketing, and Sales Strategy SIPs were finalized to coordinate drug development, clinical trial preparation, execution and follow-up publication, Marketing content generation, Sales training, and regulatory application activities to achieve market approval.

Once the pharmaceutical company receives regulatory approval to market the immunotherapy drug for cervical cancer patients, collects and publishes premarket (and postmarket) clinical data in reputable peer-reviewed journals, and key oncology opinion leaders present that data at prestigious gynecological oncology conferences, the company is able to prove that the treatment is safe and effective with a broad cervical cancer patient population. Now, oncologists in other cancer areas are demanding that their patients be permitted to use the immunotherapy treatment. The pharmaceutical company's Regulatory and Clinical Affairs teams submit additional regulatory applications to receive approval to market the treatment to cancer patients in other patient segments, such as skin cancer, lung cancer, stomach cancer, and so on, that are affected by the same protein pathway.

With new regulatory approvals that authorize the pharmaceutical company to market the immunotherapy drug to additional patients and providers, the company prepares to expand into additional market segments by initiating projects with Marketing to create new advertising and promotional marketing materials (such as website updates, direct to consumer radio and television ads, magazine ads), hiring additional Sales personnel, training their existing sales force, partnering with third-party sales forces that specialize in cancer treatments, and partnering with specialized distributors that have existing relationships with the top cancer treatment facilities and oncologists in the country.

7.5 Postlaunch: Evaluating Effectiveness and Planning for Further Expansion

Connecting with customers postcommercial launch is essential for establishing a strong foothold in the market and capturing the desired market share. To connect with customers, companies can utilize the various channels that were identified during business model and marketing strategic planning (e.g., face-to-face interactions, company website, social media platforms, customer service hotline, e-mail, letters, etc.). When given the opportunity, customers can provide invaluable data about how your product or service is being used, how easily customers are able to procure your product or service, how reliable your product or service is and its overall performance, and so on. This is referred to as Postlaunch Customer Surveillance.[24] Customer data can

24 Not to be confused with Postmarket Surveillance that is required by the regulatory agencies (see Chapter 12).

inform your company of new use cases (i.e., new ways how customers are using your product or service) and market opportunities, whether design/feature changes need to be made for the next product/service iteration, whether customers are having difficulties that the development team did not foresee, or even how the company can improve its operations prior to implementing market expansion plans.

During postcommercial launch, companies[25] need to be prepared with a strategy for collecting data, reviewing and analyzing data, summarizing that data, and then making informed, data-driven business decisions for the benefit of the company (refer to Figure 7.7). In other words, does the data indicate that business strategies or future product/service development campaigns should be modified or kept the same?

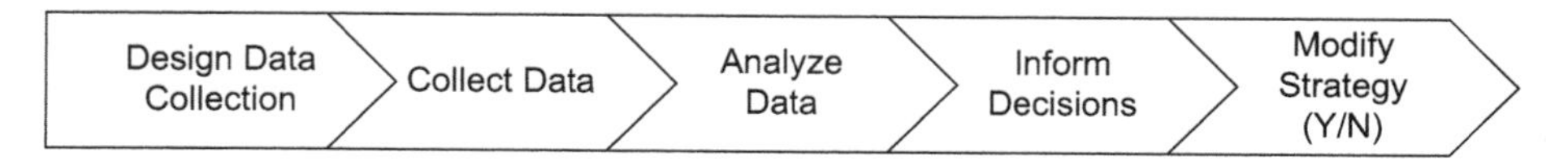

Figure 7.7: Postlaunch Surveillance and data analytics process flow.

This exercise in customer surveillance and data analytics is only as good in quality as the data itself – if the data collected is of poor quality or quantity, is incomplete, or is based on faulty assumptions, then leadership will not be able to answer key strategic questions, such as:

- How effective is our marketing and sales strategy?
- Do we, as a company, need to change our technology development strategy? If so, what resources will be required, and do we have the internal skills that are needed to make a change? Can we determine the next product line?
- Do we need to pivot on our product's use cases and pivot on the type of customers that are associated with that use case?
- Are we optimizing our various revenue streams, and how are profit margins different across revenue streams?
- Are we ready to expand into new markets, and if so, which partnerships do we need to develop and who additionally do we need to hire?

25 Marketing and Sales leadership define the Customer Surveillance strategy. Financial and marketing data analysts within the Marketing, Sales, and Business Development teams then typically collect and analyze customer surveillance data. This analysis is then presented to Executive Leadership.

7.5.1 Postlaunch Surveillance and Analytics Strategy

A postlaunch customer surveillance and data analytics strategy relies on good *people, processes*, and *tools*. A surveillance and data analytics team should be comprised of individuals (1) with good customer service, organizational, and communication skills, exceptional data analysis and presentation skills, (2) who are adept and effective problem solvers and researchers, and (3) who are attentive to detail. Marketing and Sales Data Analysts, Marketing Product Managers, and Marketing and Sales Leadership are expected to analyze the data in different ways to extrapolate meaningful information and patterns, which is sometimes lost in the details and numbers.

For example, by studying the ways in which customers interact with your company digitally via the company website, company Data Analysts can determine how the website layout, that is, the user experience and user interface, can be improved to make it easier for customers to locate and go to the company webpages that matter most to them. Or, in another example, Data Analysts can determine which resources (e.g., articles, blogs, product information brochures, etc.) are being downloaded frequently, indicating that more customers may be interested in purchasing those products or services, which may prompt the company to launch a targeted market campaign.

Customer surveillance comprises the development of the surveillance process, the execution of the surveillance, and the analysis of the surveillance data. Customer surveillance processes should be easy to follow and provide instructions for the different types of surveillance monitoring activities, frequency of monitoring, what data sets are to be collected, and how to collect and package surveillance data to make it easier for the Data Analytics team to analyze. Customer surveillance processes should also include instructions for how to identify and escalate customer complaints, when further investigation into the customer's complaint is required, and/or when the complaint poses a potential commercial or regulatory risk to the company.

Data analytics procedures can refer to standard methods for determining sample size, statistical data analysis techniques[26] (including confidence and reliability calculations), and presentation techniques, but these procedures should be flexible enough to allow for creative freedom in determining the best way to analyze and present the data.

Customer surveillance and data analytics tools, for example, such as SAP, Google Analytics, IBM's Watson, SurveyMonkey, and so on, need to be robust, easy to use, and should be verified that they meet company requirements and validated that they deliver the benefits that are expected by the users of the tools. These tools are no good to the company if the tools are too expensive, if the tools do not deliver

26 Statistical analysis terminology is described in Chapter 2.

a good return on investment (ROI), if there are bugs in the data gathering and analysis software, if the tools are inoperable due to frequent maintenance and upgrades, etc. A customer surveillance and data analysis strategy can also rely on simple, manual tools to get the job done. For example, customer surveillance information can be documented on paper forms and later transferred to a database for analysis and discussion.

Finally, a postlaunch customer surveillance and data analytics strategy can be used to positively reinforce company decisions or identify opportunities for improvement.

Example

A medical device company ensures that Sales representatives are present at every new hospital that uses its medical device for the first few surgical cases, to ensure that physicians and the surgical teams are comfortable with using the medical device and observing whether they are using the medical device in new ways that were previously not contemplated by the company. However, for one particular surgical case, the device behaves differently than intended, and so the Sales representative collects the following information about the event and the surgical case, such as:

- The date of event.
- Hospital information.
- Physician information.
- Medical device product information.
- Detailed information about the event.
  - The point at which the event occurred relative to the surgical procedure.
  - What additional surgical accessory tools were used during the surgical procedure.
- A statement from the physician (if possible).
- Whether the patient's health and safety were impacted and how.

The medical device Sales representative arranges for the device to be returned to the medical device company so that Quality and device failure engineers can perform a materials and chemical analysis and mechanical failure investigation, to determine the root cause of the event. Quality can also refer to the company's risk management procedures to determine what are the risks, as they relate to potential patient and medical device user harms, and what is the likelihood of that risk reoccurring and severity of impact when it does occur. The company issues a message to all Sales representatives to watch out for this type of behavior and to alert the company immediately if the event happens again. Before closing the report, company regulatory leadership determines whether the event must be reported to regulatory agencies, depending on whether the patient's health and safety were jeopardized, which was directly or indirectly attributed to the behavior of the device.

With this same example, the physician could also give input to the Sales representative about performance characteristics that he or she likes about the medical device and offer to provide testimony at a future medical conference or contribute to a future medical publication (i.e., marketing support). This positive feedback can be captured by the medical device Sales representative and communicated back to company Sales leadership.

7.5.2 Postlaunch Surveillance and Analytics Implementation

To implement a postcommercial launch customer surveillance and data analytics strategy, identify the communication channels you have with customers (prior to customer engagement, during customer engagement, and following customer engagement) to gather data about their needs and requirements, likes, dislikes, and habits while using your product or service. Once the communication channels have been identified, then determine which approaches will provide the most useful data for the company resources available (e.g., time, personnel, cost, etc.).

Face-to-face customer surveillance provides companies with a wealth of information that can be gained through conversation and by discussing open-ended questions. Companies can interact face-to-face with customers by exhibiting at conferences, conventions, and trade shows, by arranging sales calls (for instance, on the phone, through video conferencing, or in-person visits), by sponsoring and conducting training events, by designing and completing interviews with individuals and with groups, and so on. Face-to-face customer surveillance is a great opportunity to collect feedback from early adopters, place orders directly with customers and distributors, talk to key opinion leaders and subject matter experts about how differently your product or service can be used, designed, or marketed, and promote organic growth through word-of-mouth and referrals. Remember that all company marketing materials and personnel outward appearances should reflect the brand of the company. Though highly effective, face-to-face customer surveillance is resource intensive.

Digital customer surveillance provides companies with a lot of information quickly; the hard part is knowing which information to act upon. Companies employ teams that are solely dedicated to digital customer surveillance. For example, companies can:

- Track how customers are navigating through the company website pages and social media pages, to determine what content to promote.
- Issue customer surveys at important steps of the digital sales or digital customer service process.
- Receive complaints or product return requests via the company website or via a customer service e-mail address.
- Read reviews, testimonials, and customer feedback on message boards and consumer websites, to determine if a product recall may be needed or whether a design or manufacturing change is required.
- Monitor regulatory websites to ensure products are not causing safety issues.
- Monitor e-commerce websites to see which products are top selling, and which products need additional marketing or sales support.
- Analyze internal and external sales databases to determine which customers are repeat customers and which products are being reordered, to anticipate future demand, and so on.

To summarize, postlaunch customer surveillance data and analytics can be used to verify the effectiveness of your marketing, sales, and commercial launch strategies, determine when the strategy needs to be modified and how (usually kicking off an internal research project to determine which resources, such as personnel or funds, are required to implement the strategic change), identify potential revenue streams, and confirm that Operations is ready to handle additional customer demand, before expanding into new markets.

This chapter discussed several key aspects for launching your innovation to market by describing the strategy and implementation of hiring resources, managing vendors, designing marketing and sales operations, and analyzing postlaunch surveillance data. Chapter 8 discusses implementation in more detail by describing best practices in the management of technology projects.

8 Technology Project Management

Abstract: Technology project management is the process of managing and delivering successful projects – projects that implement well-defined objectives and acceptance criteria in the shortest amount of time with the fewest number of dollars, and resources spent in a way that is understood by the organization and any customers who receive the outputs of the project. Components of successfully managed projects include: defining strategy, managing scope, mitigating risk, executing proven technology management processes, and applying project management tools and lessons learned from previous projects, which all require continuous, effective integration and communication of results and outcomes. This chapter explores a variety of useful tools that can be used to effectively manage the critical aspects of technology development projects, such as utilizing statements of work (SOW), project charters, phase gates, and change review boards for managing the scope of the project; leveraging Risks, Assumptions, Dependencies, Issues, and Opportunities (RADIO) project elements, storyboarding, risk management governance, and lessons learned from previous projects for handling risk and scenario planning; and creating a communication plan that utilizes communication tools, such as SCRUM boards, project scorecards, dashboards, and roadmaps to facilitate communication between the stakeholders.

Keywords: Project management, scope, statement of work (SOW), project charter, phase gates, change management, change review board, scope management, risk, risk management, RADIO, storyboarding, risk management governance, lessons learned, stakeholder communication, SCRUM boards, project scorecards, dashboards, roadmaps.

8.1 Introduction

A technology project typically designs, develops, and launches a new technology or innovation. In terms of best practices, the project needs to align with and support an organization's strategy and vision.[27] The strategy and vision are created and led by senior leadership within the organization. A technology project manager is tasked with implementing that vision, that is, it is the "glue" or driving force that keeps a project from failing due to schedule delays, budget overruns, changing

27 If a project does not align with the company's vision or project changes no longer align, then project management should communicate upwards to senior leadership and resolve that discrepancy.

https://doi.org/10.1515/9783110521900-009

objectives, poor communication, lack of quality, and so on. The technology project manager is the day-to-day leader of the project.

The first rule of project management is to not let the management of the project become a project itself. The life cycle stage (to reiterate, the company life cycle stages include the due diligence stage, startup stage, growth stage, expansion stage, and maturity or exit stage) and size of the company need to support the project implementation strategy.

A technology project manager needs to balance the priorities of multiple stakeholders, all of whom have various interests and can influence the project outcome, and a technology project manager needs to bring together the work outputs of multiple project work streams. For instance, during a typical day in the life of a technology project manager, a number of things can go wrong, such as:

- Midway through the project, timelines can slip.
- Key resources (people) on the project are being asked to decrease their availability to join other projects, and these key resources need to transfer their work to less experienced resources who require intensive training.
- A manager from a different business function is asking for a change to one of his or her deliverables that the project team has already completed for that function.
- Technology service providers and vendors are slow in submitting quotes for services, leading to schedule delays.
- One of the components used to build the technology is currently only sourced from a single supplier, and there are rumors that the supplier is behind on orders.
- Project team members are constantly interrupting you with questions, asking about timelines, deadlines, and who should be working on what.
- Managers from other business functions are already voicing their disapproval about the changes that the project is implementing.
- Marketing is requesting additional Voice of the Customer (VOC) data to support and verify the user needs that the new technology will be addressing.
- Technology requirements keep changing.
- The current technology prototypes are clunky and keep breaking.
- No one knows when the project will be finished!

Sounds daunting, right? This chapter discusses concepts and tips for managing and delivering successful projects – projects that implement well-defined objectives and acceptance criteria in the shortest amount of time with the fewest number of dollars and resources spent in a way that is understood by the organization and any customers who receive the outputs of the project. Figure 8.1 summarizes important components of successfully managed projects: defining strategy, managing scope, mitigating risk, executing proven technology management processes, and applying project management tools and lessons learned from previous

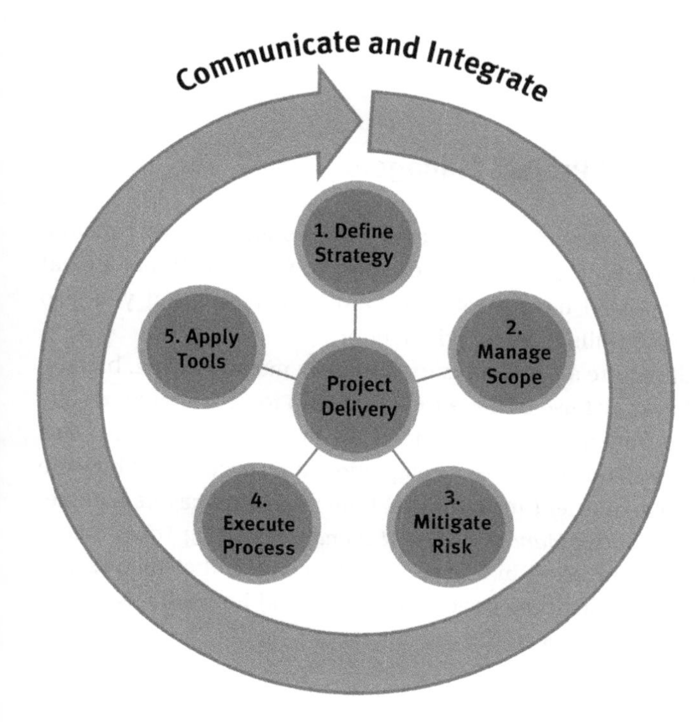

Figure 8.1: Important components of successfully managed projects.

projects. These all require continuous, effective integration and communication of results and outcomes.

These components of successfully managed projects (meaning strategy, processes, and tools) will be discussed in the following sections:

- Tools for "First-Time" Project Managers
  - Scope Management
  - Risk Management and Scenario Planning
  - Communication
- Managing Technical and Strategic Risk and Creating Realistic Timelines
- Effective Technology Project Management
- Change Management
- Problem-Solving Techniques
- Brainstorming (Ideation) Techniques

The content that follows is an overview of best practices for project management. These best practices are used by large, successful companies. Smaller companies and startups may not have the resources or need for every topic discussed in the following sections, but knowledge about the processes and tools that are available can be useful for any company of any size.

8.2 Tools for "First-Time" Project Managers

It is another typical day at work when your manager stops by and tells you about a "high visibility" project and asks you to lead it. You think to yourself, "What a great opportunity!" and "This is what I have worked so hard to achieve!" But you also wonder whether you have the skills to lead such an effort.

A complex project will require a set of tools to effectively manage scope, budget, resources, and quality. The correct use of these tools will help to deliver great results to the team and your organization. What are some of the project management tools you can use to improve your chances of project completion and success, and how are they used? Useful project management tools for first-time project managers include: *1) Scope Management Tools, 2) Risk Management and Scenario Planning Tools, and 3) Communication Tools*, which are listed in Figure 8.2. The content that follows will serve as a good introduction for first-time project managers, and hopefully as a good recap for more seasoned and experienced project managers.

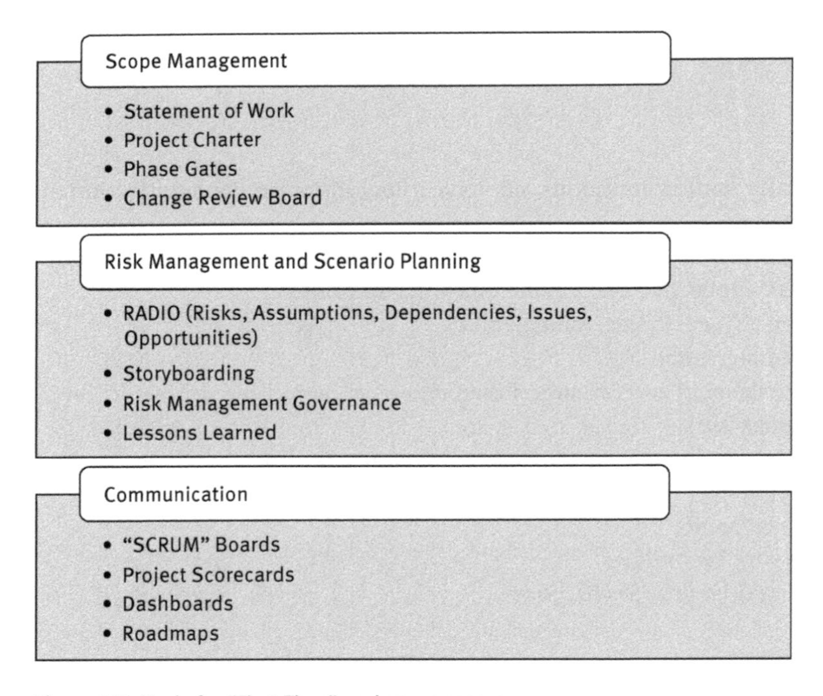

Figure 8.2: Tools for "First-Time" project managers.

8.2.1 Scope Management

Project scope defines what will be accomplished after completing the project, which means identifying what requirements will be met and how they will be met. For example, if a project is initiated to build a house, then the scope would include the design of the structure, the foundation that would be laid, the number of floors and types of rooms to build, and the building materials that are to be used.

Projects fail because: (1) project scope is not defined and agreed upon by all major stakeholders in the beginning of the project (stakeholders have different levels of influence in a project, and not all stakeholders are equal) and/or (2) scope boundaries are not respected throughout the project. When projects become too unwieldy to manage, then the end goals are always changing. When this happens, project timelines lengthen, project budgets widen, and stakeholder and organizational tolerances shorten.

There are several tools project managers can use to define, unify, and communicate with project stakeholders about scope: Statement of Work (SOW), Project Charter, Phase Gates, and Change Review Boards.

8.2.1.1 Statement of Work

A SOW is a contract or agreement between the project manager, the project implementation team, and the project sponsor(s). A project sponsor approves the budget and resources that are necessary to implement the project.

Note: In smaller companies, the project leader and the project sponsor may be the same person.

A SOW should be prepared, reviewed with the project team, and approved prior to initiating any project work. A SOW can include the following content (this is not an exhaustive list):

- A brief description of the project background (purpose, history, business case for change) so that all readers understand why the project is being initiated.
- Project Scope (what is in scope and out of scope) and applicable documents or references.
- Goals and/or objectives (should be **S**pecific, **M**easurable, **A**ctionable, **R**ealistic, and **T**ime-bound or SMART).
- Project Requirements (can be categorized by priority, such as "must have" versus "nice to have").
- Deliverables (can be organized by project work stream or by project phase).
- Additional project needs (for instance, physical resources/equipment, human resources).
- Anticipated period of performance or calendar length of project.

- Project Entry Criteria (such as an approved SOW and approved budget).
- Project Exit Criteria (such as sponsor acknowledgement of completed deliverables).
- Project Manager-specific deliverables (e.g., project plans, tracking of action items, and dashboards/scorecards).
- Project Sponsor-specific deliverables (including access to training, tools, timely review and feedback).
- Project Budget.
- Project Change Requests (Section 8.2.1.4 discusses Project Change Requests in more detail).

The more time spent up-front defining the SOW, the less headaches and finger-pointing there will be throughout the project.

8.2.1.2 Project Charters

At a high level, project charters summarize the framework for new projects (identifying the who, what, when, where, and what-ifs of the project) and are reviewed by the project leader, project implementation team, and the project sponsor. By planning ahead and being prepared, the company can prepare a project charter that encourages structure throughout the planning process. Project charters can include the following sections:

1. What is in scope and out of scope for the project.
   - These are important distinctions for stakeholders.
   - Clarity of this content can prevent future arguments, especially between the project delivery team and the project customers.
2. High-level project summary and purpose or justification.
3. Objectives for what the project should accomplish (also known as deliverables) – These should be specific, measurable, realistic, and time bound.
   - They define what the end of the project "should look and feel like."
4. Team member roles and descriptions for each role – This is not trivial and a very important section to "put down on paper."
5. High-level timeline of events or milestones – The timeline could be divided into phases and should easily be understood by others not involved with the project.
6. Team-generated RADIO list (identifying any potential Risks, Assumptions or Constraints, Dependencies (key handoffs), Issues, and Opportunities).
7. Budget summary – Estimated project spending should refer back to the high-level timeline, that is, the amount of funding needed to reach each milestone, or complete each project phase.
8. List of key stakeholders (who are impacted by the project and/or needs to provide input).

In summary, the project charter is a useful roadmap that guides the creation of more detailed project plans. Project charters should be a collection of the project implementation team's opinions. For instance, if the project implementation team is involved with the creation of the project charter, this will drive team ownership and accountability for the project's success.

8.2.1.3 Phase Gates

To guide and influence scope throughout the project, project teams can participate in project reviews with key project stakeholders and project sponsors to ensure that deliverables are being shaped and completed according to the SOW, that is, to ensure there are no "bad surprises" at project completion. These project reviews are known as Phase Gates.

A Phase Gate can take the form of a status update presentation, or can be more interactive like a Table Top Workshop (see Section 8.5.5 for more information on Table Top Workshops) or a Technology Development Life Cycle Design Review (refer to Chapter 10 for more on Design Reviews). Phase Gates can be scheduled after critical milestones are achieved or on a recurring basis (for instance, quarterly).

Attendees of the Phase Gate can identify risks or opportunities (see more on Risk Management in Section 8.2.2) during these project reviews and can help steer the team to more favorable outcomes. Usually a project team cannot move forward, until Phase Gate follow-up items or action items are addressed with a satisfactory level of detail and planning.

8.2.1.4 Change Review Boards

It is inevitable that there will be change requests on a project, for example, a request to include something new in the final output that impacts the timeline; a request to switch vendors, a request for additional funds or resources, a request for different solution requirements, and so on. A Change Review Board is simply a means to formalize, document, communicate, and manage these requests for change in a project and get stakeholders to buy into the change. Change Review Board members typically consist of stakeholders who have great influence and/or interest in the project outcome. Within large companies that are implementing multiple projects, a Change Review Board may meet monthly; for smaller companies, Board Meetings may take the place of Change Review Boards.

All Change Review Board topics that are discussed at meetings should pass a defined set of change criteria or thresholds; otherwise, unimportant topics will consume the Change Review Board's time and energy unnecessarily. Change criteria can be project specific, such as a budget change of "x%" or timeline impact of "x" months, and can be defined by the Change Review Board. Without the Change Review Board, there is a risk that large scope changes could be implemented without accounting for downstream impacts to the project in question, other projects that may be impacted, and the impact on the overall organization.

8.2.2 Risk Management and Scenario Planning

It is difficult to plan different project scenarios without considering the impact of risks. Risks can either be helpful (opportunities) or harmful (threats) to your project. As a project manager, your goal is to increase the probability of opportunities occurring and decrease the probability of threats occurring, especially those threats that impact the project budget, schedule, scope, and/or quality. Consider the following cause-and-effect sequence, depicted in Figure 8.3, while predicting, preparing for, and mitigating project risks:

- The project team thinks about possible sequences of events that can lead to a situation that becomes a project risk.
- Possible project risks are analyzed according to their severity of impact to the project and the duration of effect on the project.
- The project team then determines possible risk mitigations or controls to implement that can decrease the probability of the threat risk occurring or can reduce the duration of the risk's effect.
- The project team also determines contingency plans for if the risk mitigation is not successful.
- Implemented risk mitigations will create new sequences of events that may cause future project risks.

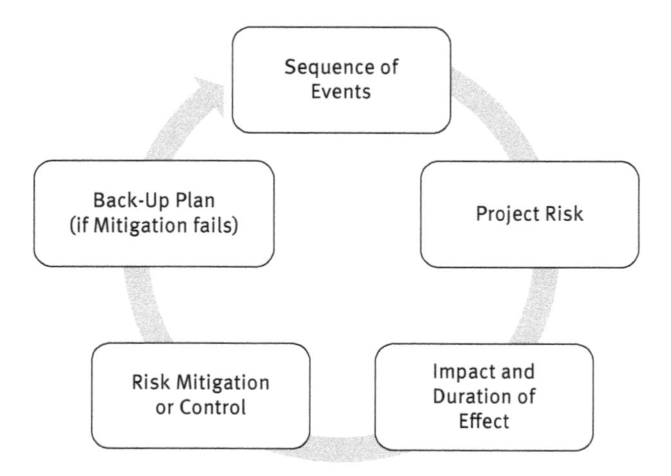

Figure 8.3: Cause-and-effect sequence of risk and mitigation events.

Risk planning can occur while creating the project communication plan and project management plan. To simplify the listing and organizing of risks, risks can be categorized by the different project work streams or major groupings of tasks that are listed in the project management plan, anything that may cause an impact, or would be impacted, by the risk event.

Risks can be caused by uncertain assumptions, complex or uncontrollable project dependencies, poorly constructed processes, conflicting requirements between stakeholders, and even by the residual or secondary effects from the mitigations that are put in place to prepare for risks.

Risk mitigations will vary, depending on whether the risks are opportunities or threats. Opportunities should be explored and developed, whereas threats should be prevented or mitigated (anything that will lessen the effects of the threat). Alternatively, project managers can decide to accept both kinds of risks.

An example of a risk opportunity could be that the project's contract manufacturer is able to shorten the timeline by adding additional resources and approving overtime hours for the operators and technicians. This will increase the spending burn rate for the project, but the benefits of completing the project early outweigh the risks of spending available funds in a shorter amount of time.

An example of a risk threat could be that a supplier of critical electrical components for your product is facing a shortage of these components, due to an increase in global demand. This threat can be mitigated by requesting the supplier to seek alternative components that function and perform in the same way, and by securing quotations and contracts/agreements with multiple electrical component suppliers, if necessary.

Project managers can use the following tools to plan for and manage risk events:

- Risks, Assumptions, Dependencies, Issues, and Opportunities (RADIO),
- Storyboarding,
- Risk Management Governance, and
- Lessons Learned.

Different formats can be useful when utilizing these risk management tools, for example, large posters that document RADIO topics are a great way to organize and run meetings, spreadsheets can organize and manipulate lots of RADIO-related data and information, and stand-alone documents, such as storyboards and lessons learned, can be created as risk communication tools.

8.2.2.1 RADIO

RADIO is an acronym that stands for Risks, Assumptions, Dependencies, Issues, and Opportunities. These terms are defined and described in Figure 8.4.

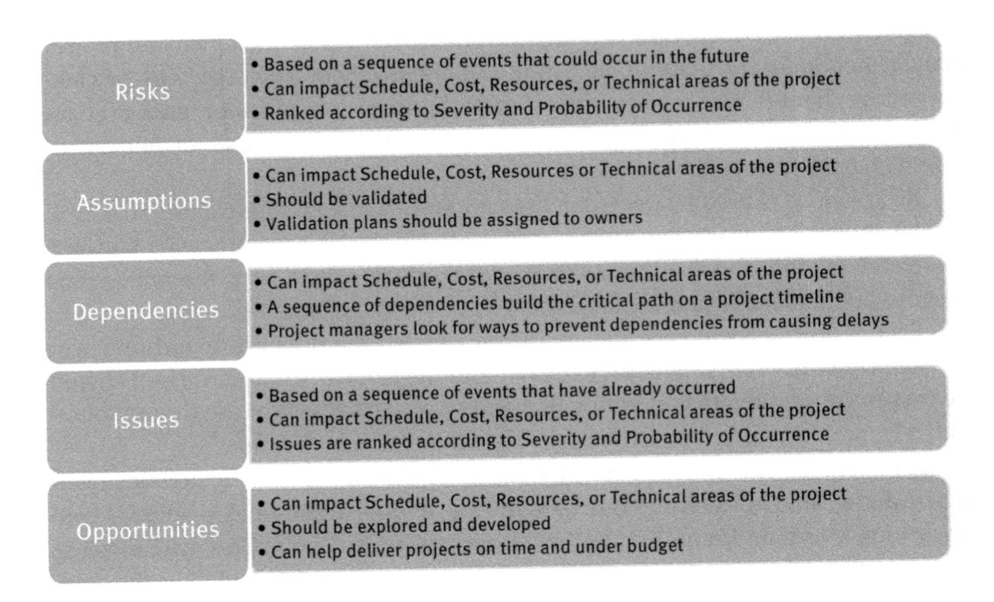

Figure 8.4: RADIO Terminology.

RADIO elements can come up in any project meeting or conversation, so project managers should get in the habit of listening for and documenting these elements. In the project meetings that you facilitate, keep separate posters as visual aids to document the individual RADIO elements that are discussed. Document these elements on the posters in front of the project team and obtain their approval on the description, next steps, and owner(s) associated with each element.

By keeping a checklist or register of RADIO items, project managers can be better prepared for project risk management meetings. Typically, project risk management meetings focus on the risks that carry the highest probability of occurrence with the highest impact (positive or negative) to the project. Many risk and issue management chart templates can be found on the internet, if your organization is not already using a template.

8.2.2.2 Storyboarding

A great way to identify project risks and to think through the major or minor effects of executed risk mitigations, is to storyboard, or use visioning techniques to map out project scenarios from start to finish. The storyboard can focus on an individual or team's experience as they go through the project steps, tasks, decisions, outcomes, and exchanges with others. Or the focus can be on inanimate project objects, such as data or flow of goods/materials or services. For example, Figure 8.5 depicts a simple storyboard for analyzing and selecting a manufacturing process

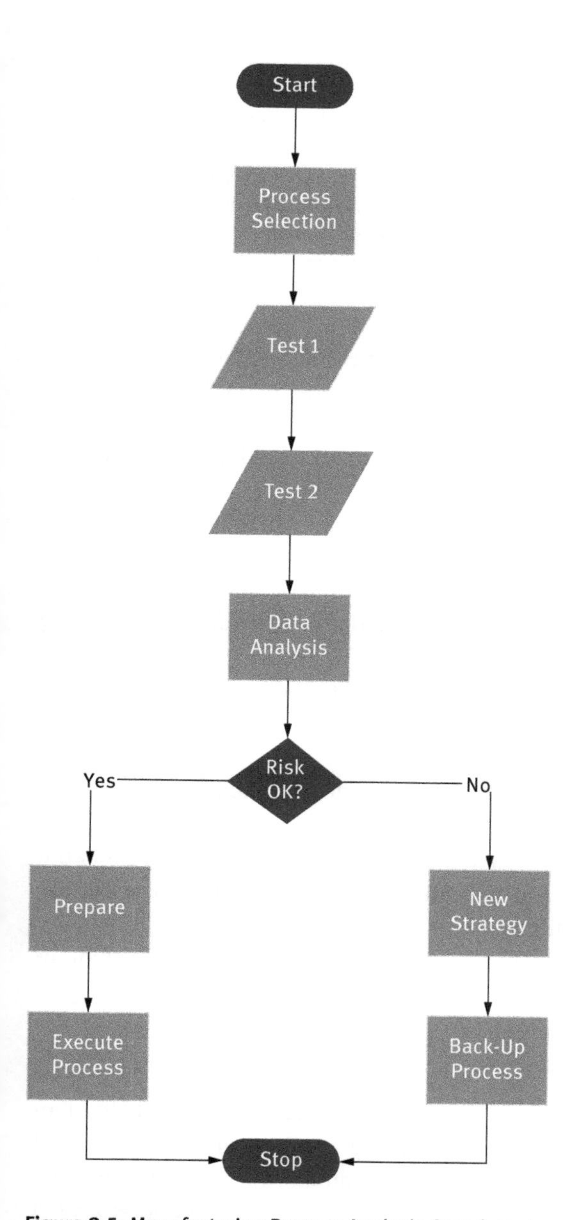

Figure 8.5: Manufacturing Process Analysis Storyboard.

that will be used to mitigate a risk associated with unsuccessful manufacturing lot release test data.

Storyboarding is a great visual and communication tool for teams: by seeing scenarios unfold, teams can identify potential RADIO elements and their alternatives. Remember, storyboard visuals do not need to be complex; most of us are not great artists!

Example
Walt Disney, the entrepreneur, animator, voice actor, and film producer, excelled at utilizing storyboard techniques during the story development process, to engage teams to work together and develop content for his movies by building off of one another's ideas. Story scenes were quickly sketched and strung up across a wall or bulletin board, where they could be modified or switched out easily for alternative content. The story literally unfolded before everyone's eyes.

8.2.2.3 Risk Management Governance

Risk management meetings can produce a lot of aggravation and very few results, when there is no governance or process in place to guide the team. Before incorporating project risk management practices into your organization or company, think about answers to the following questions (although not all questions may apply) to clarify outcomes and team expectations:

- What is the group's risk threshold or what is the company's tolerance to risk?
- Who provides expertise with regards to risk identification and monitoring?
- How often should risk management be monitored?
- How are change requests to address risks managed and integrated into existing projects/programs?
- There should be contingency reserves for budget and schedule impacts. What is the strategy for allocation, for example, expected monetary value analysis, and scenario planning?
- What are the authority levels for risk management decision-making, and which roles will have these responsibilities?
  - These authority levels and responsibilities may depend upon the probability and impact levels from the risk identification register, or the contingency reserves that will be allocated.
- What stakeholders need to be involved in this process, and what is the communication plan?

8.2.2.4 Lessons Learned

Resourceful project managers refer to lessons learned during other projects to identify patterns of risks and/or issues that may similarly apply to their projects, and what mitigation actions can be taken. An example template for capturing lessons learned is shown in Figure 8.6. Team member input can be gathered and analyzed according to whether a project outcome was successful or not successful, and whether this outcome was planned or unplanned. Once an outline is established, a full report detailing the trials and tribulations of the project can be established. A matrix or checklist of RADIO items can also be reviewed for lessons learned. These types of resources from previous projects should be used by project managers as "watch lists" – study the early signs and symptoms of problems and put a plan in place to mitigate and manage them before they occur.

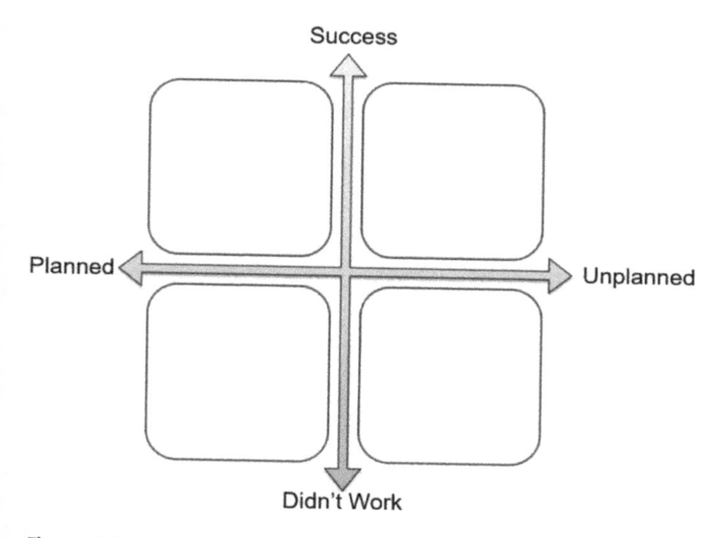

Figure 8.6: Lessons learned template.

8.2.3 Project Communication

Some of the most important roles as a project manager are to see the bigger picture and all the moving parts within a project and to explain how the parts of a project fit together, all while communicating different messages with different themes or emphases to audiences of different influence and interest in the project with different backgrounds. This requires great communication skills and the use of an assortment of communication tools.

Communication consists of three parts: the sender, the message, and the receiver of the message. As depicted in Figure 8.7, in addition to communicating the message, the sender needs to listen just as intently as the receiver, to confirm that the receiver has understood the message.

As a project manager, you will not always have the benefit of (1) communicating your messages about the project directly and/or in person to receivers, (2) studying receiver reactions to the message, and (3) ensuring receivers understand, clarifying your message if necessary. For example, project meeting notes that you complete may get forwarded to project sponsors and senior stakeholders, so project meeting notes need to include enough information and be written well, so that they are stand-alone communication tools. Refer to the example project meeting notes template in Figure 8.8 (an additional full-page template is available online at www.keystonescientific.net/resources.html). Additionally, these messages can provide a communication history that can either be used as reference to past conversations, or, in a worst-case scenario, as legal documents.

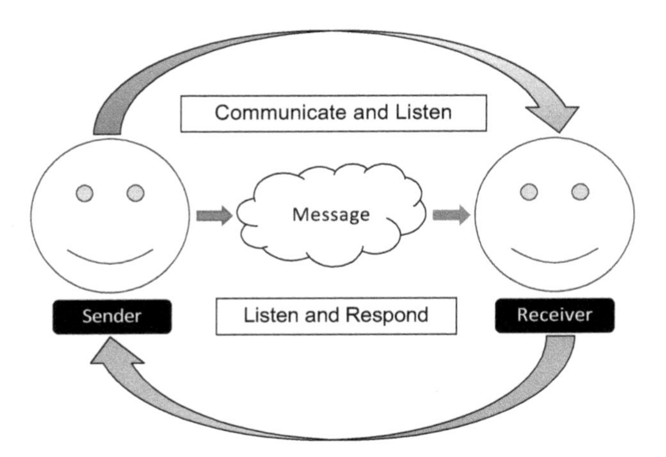

Figure 8.7: The three parts of every communication – sender, message, and receiver.

Or in another example, your presentation slides may be used as reference materials in future meetings, without you being present. These project messages, therefore, should be stand-alone from the project manager, and the intent should be clear. Visual or pictorial messages, when used correctly, can stand alone and can be used as tools to convey a vast amount of project information. Similar in intent as infographics, these communication tools tell a story about the progress and status of the project. We have all heard or used the idiom "A picture is worth one thousand words," and the same is true for visual project communication tools. These tools are absolutely essential in today's global economy, where most projects involve many people, both internal and external to your organization, and not everyone is located in the same time zone or speaks the same language fluently.

8.2.3.1 Communication Planning

A communication plan is an engagement strategy for how you, as a project manager, will communicate to stakeholders during the course of a project. Who will receive project communications, when will project communications get distributed, and how? A communication plan is created during the early stages of project planning, meaning that a communication plan is created while developing the project management plan, project risk management plan, and project quality management plan.

Before getting started on creating project communication tools, think about the different groups of project stakeholders that will be receiving and interpreting the messages that are conveyed. Not every stakeholder group or individual has the same level of interest nor the same level of influence in your project. Modify your communication plan according to these differences in stakeholder needs or requirements.

Weekly Status Meeting
DATE

Attendees:

Action Items:

Action Description	Owner	Completion Date

Parking Lot:

Item Description	Owner	Follow-Up Notes

RADIO (Risks, Assumptions, Dependencies, Issues, and Opportunities):

Item Description	Owner	Mitigation / Next Steps

Meeting Minutes:

Figure 8.8: Project meeting notes template.

Decide what project information needs to be conveyed to which stakeholders and at what levels of detail and choose which communication tools is best for a given stakeholder. Refer to the example communication plan template in Figure 8.9.

Another very important aspect of communication planning is Change Management – preparing your stakeholders to receive the message and ensure that it "resonates," once the message is delivered. Change management, as it pertains to project management, is covered in Section 8.5.

Once your communication plan and change management strategy are solidified, you can begin to think about the type, intent, design/format, usage, and follow-up for the project communication tools that closely fit or complement your project. Project managers can use the following tools for communication planning:
- SCRUM Boards.
- Scorecards.
- Dashboards.
- Roadmaps.

Descriptions of these project communication tools are found in the following section.

	Communication #1 – Stakeholder A	Communication #1 – Stakeholder B	Communication #2
What triggers the communication, for example, project milestone or calendar date?			
Who is the audience?			
What or who influences/ impacts the audience?			
What does the audience need to know?			
How does the audience prefer to receive the communication?			
How can you verify the communication is effective?			

Figure 8.9: Communication plan template.

8.2.3.2 SCRUM Boards

When rugby players are about to restart play, for example, after a penalty is called, they form a huddle and point their heads down, moving around to try and gain possession of the ball. This is known as scrum, which is short for scrummage. When project teams begin a new day, they can also huddle and attend SCRUM[28] meetings that are centered around a tool known as a SCRUM board, to discuss what needs to be accomplished on that particular day. In SCRUM meetings, project teams meet on a daily basis to communicate status updates and alert the team, if there are difficulties that the project manager should address. Although the SCRUM process is most frequently used in software development, SCRUM boards are used by many project managers on a daily basis to convey status of programs involving multiple projects or work streams with overlapping dependencies.

A SCRUM board is a fancy team "to-do" list for project stakeholders, usually divided into individual project work streams or individual projects as part of a greater program. Project managers use different, color-coded sticky (adhesive-backed) notes to identify tasks assigned to specific individuals or teams. The list of tasks on the SCRUM board include which tasks are being worked on that day or that week, which tasks have been completed since the last SCRUM board discussion (usually brief, fast-paced discussion held either daily or biweekly), which tasks are

---

28 SCRUM is a subset of Agile that emphasizes brief and concentrated time periods of project team work, or "work sprints", to achieve a lot of progress on complex projects in a short amount of time. Agile began in 2001 in Utah, USA, by software developers to prioritize a backlog of tasks.

considered Works in Progress (WIP), and which tasks are experiencing roadblocks, due to a lack of funds, resources, delays, process inefficiencies, and so on. Roadblocks on the SCRUM board are usually identified with an additional symbol, such as a red arrow or red star. Refer to the example SCRUM board template in Figure 8.10 (an additional full-page template is available online at www.keystonescientific.net/resources.html).

"Engineering Innovation" Book Project					
	Research	Writing	Editing	Marketing	Business Development
Daily Tasks to Complete					
Weekly Tasks to Complete					
Works in Progress					
Completed					

Team
Jennifer
Amber
Benjamin

★ = Roadblock

Figure 8.10: SCRUM board template.

Clients and stakeholders like SCRUM boards for their transparency. Project managers like SCRUM boards because they are an easy tool for visualizing:
1. Overutilized or underutilized resources.
2. Current roadblocks (issues) or pending roadblocks (risks) that need to be eliminated by the project manager.
3. The moving parts and interdependencies that need to be monitored by the project manager.

8.2.3.3 Scorecards

Project scorecards are frequently used to compare status of the critical path on the current project plan versus the critical path from earlier versions of the project plan. The critical path of the project plan is the shortest sequence of tasks that, when completed, will indicate the completion of the project. Any delay in completion of a critical path task will delay the final completion date of the project. As a project manager, you frequently receive project updates from stakeholders and then update the critical path timeline on your project plan.

For example, the following data points about the critical path can be recorded in a scorecard table and shared with senior-level stakeholders. The background in each cell can be highlighted as red (not good), yellow (at risk), or green (good), depending on the agreed-upon definition for each:
- Total number of critical path milestones.
- Number of critical path milestones that are on track.
- Number of critical path milestones that are at risk for being delayed.
- Number of critical path milestones that are currently delayed (1–2 weeks).
- Number of critical path milestones that are currently delayed (greater than 2 weeks).

An example project scorecard is featured in Figure 8.11.

Project Schedule Critical Path Scorecard

Scorecard Data	Status	Resolution Plan/ Comments
Total # of critical path milestones		Goal is to remove tasks from the critical path
# of critical path milestones that are on track		Overall status of critical path
# of critical path milestones that are at risk for being delayed		Maintain frequent communication with suppliers
# of critical path milestones that are currently delayed (1–2 weeks)		Stay on target!
# of critical path milestones that are currently delayed (greater than 2 weeks)		Additional resources are needed to bring in task(s)
Project end date		Project end date keeps slipping

Figure 8.11: Example scorecard for reporting status of project schedule critical path.

8.2.3.4 Dashboards

Project dashboards present lots of information in an organized and succinct way, especially information that is quantitative (e.g., numbers, percentages, calendar dates, progress bars, and status indicators). Refer to the example dashboard template in Figures 8.12 and 8.13. Project managers carry dashboards around with them into meetings because they quickly convey the "pulse" of the project, and project managers keep stakeholders informed with the latest data by updating and using dashboards. Dashboards help project managers hone in on the Key Performance Indicators (KPIs) and Return on

Name of Project
Name of Work Stream or Sub-team
Date

Status Item	*Current Status*	*Prior Status*
Overall Project Status	Green	Green
Technical	Green	Green
Schedule	Green	Green
Cost/Resources	Ahead Budget	Ahead Budget

On Track = Green, At Risk = Yellow, Delayed = Red)

Top Program Issues and Resolution Plan

Past Month Accomplishments
•

Tasks Next 1 Month
❑

Figure 8.12: Example project dashboard template.

Investment (ROI) attributes that project sponsors value. KPIs and ROI attributes are used to define and measure what project "success" looks like and should align with the organization's vision and mission.

8.2.3.5 Roadmaps

Project roadmaps are visual depictions of the High Level or Level 1 views within a project plan or Gantt chart. After a Gantt chart is created, subtasks can be rolled up into summary tasks, so that only the summary tasks are presented in the Level 1 view. Roadmaps are presentation tools that depict the Level 1 view of a Gantt chart. Roadmaps are great visual aids for all project stakeholders because roadmaps show the sequence, interconnectivity, and timing of key milestones and summary tasks along the critical path, and/or a roll-up of subprojects or work streams in a larger program. Roadmap timelines can be divided into months, yearly quarters, or even years, depending on the size of the project. An example roadmap is shown in Figure 8.14 (additional information on the FDA regulatory agency and medical devices is provided in Chapter 12). By using a project roadmap, every stakeholder can see how their individual "piece" fits into the broader "puzzle" of the project.

One Month Look Ahead:

Critical Tasks	Schedule Dates

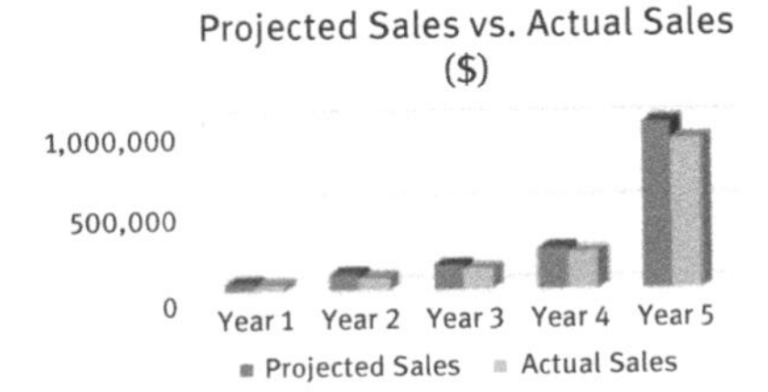

Insert Visual Data Here – Test Results, Screenshots, Pictures, and so on.

Figure 8.13: Example project dashboard template.

8.3 Managing Technical and Strategic Risk and Creating Realistic Project Timelines

Closely managing a project's technical risks and strategically de-risking the development of the technology improves the probability that the project will finish and accomplish its goals within the anticipated project timeline. However, the timeline that is created needs to realistically reflect the known and likely technology design and development risks. This section discusses how an organization can be more strategic with technology and innovation development projects, and includes helpful hints for creating realistic project timelines.

8.3.1 Managing Technical Risk

As a technology project manager, guide your team to simplify the engineering processes, technical solutions and, ultimately, the final outputs of the project. It is easy to overcomplicate technology development. Overly complicated projects,

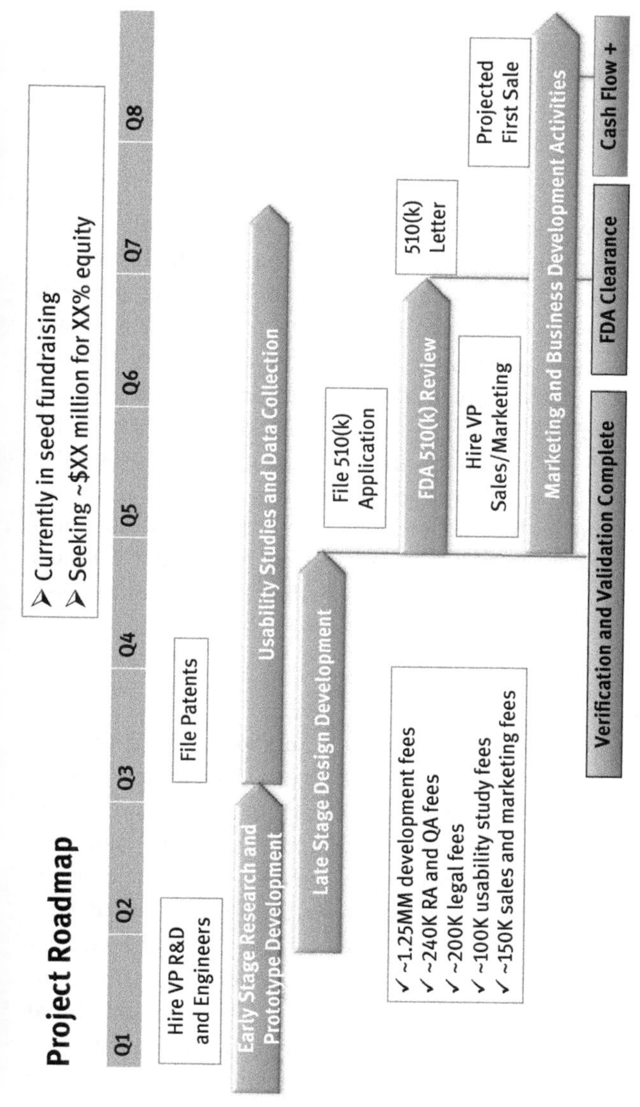

Figure 8.14: Example startup company roadmap for an FDA regulated class II medical device.

scope creep (where the scope of a project continues to grow, usually due to poorly defined project charters or statements of work [SOW]), and issues with vendors are the common causes of delay, during the development and commercial launch of new technologies.

Simplify, simplify, simplify!

It is always better to simplify an idea and then stick with that simplified plan, to ensure a quicker path to market. The more changes that are made during development, such as changes that impact materials, manufacturing process, quality, and testing, the longer it will take to launch the product. The longer this development project runway lengthens, the higher the probability that issues will occur, the higher the probability that a competitor will beat you to market, and the higher the probability that you will lose project team members. More "bells and whistles" on a product are not always better, especially when launching the first iteration of a product. Enhancements and improvements can always be introduced in future iterations, after the initial commercial launch, where customer feedback can be acquired to highlight (1) issues with the current product, and (2) key features that customers will want to see in the next version.

Listen, Research, and Analyze

When technical challenges do arise, solicit feedback and expertise from outside the team, both internal and external to the company. Sometimes a fresh pair of eyes during a "cold eyes" review or during a technology design review meeting can make all the difference. In these discussions, utilize the data collected from prototype builds and tests, models, and so on, to drive decision-making. When making decisions, do not overlook how decisions about design and functionality are impacted by the manufacturing process. Risk of technical failure is increased, if the team does not think about how manufacturing conditions, such as pressure, temperature, and energy, can alter the functional performance of the product. For example, a product's material properties (for instance, ductility, hardness, strength, resistance to corrosion or degradation) can be altered from the original material selection, if the material is exposed to a process condition that is above or below its limits. An unexpected change of material property may lead to an unacceptable manufacturing process verification or acceptance test result.

Lead the Extended Team

A technology development project typically involves many external vendors and service providers. Engage with these project partners frequently:

- Seek to **clarify** requirements and intended use(s) of the technology being developed,
  - *Note: Sometimes a vendor's solution or design choice may be faulty, if they are not fully aware of the intended uses of your technology (especially with regard to how a component is built and processed).*
- **Drive** them to perform,
- **Monitor** them for quality,
- **Control** their costs,
- **Assist** them with implementing technical risk mitigations, and
- Hold them **accountable** for their deliverables.

8.3.2 Managing Strategic Risk

Besides the complexities surrounding the technology development itself, strategic risk cannot be ignored.

As discussed in Chapter 1, conduct market research to establish a strong, justifiable business case. Know which intellectual property is already protected and where there is opportunity to file patents containing new ideas and claims. Be aware of prospective enabling and interoperable technologies that can provide a complementary feature to your technology or can communicate with your technology – think wearable devices, electronic healthcare systems, and be aware of major development milestones for competing technologies.

By keeping an eye on the developmental progress of your competitors, you can gauge whether you need to adjust your development strategy to ensure you can beat your competitors to market. If beating your competition to market is critical for your commercial launch plans, then you can make project-related decisions to shorten the critical path, including decisions to:

- Add resources to the project.
- Ensure there are backup plans and backup suppliers or contractors.
- Analyze risk versus reward or cost versus benefit.
- Know where to focus your time.
- Streamline deliverable reviews.
- Improve team productivity by colocating team members for periods at a time.
- Minimize the number of unimportant/non-value-added meetings.
- Make sure you have the right people for the job.

As discussed in Chapter 2, increase your probability of commercial success by obtaining customer feedback (VOC feedback) from initial technology concept through commercial launch and even postlaunch, by testing product positioning and branding/messaging options within target market segments. Without this feedback, your development team will not have its finger on the "pulse" of customer wants and needs, and your solution will not provide value.

If budget becomes an issue and you are employed at a startup, seek alternative sources of funding when necessary, such as grants, competitions, crowdsourcing, friends and family contributions, and even lucrative partnerships (all discussed in Chapter 6), and licensing agreements (discussed in Chapter 5).

If budget becomes an issue and you are employed at a larger company and are managing a pipeline of technology development projects (versus a single project), maintain a balanced portfolio of high-risk, high-reward commercial opportunities with low-risk, lower-return commercial opportunities. As a manager, ensure your projects have secured enough funding and company support, especially when there are changes made to company leadership or project executive sponsorship. Seek business allies and coaches who will advocate for your projects' business cases.

Risk-conscious leaders also create sustainability plans, if ever there is an emergency due to sickness or death, natural disasters, fires, computer viruses, and so on, which impacts the progress of a project. For example, all key project documents can be stored on a cloud-based file-sharing platform, and multiple individuals should be trained and capable of performing a particular role on the project.

8.3.3 Creating Realistic Project Timelines

Solid (1) information, (2) preparation and planning, (3) communication, and (4) processes will support the creation and maintenance of realistic project timelines.

Information

A timeline is only as good as the quality of information that you have. Therefore, carefully think through task sequences, task durations, and task dependencies, especially for tasks that appear on the project's critical path. To verify the plan that is created, seek multiple opinions within the organization, especially from those with lots of technology implementation experience. Get feedback on the practicality of the schedule and subject-matter input on the technical difficulty of the project deliverables. Be realistic with what can be accomplished – do not overextend the team.

Preparation and Planning

There will always be a degree of uncertainty in the project timeline because technology development is complex and can be unpredictable, so project managers should create **base**-case plans (with everything running smoothly according to plan) and

best-case plans (with deliverables and milestones being met ahead of team expectations). All plans, however, should include contingencies for schedule delays and technical issues, because there may still be setbacks. Project managers can then look for opportunities to shorten timelines, for example, by mitigating project risks that could negatively impact the project timeline.

Planning should also account for the following:
- Periods of lower employee productivity and efficiency, for example, during the end-of-year holidays, summer Fridays, just after a huge milestone is achieved or a significant deadline is met, and so on.
- Periods when vendors will be in high demand for their services (check ahead for availability of test or laboratory equipment, availability of manufacturing and processing resources, etc.).
- Logistical requirements, such as the time required for shipping goods and personnel travel time.
- Time required for the review and approval of key deliverables.

Communication

Effectively communicate and listen to stakeholders and project sponsors, to minimize lack of clarity on the project. Get clear consensus on scope, deliverables, and project goals, and what endpoints need to be achieved to know that the project is successful. Project risks and issues must be freely communicated.

To address project risks that are outside the direct control of the project implementation team, communicate often with service providers and vendors. For example, when vendors provide price quotes for parts or services that directly impact the project timeline, it is okay to push back and ask for shorter durations and earlier delivery dates (within reason). If a vendor requires a long duration to complete a task or deliverable, touch base with the vendor frequently and discuss opportunities to shorten durations – problem solve and find solutions *together*.

Also, communicate clearly and manage the organization's expectations, which can be difficult. Depending on the project, to set realistic expectations, a project manager may incorporate a buffer of time into the project schedule to compensate for delays and technical issues. With this buffer, if a project is completed early, then the project manager is viewed as dependable and successful, whereas without this buffer, if a project is behind schedule, then the project manager is viewed negatively.

Processes

For most companies, there is a huge dependency on the services of vendors and consultants, including design, development, manufacturing, testing services, and so on, to complete technology projects, and there may be a large number of suppliers within the technology supply chain. Therefore, companies need to have quality-driven, risk-based processes for managing vendors and suppliers and avoid having only one option available to get work accomplished. These controls are described in more detail in Chapter 11.

Do not forsake quality for project speed – if it is necessary to collect more data before making strategic project or purchasing decisions, then take the time to do so. If your team is not working smarter, everyone ends up working harder.

8.4 Effective Technology Management

Within a large organization, technology management across a portfolio of projects should be focused and structured, using proven program management methodologies that are designed with streamlined processes in mind. This will improve the organization's likelihood of capitalizing the most return from its investments. In the following sections are some of the guidelines that large organizations can follow to be more effective at technology management.

8.4.1 Establish Central Governance

Central governance or oversight means that funds and resources can be allocated both vertically and horizontally across organizational functions by a central governing body. The central governing body includes representatives from across the organization. In this way, the invisible or visible barriers and politics that separate functions and create organizational "silos" can be overcome. Resources with specific backgrounds and skill sets can be matched to specific program needs. Also, processes to fund and approve programs, initiate programs, and close out programs can be standardized. Requesting and implementing changes that measurably impact program budgets, timelines, resources, and outcomes can also be monitored and approved via central governance.

8.4.2 Utilize a Project Management Office

A centralized Project Management Office (PMO) includes a hierarchy of project management personnel, for example, program managers (who manage a portfolio of projects), senior project managers and project managers (who manage single projects), and project management assistants. A PMO is one classic signature of a matrix organization.

A matrix organizational structure typically does not appear in startups; in startups, a development engineer or Chief Technical Officer/Chief Operations Officer may be assigned to be the project manager. In matrix organizations, individuals on temporary project teams report in to the project manager, shown as a dotted line on the organizational chart, and also report to a manager or supervisor, usually within their own functional area, shown as a solid line.

Referring to an example project organizational chart shown in Figure 8.15, the functions that make up the project implementation team – marketing, product development, quality assurance, manufacturing process engineering, and so forth – report in to the project manager, shown as a dotted line. The project assistant directly reports to the project manager via a solid line.

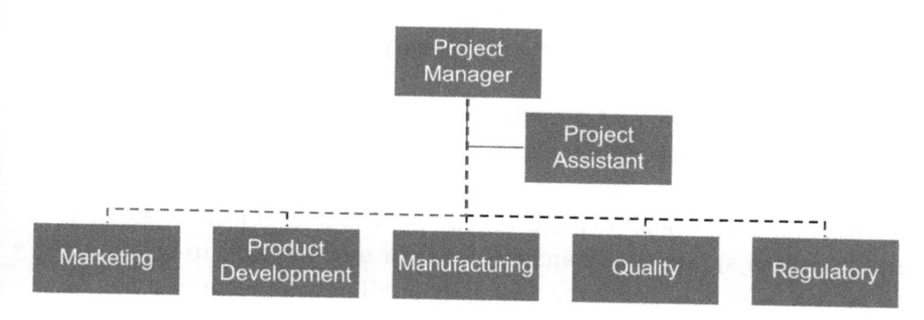

Figure 8.15: Example PMO organizational chart.

A PMO provides a structure for how projects and programs are designed, planned, and executed. Good PMOs provide training and professional development opportunities for their project managers and share lessons learned from previous projects. PMOs usually operate according to a PMO "Playbook" or a collection of best practices on topics such as financial management, communications and change management, risk management, and so on. PMO project managers work for the benefit of an entire organization and not just an individual function, and therefore have an interdepartmental perspective of each project and its required resources.

8.4.3 Design an Employee Performance Incentive Program

Employees should be incentivized to perform well on projects and be rewarded accordingly (through bonus payments, additional paid time off, and employee recognition awards) when project outcomes are positive. Put another way, employee performance reviews, which result in employee ratings and rankings, could be directly linked to project outcomes. Managers should also encourage employees to write their own performance goals – both professional and personal – to align with project objectives.

For example, an organization may be seeking an ISO certification to remain ISO compliant. The quality engineers on the project at the same time may be seeking quality certification through a professional society, such as the American Society for Quality. It would benefit the quality engineers to learn the ISO certification process by working towards achieving ISO certification for the organization. After the project objective is met, the quality engineers can apply that experience and achievement toward their professional certification.

8.4.4 Execute Best Practices in Project Management

To close out this section on effective technology management, these general project management principles will help ensure a technology project is completed on time and within budget.

- Define **S**pecific, **M**easurable, **A**ctionable, **R**ealistic, and **T**ime-bound (SMART) incoming and exit project criteria which align with an organization's mission, vision, and strategic goals. These criteria should clearly describe what project success will look like, when the project is complete and its objectives are met.
- Secure the written approval of managers for their direct reports to be allocated to your project team and secure the necessary funding before a project begins.
- Implement risk-based project management. Identify and mitigate risks as they arise, and determine the financial, timeline, and resource impact of project decisions.
- Follow a stakeholder communication plan. Sometimes, individuals with very little interest may hold a considerable influence over the project and vice versa. Prepare a stakeholder communication plan that aligns with their needs. For an example of stakeholder communication plan template, see Figure 8.9.
- Include organizational and change management activities in the project execution, for instance, developing a strong case for the technology change, aligning business processes with technology requirements, defining employee accountabilities and responsibilities regarding technology usage, planning a road show to showcase the new technology, designing organizational training for the new technology, and so on. Refer to Section 8.5 to learn more about change management activities.
- Project teams should be cross-functional and colocated together during important project meetings or milestones, where possible.

8.5 Change Management

Project managers: a very important aspect of communication planning and project implementation success is change management – preparing your stakeholders to receive the change, ensure that it positively "resonates," and that it is successfully adopted by the organization once the change is delivered. Change management can be performed as a series of steps, each of which is discussed in more detail as follows:

Step 1: Assess and Understand the Current State of the Project
Step 2: Research, Analyze, and Agree to the Problem
Step 3: Communicate a Vision
Step 4: Design and Implement a Project to Deliver the Change
Step 5: Assess how the Change is Impacting the Organization

8.5.1 Step 1: Assess and Understand the Current State of the Project

To do this, you can interview multiple stakeholders within the organization at different management levels and make sure various functions or departments are included. Discuss the following topics:

- What are the stakeholder pain points?
- What are the current process roadblocks or bottlenecks?
- Where is productivity lacking?
- Where and how is the organization not meeting metrics or goals?
- When and how are projects stalling?

8.5.2 Step 2: Research, Analyze, and Agree to the Problem

With stakeholders, utilize tools such as documenting the strengths, weaknesses, opportunities, and threats elements of SWOT analysis (a sample SWOT analysis can be found in Chapters 1 and 4); Fishbone/Ishikawa Diagrams; and the five WHY'S to understand the root cause of the current problems. An understanding of the current state needs to be based on fact, not opinion. Facts will help you establish an argument or case that change is needed, and facts will help you build consensus for the change with disparate groups of stakeholders. Example of five WHY's and Fishbone root cause exercises are depicted in Figures 8.16 and 8.17.

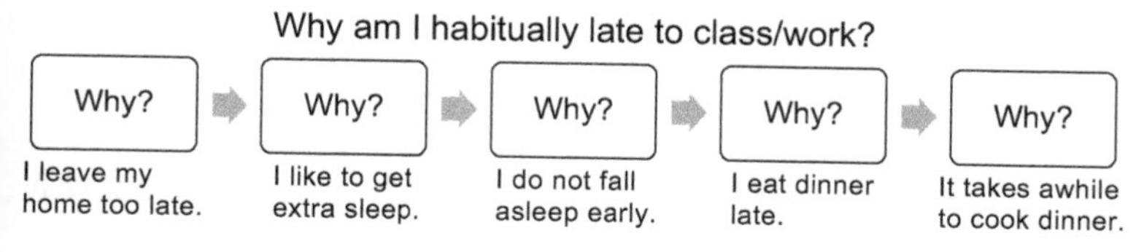

Figure 8.16: Diagram showing an example of five WHY's.

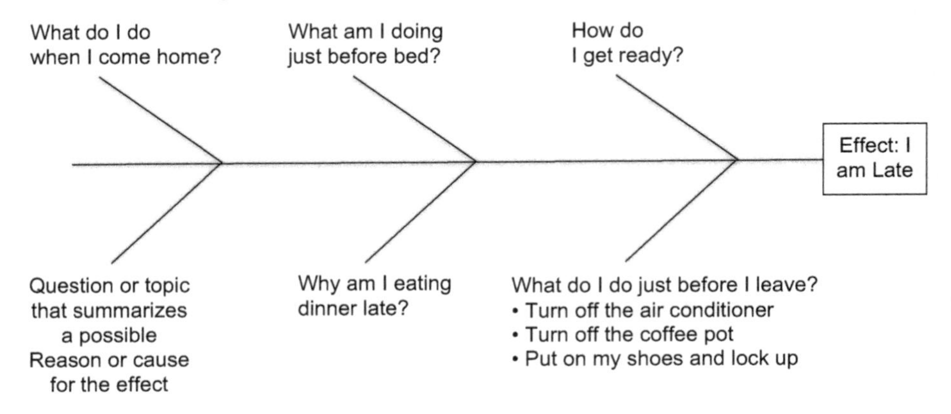

Figure 8.17: Diagram Showing an example of Fishbone.

Note: If you are stuck on collecting answers as part of your research, are there opportunities to conduct research on other best practices, industry benchmarks, and so on? Ask yourself: How is this problem solved in another industry or sector?

8.5.3 Step 3: Communicate a Vision

Inspire project stakeholders by depicting a clear vision for the future. Use storyboards, pictures, charts, and diagrams, where possible, to describe the future state of the business after change is implemented. Your vision can describe:
- How teams will work together.
- How intra- and inter-function/departmental collaborations and partnerships will occur.
- How the organization will thrive financially (by way of revenue growth, reduction in spending, or an increase in ROI).
- How policies will support new processes.

Remember, the vision for the future must be shared with all key leaders and be in line with the company's mission and strategic plans.

8.5.4 Step 4: Design and Implement a Project to Deliver the Change

As you prepare the project charter and plans to implement the change, begin with the end in mind.
- Know the key strategic implications and business challenges.

RACI Matrix

	Concept	Design	Development	Commercialization
Business Development	I	I	I	R
Engineering	R	R	R	C
Finance and Payroll	I	I	I	I
Human Resources	I	I	I	I
Legal Affairs	C	C	C	C
Manufacturing	C	C	R	R
Marketing	C	C	C	R
Quality Assurance	C	C	C	C
Regulatory Affairs	C	C	C	C

Figure 8.18: Example of a RACI matrix.

- Think about how business processes and organizational roles and structure will be impacted, and identify risks to be mitigated.
- Prepare a detailed implementation strategy that includes stakeholder communication and rollout plans, such as organizational "road shows" and presentations, various methods of training, and so on.
- Recommend process and organizational KPIs, for example, quality metrics, revenue and profit targets, employee satisfaction, and additionally, roles and responsibility frameworks, where possible.
- Create RACI (Responsible, Accountable, Consulted, Informed) matrices that communicate which functions or roles will be performing tasks in the future state and which functions or roles will be approving or contributing to tasks. Refer to the example RACI matrix in Figure 8.18. In Figure 8.18, "R" stands for Responsible, "A" stands for Accountable, "C" stands for Consulted, and "I" stands for Informed. The information in RACI matrices are used by leaders to understand task workloads, task ownership and governance, and can be used to create organizational charts.

Regarding change, project stakeholders will want to know:
- What does success look like? Or, what is the culture of accountability that the change will create?

- Which organizational roles/stakeholders need to be engaged throughout the change implementation?
- How will progress be communicated?
- Does governance need to be created to monitor this change and other changes in the future?

8.5.5 Step 5: Assess How the Change is Impacting the Organization

Determining change impact is all about awareness and continuous improvement. After the change is implemented, find out what is working and not working – do your own version of learning to validate that the changes are working. How do stakeholder moods/behaviors/actions change over time? If necessary, adjust and check again. This aspect of change management never ends – continue to engage with stakeholders and get their feedback.

8.5.6 Table Top Workshops – A Live Example of Executing Change Management

Table Top workshops are particularly useful change management tools that inform and visualize process or system changes with a live audience. In Table Top workshops, participants are assigned a process or system role and walk through or act out predefined scenarios with the purpose of exploring new customer requirements, reviewing proposed business rules, identifying and mitigating risks, making decisions, and aligning different groups of stakeholders. The workshop participants have the opportunity to be informed of the change, refine the change, acclimate to the idea of the change, and gradually "buy-in" to the change, once they can visualize how the change will improve their current situation.

Table Top workshop participants can then serve as change advocates, adopters, and communicators of the change throughout the organization, increasing the probability of a successful change implementation. These participants can be the go-to experts and serve as the project manager's "eyes and ears" on the ground within the organization, helping other stakeholders throughout the transition and alerting the project manager of implementation difficulties or errors.

Table Top workshops can apply to many types of projects and processes, for example, customer service/engagement, knowledge and information management, and product ideation, . For example, Table Top workshops can focus on an information technology (IT) application and software deployments, usually to support new business processes resulting from a merger or new regulatory requirement. Our advice is to not let Table Top workshop scenarios become too technical in nature, or else your audience may get bogged down in the details; utilize pictures, process flows, and group exercises as much as possible to keep the workshop engaging.

8.5.7 Final Thoughts on Change Management

Remember, change can be viewed as scary or unsettling, and it is your job as project manager to inform your organization of the changes that will occur as a result of the project implementation and to find better ways of guiding stakeholders throughout the transition.

Great project managers go out of their way to establish trust and develop relationships across their organization. Showing genuine empathy and seeking to understand people's concerns and understanding of the change can alert project managers of impending risks and potential solutions. Keep asking yourself the following questions and continue to seek answers: how will the change impact stakeholders, how can the change benefit stakeholders, and how can you help transition stakeholders?

8.6 Problem-Solving Techniques

Problem solving, if nothing else, tests your resourcefulness and patience as a project manager. Remember, problems always seem a lot scarier and, perhaps, insurmountable, before you begin to understand them, so relax and take it one step at a time. A clear mind and a calm disposition go a long way in problem solving, and never forget to utilize the resources and strengths of your team in assisting you.

8.6.1 What is the Problem?

Before attempting to solve any technical or nontechnical problem that may arise on a project, first focus on breaking down the problem into manageable pieces of information, and continue to research and ask questions so that the root cause or causes of the problem are understood, identified, and described in enough detail.

Sometimes, the root cause of the problem may not be the most obvious choice or may not be supported by the largest body of evidence, so be careful to not jump to the most likely candidate. Think about all of the different types of possible failure modes or types and sequences of events that could lead to the problem. Break down the problem into mini-problems, each with its own cause and effect, or each with its own possible input and output. You can utilize visual thinking techniques such as the five WHY's and Ishikawa/Fishbone diagrams (see Figures 8.16 and 8.17) to help conceptualize the possible root causes of the problem.

Not every person will see the problem the way that you do and not every team or function within the organization operates like yours, so try thinking about the problem from the viewpoints of different stakeholders, for instance:

- "How would the technology product manager gather insight from customers about this problem?"
- "How would Quality search for records of evidence of this problem?"
- "How would Manufacturing try and recreate the problem?"

Additionally, think about how this problem arises in other industries or sectors – what research has already been collected, analyzed, and published? This is where the power of an extensive personal and professional network of contacts external to your organization comes in handy.

If the problem is too complex for one person to solve, form a team that is specifically tasked to solve the problem, and ensure the team is diverse. A team that is diverse in age, gender, ethnicity, work experiences, career backgrounds, and so on is very important in generating ideas and sharing insight. If the team gets stuck, take a step back and review the methodologies that are being used to solve the problem. Sometimes, an alternate problem-solving approach can force the team to see the problem in new ways. Being persistent in your efforts to frame the problem is key.

8.6.2 Finding Solutions

Solutions come from knowledge. To come up with potential solutions, you need to think broadly and rely on your own (and others') knowledge, expertise, and insight to generate ideas. Do not assume that the knowledge you have is also shared by the individuals on your team just because you all are part of the same project or organization.

If you don't possess the knowledge, then go out and get it. Do your own research – read literature, review websites, talk to experts, even attend conferences and workshops as necessary. If the problem is technical in nature, think about the technology requirements and re-review internal and external stakeholder wants and needs.

Once you have the knowledge, generate a list of possible solutions by organizing and facilitating brainstorming workshops with the team. Brainstorming facilitation is discussed in Section 8.7. After brainstorming, review the results with the team by discussing the following questions:

- How was this problem solved in the past?
- How is this problem solved by other organizations in this sector and other sectors?
- What are the technical and economic challenges with implementing a chosen solution?
- Can a proof-of-concept (POC) prototype or model be created quickly, or can a simple test be performed to test a chosen solution?

- Can off-the-shelf statistical tools be used to extrapolate data to analyze a chosen solution?

Then, review the list of possible solutions with individuals who are not involved in the brainstorming workshops – get a "cold eyes review" of the list of possible solutions. Independent feedback, especially from senior subject-matter experts within the organization, can help validate a proposed solution and secure upper management approval of a decision.

To select a solution, begin with the end user or customer in mind. Without understanding how solutions will be accepted, how will you know whether one solution should be favored over another? Clearly defined acceptance criteria or project metrics that are **S**pecific, **M**easurable, **A**ctionable, **R**ealistic, and **T**ime-bound (SMART) make it easier for teams to rank order the different solution options. Once the solution options are ranked in order of preference, be comfortable in knowing that you have performed the research, gathered the knowledge, brainstormed potential solutions, generated new ideas from brainstorming results, and are now making an informed decision to select the right solution with the information you currently have.

8.7 Brainstorming (Ideation) Techniques

How do teams get the most value from brainstorming meetings to generate a broad list of prospective solutions? You are asked to lead a team brainstorming meeting to develop a strategy or propose solutions to an issue – now what? Do not be fooled, brainstorms are not typical "informational sharing" or "team update" meetings. Organizations can waste a lot of resources in brainstorming meetings because of (1) inadequate preparation, (2) poor meeting facilitation techniques, (3) limited analysis of the output data, and/or (4) lack of follow-up on next steps. A brainstorming meeting is an investment in your organization's future, so treat these brainstorming meetings as a valuable source of knowledge capital. Knowledge capital can be the "asset" difference between high-performing and low-performing organizations.

To extract more value from team brainstorming meetings, do your homework in advance of the meeting, use effective facilitation techniques, do not settle for just a list of ideas (take things a step further), and follow through by being proactive.

8.7.1 Do your Homework

- Select a meeting location that the team will not expect. Examples of nontypical meeting locations include a cafeteria or restaurant (during off-peak hours), a shared lab space, outdoors, a non-sterile production or packaging area, and so on.

- Arrange seating so that team members may get up quickly and move easily around the meeting area.
- Assign seating (1) if you know certain team members do not get along, or (2) to encourage team members from different functions/departments to work together to come up with new ideas.
- Come to the brainstorming meeting prepared with examples or prior cases, prototypes, models, and/or props, if necessary, to jump-start the discussion and promote creative idea flow.
- Don't forget to bring materials so the team can document their ideas, for instance, multicolored markers and sticky notes and "flip-chart size" posters.

Tip: Posters that contain adhesive strips on the back come in handy.

Prepare an agenda for the meeting by:
- Writing a list of open-ended questions so the team's answers will gradually build upon each other.
- Assigning a time limit for team members to record and discuss their ideas for each question.
- Including time for idea/output analysis or schedule a follow-up meeting to complete and review this analysis with the team.
- Including time at the very end of the brainstorming meeting to discuss next steps and assign actions.

8.7.2 Facilitation Techniques

As a facilitator, you are in charge of the flow and outcomes of the workshop or meeting. Set the tone of the meeting through introductions, icebreaker or getting-to-know-you activities,[29] and outlining the agenda. Throughout the meeting, pay attention to the clock and make sure the team is addressing each question or at least the important ones on the agenda.

Do not let your own views influence idea generation or follow-up discussion. Your job is to ask questions, not to answer them! Also, you can move around the room, as you ask questions, to raise the energy level of the discussion and to engage with all members of the team. It is okay to single out individuals with questions, if their ideas are not being discussed or if they seem distracted.

29 Icebreaker activities are used by facilitators to get their audiences quickly engaged in the meeting discussions by shifting their focus at the start of the meeting from other, outside work to what needs to be accomplished, and to allow new audiences to get to know each other. Icebreaker activities should be entertaining and they can be funny, depending on the mood of the audience that the facilitator is trying to achieve.

For each question, ask all team members to document their own ideas on the multicolored sticky notes, prior to discussing with the team. Ensure that each team member uses a different color marker or sticky note, so that you can follow-up afterwards with individuals, if you have questions about their idea.

It is helpful to designate separate posters to document all ideas or comments for a given question. If there is plenty of room, encourage participants to walk around the meeting area and post their sticky notes on the separate posters which pertain to their answers. You can utilize different posters to capture Risks, Assumptions, Dependencies, Issues, and Opportunities (RADIO) (see Section 8.2.2.1) and "Parking Lot" ideas or comments; these suggestions answer different questions and should not distract the team. Parking Lots are topics that are meant for follow-up discussion and/or action – they are captured on the poster but are not meant to be discussed further in that particular meeting. RADIO and Parking Lot comments can be assigned action items and action owners to be addressed at a later time. Refer to the example brainstorming poster set in Figure 8.19.

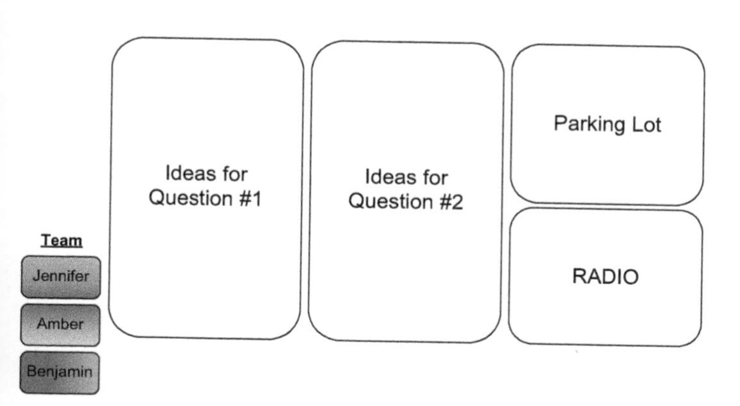

Figure 8.19: Brainstorming poster set example.

8.7.3 Do Not Settle with a List, Take it Further!

There are several exercises you can try with teams to invoke new ideas or ideas that complement or deepen what others have already shared. For example, start with one person communicating his or her idea with the group and keep going around the room, letting each team member voice ideas until the team runs out of ideas and/or comments. This is fruitful because it generates new ideas in the minds of the individuals in the group, while listening to the ideas of others. If the team is stuck, try asking the question in a different way.

- What would happen if we tried to make the problem worse?
  - Why exactly is that an issue?
  - And so, what could we do to avoid that issue or the events/scenarios leading up to that issue?
- How would an idealist or pessimist respond?
- How would an extrovert or introvert respond?
- How might a colleague from a different function within the organization solve this problem?
- How would your customers and users feel about this solution?
  - How would they solve the issue?

As a facilitator, continue to ask open-ended questions for clarification of the ideas being discussed. It can seem frustrating for some individuals when a facilitator keeps asking, "Why?" or, "How?" but this is how robust ideas and more focused implementation plans are formed. It is helpful to advise the team early in the meeting that you will be asking lots of questions for the sole purpose of learning more about their ideas. Sometimes, facilitators rephrase the statements made by team members to help clarify or translate the intended message into terms that the rest of the team can understand better or in terms that the team would be more receptive towards.

Ensure that side comments and questions are not missed by assigning a member of the team to take copious notes, and rotate this task from meeting to meeting. This is an important role during the meeting and follow-up deliverable after the meeting. Ideas that are not documented will most likely be forgotten.

8.7.4 Follow Through, Be Proactive!

While the team is present and energy levels are high, assign individual team members to take accountability for the different follow-up actions identified during the meeting. Have each action item owner define and commit to a completion date. As a facilitator, you can decide the best way to follow-up with each action item owner, for example, using team meetings, one-on-one discussions, and task management tools.

There are several ways to assist the team in moving the project forward after the brainstorming meeting. For example, for each question asked group similar responses and ideas together and into categories. Send a follow-up presentation or memo within one to two business days after the meeting, describing the key discussions and themes, main ideas, and action items. Describe what follows next, how the team's ideas will be used, and by whom. Team members who understand how their actions and ideas fit within the broader context and goals of the project are more likely to follow through on commitments.

Remember, brainstorming meetings should challenge the status quo and promote creativity. As a meeting leader, it is okay to take risks! Think about it: What ideas or suggestions have worked for your team brainstorming meetings in the past?

8.8 Conclusion

In the beginning of this chapter, we presented an example list of what could go wrong during a typical day in the life of a technology project manager. Following are ideas for how to address or resolve each of these issues (Figure 8.20).

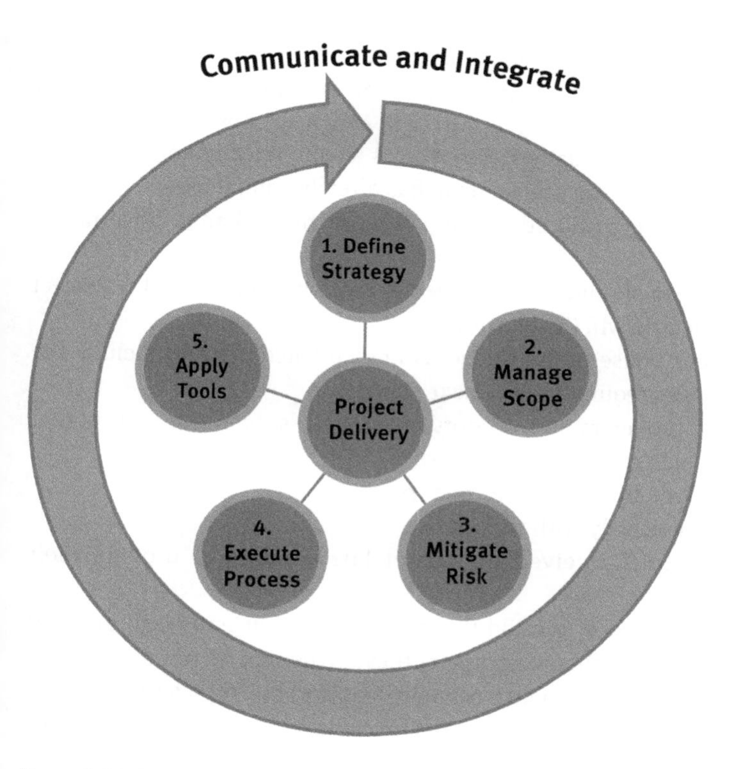

Figure 8.20: Important components of successfully managed projects.

- Midway through the project, timelines are slipping →
  - Organize and schedule the necessary meetings.
  - Discuss program risks, which could impact timelines, at the beginning of the project with the project team and update these periodically.
  - Add program risk mitigations into the project plan.

  - Work smarter, not harder – hold team reviews of key deliverables so that review and approval cycles are shortened.
  - Minimize the number of unnecessary meetings where the focus is on status updates and non-value-adding discussions, and instead focus on completing deliverables.
- Key resources on the project are being asked to decrease their availability to join other projects, and therefore, these key resources are transferring their work to less experienced resources →
  - Project resources should be negotiated with other teams and functions ahead of the project.
  - To boost project retention, link project team members' annual performance objectives with project outcomes.
  - Name key resources in the project charters and SOW so that their involvement is visible to upper levels of management.
  - If resources need to change, then develop and execute a robust transfer of knowledge plan from departing resources to new resources.
- A manager from a different business function is asking for a change to one of his or her deliverables that the project team has already completed →
  - Project scope changes should be agreed upon by a change governance committee.
  - Project scope creep should be minimized at all times, otherwise the project budget and schedule will likely overrun.
  - Once the current phase of the project is implemented, look to include the suggested changes required in the next phase.
- Technology service providers and vendors are slow in submitting their quotations for services, leading to schedule delays →
  - Request quotations from multiple vendors.
  - Communicate frequently with vendors.
  - Leverage quotations received to prompt late vendors to provide their quotations.
- One of the components used to build the technology is currently only sourced from a single supplier, and there are rumors they are behind on orders →
  - Deploy "tiger" teams or "problem-solving" teams to the supplier where necessary, for example, if they are having technical issues or if their manufacturing line is down.
  - Try and find all ways to take that supplier off the project critical path.
    - Develop alternative solutions that would not require that component.
    - If the component can be substituted with something similar, outsource that component to a different supplier.
- Project team members are constantly interrupting you with questions, asking about timelines, deadlines, and who should be working on what →

  - Utilize project management communication tools, for example, SCRUM boards, dashboards, and roadmaps.
  - Include high-level project responsibilities and dependencies in the project charter and make sure all team members have a copy.
- Managers from other business functions are already voicing their disapproval about the changes that your project is implementing →
  - Use change management techniques – communicate the vision and the need for change.
  - Schedule Table Top workshops or project roadshows at key milestones in the project with influential stakeholders to confirm that the solution meets their needs.
- Marketing is requesting additional VOC data to support user needs →
  - Conduct market surveillance while the organization is developing the case for change.
  - Survey internal and external stakeholders and subject-matter experts.
  - Analyze benchmark data from internal and external projects.
- Technology requirements keep changing →
  - Escalate change requests to the change governance committee.
    - Remember, requirements can always be implemented in a future project or the next phase of the existing project.
  - Review requirements at project phase gate or design review meetings – during concept selection and definition, once the "POC" prototype is complete, before design freeze, that is, before the design is complete and no more changes are made prior to final testing.
- The current technology prototypes are clunky and keep breaking
  - De-risk prototype development by asking independent subject-matter experts to perform a "cold eyes" review of the design at multiple stages of development.
  - Don't try to tackle too many requirements and enhancements all at once – think minimum viable product or solution.
  - Develop a works-like prototype first, work out the kinks, and then develop a works-like, looks-like prototype.
  - Utilize problem-solving and brainstorming techniques.
- No one knows when the project will be finished →
  - Establish clear exit criteria and SMART goals.
  - Develop a robust critical path on the timeline so you know which tasks and dependencies are pushing out key dates.

As a final takeaway, following are intangible learning points, or "words of wisdom," gained over the years by the authors:
- Project management skills take practice, practice, practice.
- Listen to your team members – know their strengths and weaknesses.

- Always make informed decisions.
- Don't be quick to anger or anxiety when a problem arises – problem solving requires clear thinking.
- This project, or any project, is not about you – lead and manage with humility.
- Be an anchor for the team – understand the intricacies and how things work together.
- Clearly communicate the project goals and how the work ties back to the mission and vision of the project and the organization.
- Set level expectations with stakeholders – do not over promise and under deliver.
- Things will always go wrong – it's how you plan and prepare your team for missteps that makes all the difference!

Part B: **Engineering the Innovation**

9 Needs Finding, Concept Generation, and Prototyping

Abstract: Every successful product satisfies an unmet need. This chapter explores what an unmet need is, how to convert that unmet need into exhaustive concepts to help satisfy the unmet need, and how to then convert the concepts into intellectual property. An unmet need, or "pain point," is the identification of a problem. Determining whether an unmet need is worth pursuing means determining whether there is a place in the market for your product. The exercises in this chapter begin with the needs finding approach that is commonly used in academic institutions, then shifts to industry and startup perspectives to ensure the development of an innovative product, while taking time to create and protect intellectual property. Once top product concepts have been generated through the needs finding process, then the development of Proof-of-Concept (POC) prototypes and α-prototype can help test the viability of the concepts before developing a later-stage prototype.

Keywords : Needs finding, needs gathering, unmet need, needs refinement, validation, high-level validation, in-depth validation, pain point, concept generation, need selection, functional decomposition, segmented ideation, idea aggregation, concept screening, concept validation, concept selection, selection criteria, Pugh matrix, prototyping, proof-of-concept (POC), α-prototype, design for manufacturing (DFM), design for assembly (DFA), sourcing.

Needs Finding is the process of looking at a situation and identifying an unmet need, which is a problem or "pain point" for a stakeholder. Concept Generation is a technique used to ideate solutions to an unmet need or problem, rank the solutions through the user of a decision-making process, and delivering thoroughly researched and ranked final concept selections. Together, Needs Finding and Concept Generation are the first two steps in the Product/Technology Development process for creating an innovation.

The benefit of Needs Finding and Concept Generation is that once a team has completed the processes, then exhaustive ideation has been pursued *and* all developed concepts are organized into a format that is easily translated into a provisional patent application, so that every concept surrounding each aspect of the invention is protected. This approach provides parallel market analysis in conjunction with legal protection for the intellectual property (IP) you are developing, even though a final concept may still be under development. This IP protection is necessary while (a) validating the concept and (b) trying to obtain startup investment. Needs Finding and Concept Generation combine business and technical due diligence, and the team's capabilities, to systematically assess the commercialization potential of the unmet needs and the concepts generated around them.

https://doi.org/10.1515/9783110521900-010

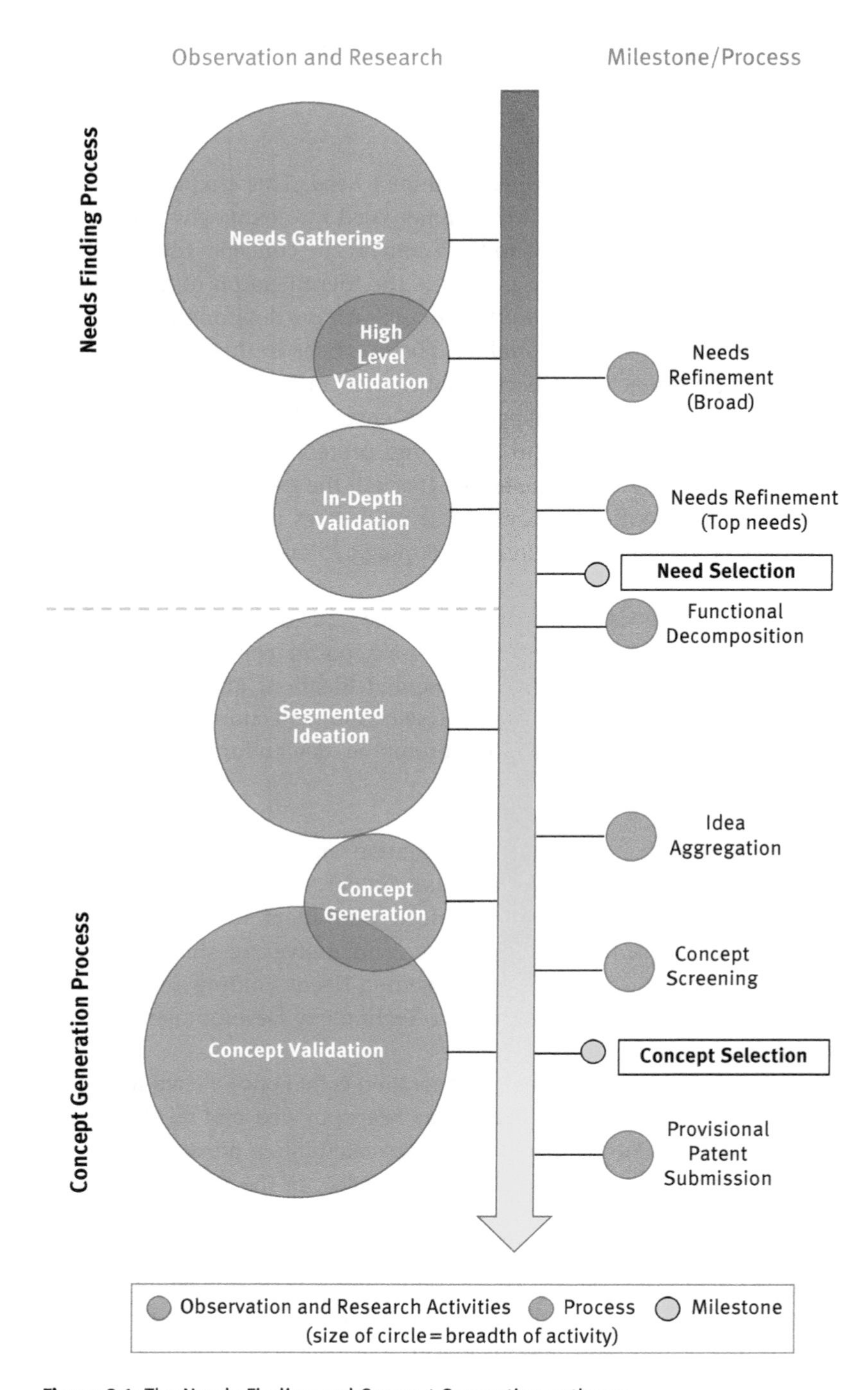

Figure 9.1: The Needs Finding and Concept Generation pathway.

As shown in Figure 9.1, there are various Observations and Research steps with associated milestones and processes that make up Needs Finding and Concept Generation. What follows is a high-level explanation of the content depicted within Figure 9.1 (circle size represents the breadth of the action; whether it be time, effort, resources or a combination of each).

The left side of Figure 9.1 illustrates the Needs Finding process. The process of gathering unmet needs involves observing customer/industry pain points, identifying an unmet need from the observation(s), then translating the unmet need into a refined Need Statement. Depending on the size and experience of the team that is completing the Needs Finding process, the number of unmet needs (i.e., Need Statements) can vary from tens to hundreds.

The process of refining Need Statements is closely followed by High-Level Validation, which involves the team conducting research surrounding the market associated with each need. The validation research is designed to help filter out the Need Statements that (a) are not true needs, (b) do not fit the team's skill sets or the company's mission, or (c) are not a good market fit. Examples of conclusions that could be drawn from High-Level Validation research might include discovering that there is a product on the market that addresses the Need Statement, realizing that your team does not have the appropriate skill sets to address the unmet need, or finding that a major player in the industry has complete control of all the products in the portfolio of existing products for a specific need. High-Level Validation is not intended to disqualify an unmet need, but rather paint a picture of the landscape around the unmet need so that the team can make better decisions.

Next, the Needs Refinement is the structured team process for organizing, comparing, and ranking the validated observations. The pared down list of unmet needs is now manageable for conducting In-Depth Validation research. In-Depth Validation research is where exhaustive market research (Chapter 1) will provide clear criteria (e.g., market size, market segmentation, market saturation, competition, technical feasibility, etc.) for determining which unmet need has the most potential (and which other unmet needs are viable alternatives). In-Depth Validation includes reaching out to stakeholders to validate the Need.

The last part of the Needs Finding process is Need Selection. Establishing a process that works for the team, and the use of decision-making tools, will allow for structured, informed team decisions that will reduce the risk that decisions will be made based on opinions or bias, rather than based on facts.

The right side of Figure 9.1 illustrates the Concept Generation process. Concept Generation assesses the currently used processes/workflows for an unmet need, then involves generating ideas around each step of the existing process to explore every alternative possible to satisfy the process. This systematic and comprehensive ideation is designed to translate into concepts that can be aggregated into a detailed provisional patent application, thus legally protecting any shifts in the scope of the product design during the first year of product development.

The Concept Generation process begins with the Functional Decomposition, which is the breakdown of the steps, also known as functions or subfunctions, of the existing processes that are associated with the unmet need.

Once the functions and subfunctions are identified, then Segmented Ideation of each function and subfunction occurs. Segmented Ideation involves segmenting the approach to generate ideas around each function or subfunction, while exploring every possible solution regardless of the technical feasibility, cost, or connection with other parts of the overall existing or developing process. The use of tools such as Mutually Exclusive and Comprehensively Exhaustive (MECE) (MECE is discussed in more detail in Table 9.1) allows team members to perform Segmented Ideation individually, and then the team performs Segmented Ideation a second time together as a group in order to explore every outcome imaginable. When generating ideas around each function or subfunction, if possible, the ideas should be captured as drawings or images (as a picture is worth a thousand words) along with other descriptors to clarify the idea.

Table 9.1: Example MECE list for physical systems.

Mechanical	Electrical	Magnetic	Optical	Radioactive
Thermal	Biological	Chemical	Acoustic	More

The ideas that are generated from the Segmented Ideation are then aggregated into a single document or spreadsheet, with the drawings or images attached (or linked) to each idea in a process known as Idea Aggregation. The team then assembles to discuss each idea and collection of ideas surrounding the individual process steps of the Functional Decomposition, with the ultimate goal of (a) generating completed concepts and (b) reducing the number of steps in the Functional Decomposition.

The removal of one or more functions or subfunctions from the existing process in Functional Decomposition may qualify as "innovation." In parallel with Concept Generation, teams often need to validate their concepts through Concept Validation. Concept Validation involves conducting market research and customer surveys on the finalized concepts. Concept Generation and Concept Validation often overlap, since it is not uncommon for concepts to need adjustment in response to the market research and customer survey data gathered during Concept Validation.

The next step is Concept Screening. Concept Screening is when the team assembles and sets to work using decision-making tools to rank and filter the generated concepts. Not unlike the Needs Refinement process, Concept Screening uses High-Level Validation to generate the acceptance criteria, with the goal of filtering out the concepts with the lowest commercialization potential.

Whittling down the concepts leads to Concept Selection. Concept Selection is the point where the team decides on a single concept, while keeping alternatives in mind. Analogous to Needs Refinement, the Concept Selection process occurs during

In-Depth Validation where the team reaches out to stakeholders to validate the concept. Note that Concept Validation continues even after the highest ranking (or scored) concept is selected. Validation should continue even after market entry.

The Provisional Patent Application is then prepared and submitted, encompassing all ideas and concepts so that any pivot during product development will be accounted for under the provisional patent application. As the proof-of-concept (POC) shifts to a prototype and then to a product, continued collaboration with your company's IP attorneys will protect your invention/innovation.

9.1 Needs Finding Process

An unmet need is the identification of a problem, or "pain point," of a stakeholder resulting from an observation. Think of the identification of an unmet need as seeing that there is a missing piece to a puzzle. Finding a viable unmet need is not easy, but an unmet need is more likely to come to light with the accumulation of experience and astute stakeholder observations within a specific industry. In most cases, the whole of the puzzle is difficult to see, and due diligence will uncover the upstream–downstream stakeholders (discussed in Chapter 2) and their needs and interests, thus helping complete the puzzle. A need is *not* a solution to a problem, but rather is a one-sentence statement of a problem. The reason for this is to ensure that you are identifying the problem – not solving it. A team first must identify the problem (or the unmet need) that must be addressed before exhaustively solving the problem, which occurs in the Concept Generation stage. The formation of an unmet need begins with observations.

9.1.1 Observations, Unmet Needs, and Needs Statements

9.1.1.1 Observations and Unmet Needs

Needs Finding is a formalized process for performing observations. Observations are the analysis of processes, procedures, actions, or occurrences from an impartial, objective point of view. *Not all observations will identify unmet needs.* When making an initial observation to gather information for determining whether there is an unmet need, the observation of a process or user workflow does not need to be all-encompassing. The goal is to conduct a cursory assessment of a process or procedure observation and translate that observation into a Need Statement (Section 10.1.1.2) that clarifies the unmet need. During observations, each team member collects, identifies, categorizes, and assesses every action of a situation. Figure 9.2 illustrates the process of getting organized and starting the Needs Finding process.

1 Identify Team Strengths

Needs Finding begins by identifying and assessing the strengths of the team and determining where conducting observations will be most fruitful. If you were tasked with

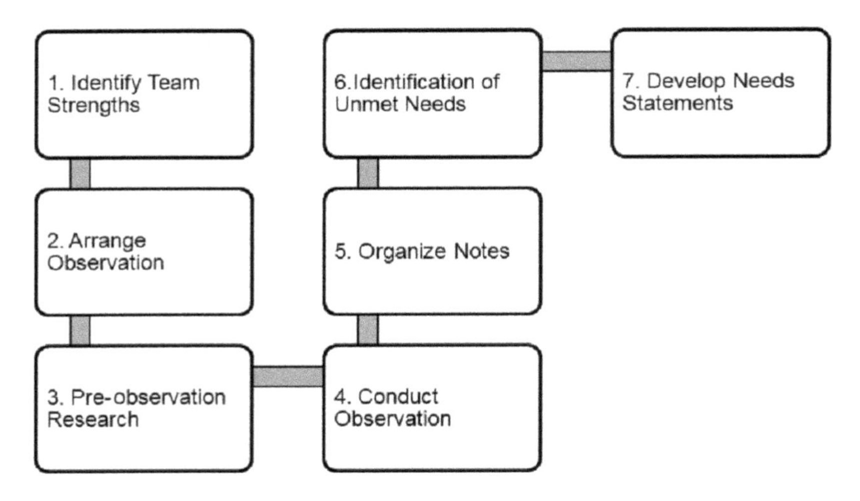

Figure 9.2: Needs Finding observation pathway.

developing a solution to a healthcare problem and your team consists of engineers and a pediatric surgeon, then the strength of the team would be in pediatrics or pediatric surgery. Additionally, once your team determines its area of focus for Needs Finding, the field expert's experience on the team (such as the pediatric surgeon) will need to align with the unmet need in order to be an effective representative for the unmet need and, later, for the product. For example, an engineer observing a pediatric surgeon may come up with some interesting findings, but when it comes time for the team to advocate for the unmet need to prospective investors, the engineer is not the best person to represent the unmet need. However, the team's pediatric surgeon is a good advocate, because fellow clinicians will find the surgeon's opinions/statements more credible. Even more compelling is the support of a key opinion leader (KOL), or revered expert in the same specific field as the unmet need (discussed in Chapter 2).

2 Arrange Observation

Organize who and where you would like your team to observe or shadow. If you are looking for unmet needs in the pediatric surgery space, arrange to shadow multiple pediatric surgeons in different types of hospitals and practices at different times or during different shifts. The quantity of observations will depend on the team and the number of observation opportunities that are available.

3 Pre-observation Research

If possible, before each scheduled observation, team members should take the time to review the procedure(s) being observed. Relevant videos, standard operating procedures, and tutorials are often available on the Internet, or there may be other background material available on the area of observation (such as background information on a physician, operating procedure, manufacturing line, and process) that can be reviewed prior to conducting observations. Most importantly, try to learn the vocabulary and parlance that pediatric surgeons use. Every field and profession has its own jargon and dialect. Understanding the jargon and dialect of a particular field is a huge learning curve, but knowing and understanding the meaning of industry-specific terms is important to getting the most out of your observations.

4 Conduct Observations

Before beginning observations (i.e., shadowing), start a notebook to record your observations. In the notebook:

- Write down a brief synopsis of the observation, procedure, shift, etc.
- Write down the location(s) of the procedure.
- Draw a picture of the room(s) (or site) and all of the equipment in the location.
- Observe the 4Ps: People, Places, Procedures, and Products (see Figure 9.3 for an example mapping of the 4Ps).
- Note how many people you observe during the procedure or shift: their names, titles, job/role, location within the room, and the activities each person engages in during the procedure or shift.
- Write down every observed step of the procedure or activity, what time it occurred, who performed it, and if there were any events that occurred as a result.
- Take note of the products (for instance, equipment, disposables, accessories, tools) used during the observation.
- If possible, include the serial number, lot number, product ID, and so on of any key products used during the observation.

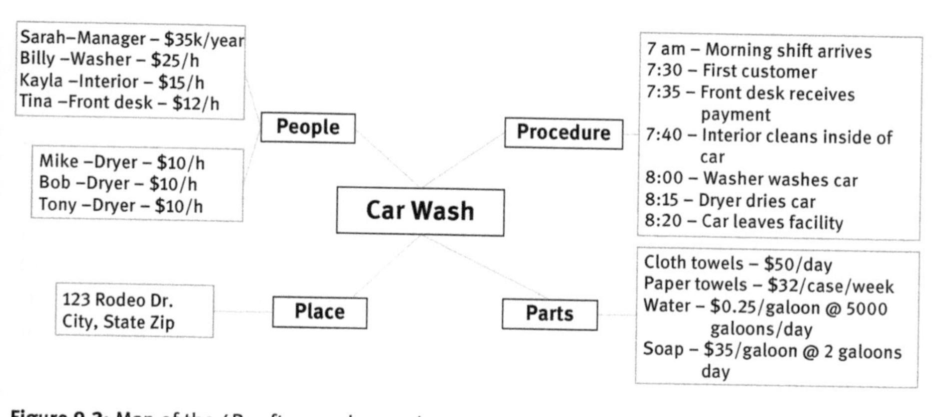

Figure 9.3: Map of the 4Ps after an observation.

5 Organize Notes

It is okay for the notes to be messy and incomplete, as long as you can decipher them later. There will be a lot of information to sift through. After each observation (or collection of observations) organize your notes into a flowchart or mind map

(Figure 9.3 is an example of a mind map). From here, you and your team can start developing Need Statements.

6 Identification of Unmet Needs
The realization that an observation or collection of observations reveals an unmet need can occur anytime during the observation, but in many cases, the identification of the unmet need will emerge during the organization of the observation notes. Unmet Needs are the initial unrefined description/summary of the problem statement.

7 Develop Need Statements
Once an unmet need is identified, then the refinement of the unmet need into a Need Statement clarifies the problem statement.

9.1.1.2 Need Statements

A Need Statement is an unmet need presented in a format that concisely clarifies (a) who the need is for, (b) the unmet need/problem, and (c) the outcome of addressing the unmet need. The simple format for a Need Statement is:

"There is a need for ______ to ______ while ______ in order to ______."

This format is intended to help the team break down the different components of the unmet need. *A Need Statement does not include a solution within the need statement.*

Example
"There is a need for car wash companies to lower their overhead, while maintaining customer satisfaction, in order to sustain profitability."

Once developed and refined, Need Statements should then be aggregated by the entire team into a centralized list, spreadsheet, or database. The spreadsheet should contain the owner of the statement (specific identification of the individual who made the observation), the date of observation, and any other identifiers of importance. Once the team has gathered enough need statements, then it is time to filter and reduce the aggregated selection of need statements down to the unmet needs with the most commercial potential.

9.1.2 Need Statement Selection

Needs Finding should generate both high- and low-quality Need Statements (i.e., unmet needs). Needs Refinement, or Need Statement Selection, is the systematic process of determining which unmet needs have the best potential for commercialization in relation to your team's skill sets. The quality of the Needs Statement is determined through market and technical research. *Not all good unmet needs are viable market opportunities*. During the Needs Validation stage, assessment of the market will determine if the product is commercially viable. Market research on market size, market segmentation, market saturation, and competition, paired with validating the technical feasibility, team fit, regulatory hurdles, and so on, should be conducted for each Need Statement. Often, observed unmet needs may only be relevant at a single observation location or are seen because of a chance occurrence. Sometimes there is a need for a product or service, but because of the market dynamics, supply chain and distribution challenges, legislation requirements, and so on, the solution to the unmet need may not be profitable.

Example

A percutaneous endoscopic gastrostomy tube (G-tube) is a tube used for enteral feeding, when a patient needs assistance getting nutrition. The most commonly used G-tube is an inexpensive piece of silicon tubing with a retainer at one end. It is common for patients who have a G-tube to experience a dislodgement of the G-tube at some point during use of the G-tube, and so there is a need for an improved way to deliver nutritional liquids to a patient. The G-tube product is a consumable/disposable product. The current G-tubes cost manufacturers less than a quarter of one US dollar per unit to make, whereas the rest of the components of the G-tube installation kit are significantly more expensive. Since the G-tube is a low-cost consumable medical device product and since large companies control a majority of the distribution channels for these products, there is little incentive for manufacturers to develop an improved G-tube. Also, there are several factors working against new competitors to the market. For instance, it is a very competitive market (with about 3,000 US patents), and due to low manufacturing costs, it is difficult to make a new and improved G-tube that is significantly less expensive (in alignment with the make it "cheaper, better, faster" approach to new product development). Consumer (in this case doctors and hospitals) adoption of a new G-tube product can be difficult, because many G-tube products already exist on the market and are familiar to consumers.

To refine and select the need as a team, you will need to determine the selection criteria and ranking system that best fits your team's decision-making process. Decision-making tools (discussed in Section 9.3.1.1) are often used to help remove bias from team decisions. Figure 9.4 illustrates the multiple iterations of the decision-making (or brainstorming) process that will be required for Needs Refinement. This process includes conducting additional background research (due diligence), then validation of unmet needs. The centralized list, spreadsheet, or database that organizes the Needs Statements can be used as a foundation for developing templates and tools for facilitating the selection criteria and the ranking process. The brainstorming sessions are used to ensure each unmet need has good market fit, aligns with the company's strategic plans and team strengths, and has commercial potential.

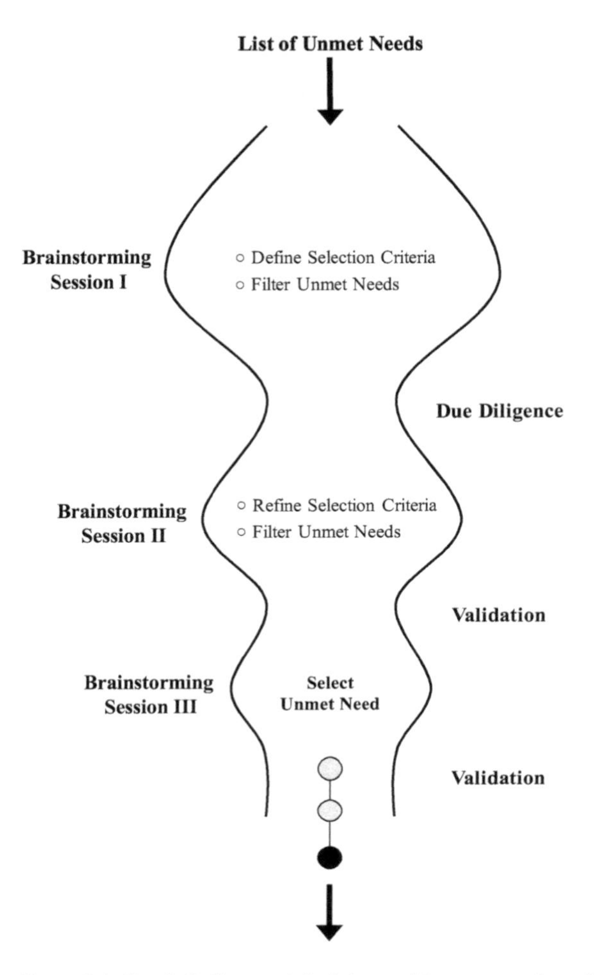

Figure 9.4: Needs Refinement decision-making process for selecting an unmet need.

During an initial decision-making session, there may be a breadth of unmet needs. In this case, utilizing fewer, directed selection criteria to help filter out a majority of the topics will be important for time management. The selection criteria should be simple and in accordance with the high-level validation that your team conducted around each need. Before each brainstorming session, update the selection criteria. The selection criteria should be designed to filter out unmet needs that:

1. Do not fit the team skill sets.
2. Do not seem to have a good fit in the market.
3. Fall into a market where there is a lot of competition.
4. Are not technically feasible to implement in view of the currently available technology.

Once the selection of needs is reduced to about 10–15 topics, selection criteria refinement concurrent with in-depth validation through stakeholder interviews will facilitate the next decision-making session. Repeat the decision-making, criteria refinement, and In-Depth Validation processes until a top unmet need and several alternatives unmet needs are agreed upon. The identification of questions or concerns surrounding each Needs Statements will identify the due diligence and validation processes necessary to prepare for the next brainstorming session.

9.2 Concept Generation Process

Once the team chooses an unmet need (and alternative unmet needs are identified, in case continued In-Depth Validation uncovers disqualifying information), then it is time for Concept Generation. Concept Generation is a system designed to exhaustively and categorically identify solutions for each stage of a process, then assess and aggregate the independent ideas into a single cohesive solution (i.e., concept). The process of Concept Generation is a technique used to ideate solutions, rank the solutions using a decision-making process, and deliver a thoroughly researched and ranked final concept that has been analyzed from component to full concept. The structure of the analysis translates the individual ideas and full concepts into IP that can be captured in a provisional patent application (Figure 9.5).

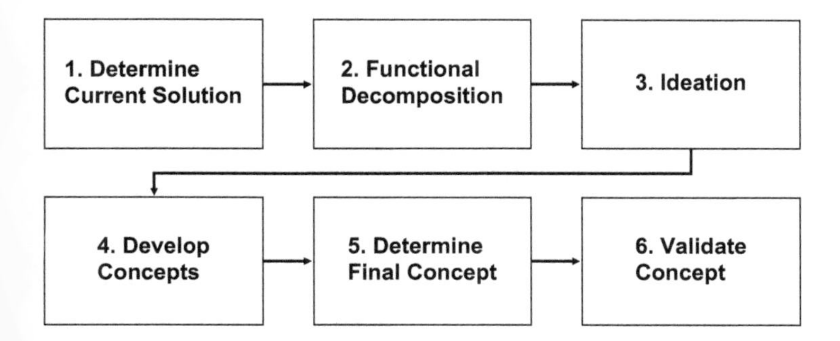

Figure 9.5: Concept Generation pathway.

1 Determine Current Solution

Even if there are not any products on the market that directly solve your unmet need, there is still at least one process or solution that exists to circumvent the unmet need. If there is a process or "gold standard," then it, and alternative processes, should be identified.

2 Functional Decomposition

Functional Decomposition is the process of analyzing a product, device, system, or procedure to break it down into its smaller, simpler parts or functions. Within a Functional Decomposition, a *function* is the simplified task or procedure, whereas the decomposition is the process of breaking down the overarching product, device, system, or procedure into functions and/or subfunctions. A Functional Decomposition begins by creating a functional decomposition diagram (i.e., a picture to help understand how all of the general tasks and subtasks fit together). The development of the diagram begins by taking the identified process (i.e., process pathways) of the current solution and breaking it down to its essential, simplified components (Figure 9.6). Since every product, device, system, and process is unique, there are varying levels of complexity. A general function is a function that is dependent on other functions and/or subfunctions. A subfunction is a function that has to work in order for a more general function to take place. Finally, a basic function is a function that has no remaining subfunctions.

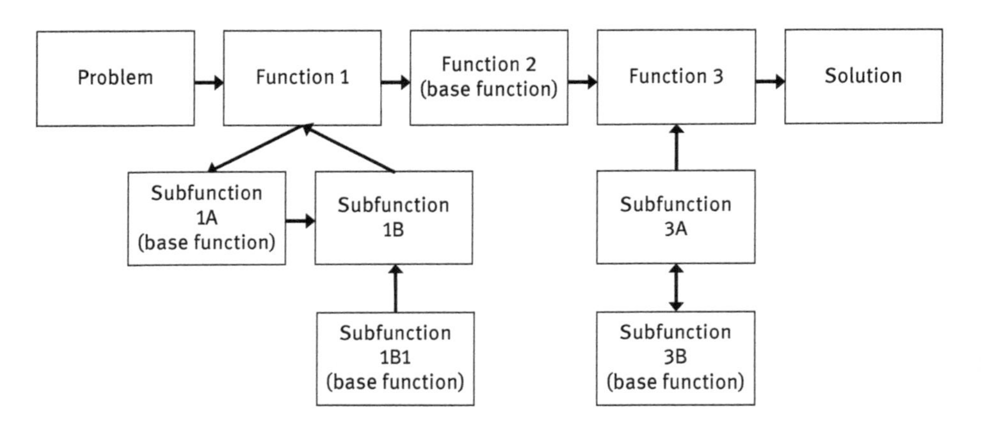

Figure 9.6: Basic example of functions, subfunctions, and base functions within a Functional Decomposition diagram.

Work as a team to determine the multiple known process pathways currently addressing the unmet need. There are usually several different process paths toward a single existing solution. Undoubtedly, there can be multiple functional decomposition diagrams for a single problem or solution.

3 Ideation

Ideation is the creative process used to exhaustively and categorically develop solutions to address a function or subfunction from the Functional Decomposition. Ideation begins by outlining a list of approaches. For example, when developing a physical manufacturing process or medical device, the use of the MECE principle (illustrated in Table 9.1) is helpful for ideation, so that all different types of ideas are represented. Different industries will use different terminology to group ideas. For example, terminology that is used in computer science may include computational techniques, computer languages, database structures, and so on.

The MECE list of approaches should be extensive and encompassing. The process for making diesel fuel is a good example. You may not have thought of including the Biological approach when assessing oil refining processes, though biodiesel is a viable internal combustion engine fuel type. The next innovative separation technique used during making fuel could be acoustic- or magnetic-based. Thinking about "outside-the-box" solutions offers a key to finding something truly innovative. Sometimes the simplest solution is the most innovative.

To maximize effectiveness of the Ideation exercise, attempt to first ideate individually, then collectively as a team. During independent ideation, each team member can retreat to a comfortable place to conduct ideation. Use a blank piece of paper or a template to record the:

- Team Member's Name
- Function Identification
- Approach (e.g., Thermal)
- Concept Description
- Drawing (if possible)

Each team member should come up with as many concepts as possible per approach. It does not matter how far-fetched or outlandish an idea is, as long as it provides a solution to perform the function (or task). Each concept should be documented on its own record and include diagrams, if possible.

Once the individual team members return from their Ideation exercises, each person should have a collection of completed templates with concepts for all of the approaches. Each concept should be logged in a central spreadsheet or database with the appropriate identifiers. The concepts should be organized by function and approach. Often during the organization process, the team will brainstorm additional ideas and idea hybrids can be added to the central concept database. Scan all of the drawings and link them to their descriptions within the database.

A team brainstorming session to establish the selection criteria and ranking system will provide insightful discussion about the different ideas that have been developed. The ranking (or scoring) of the ideas will filter out the ideas that do not seem like a good fit.

4 Developing Concepts

It is recommended to leave a couple of days in between the Ideation and Concept Generation activities because (a) decision-making processes can be labor intensive and time consuming, and (b) the extra time will allow each team member to reflect on the independent ideas and formulate them into unrefined concepts. When beginning the generation of full concepts, the team will take the ideas that scored well during Ideation and organize them together under the functions and subfunctions that were listed in the Functional Decomposition.

The established selection criteria (e.g., quality, estimated costs, feasibility, and team interest) will identify low-ranking concepts that can be set aside. Decision-making tools can be used to correlate ideas with functions and subfunctions (as illustrated in Figure 9.7) within the functional decomposition diagram. Repeat the decision-making model (as many times as needed) with the highest scored collection of ideas to arrive at a complete concept.

Innovation often occurs when a function or subfunction can be removed from a process to improve efficiency, improve profitability, reduce waste, and so on. Note that some concepts (and sometimes whole processes) are not technically possible, not cost effective, not optimal for the environment it is performed in, or may take too many resources to develop.

Turning to IP for a moment; even though the team has selected a "most viable" concept and identified a couple alternatives, an invention disclosure *encompassing all concepts* should be developed in accordance with the instructions provided by your patent attorney. Your team's organized spreadsheet with linked diagrams will then be translated from the invention disclosure into a provisional patent application by your patent attorney. This approach provides comprehensive protection for the company's IP assets, while the team is developing prototypes and testing and validating the concept. Additionally, while conducting extensive due diligence, if you find that your concept infringes on another's patent rights or you cannot get a Freedom to Operate legal opinion, then the alternative concepts included in the provisional patent application afford some "wiggle room" for shifting the development of the final concept within the first year after filing.

5 Determine Final Concept

Using decision-making tools, establish selection criteria and ranking scales to objectively rank the possible concepts that were previously generated. The ranking process will also establish alternative concepts or variations of a concept, in case the chosen concept encounters problems during concept validation.

6 Validate Concept

As discussed earlier, once the final concept is selected by the team, then the Concept Validation process should continue. When conducting Concept Validation, stakeholders will be able to help determine usability and assess human factors of the final concept. If validation of the concept uncovers a disqualifying feature, then referring back to the Concept Generation ranking can offer alternative solutions.

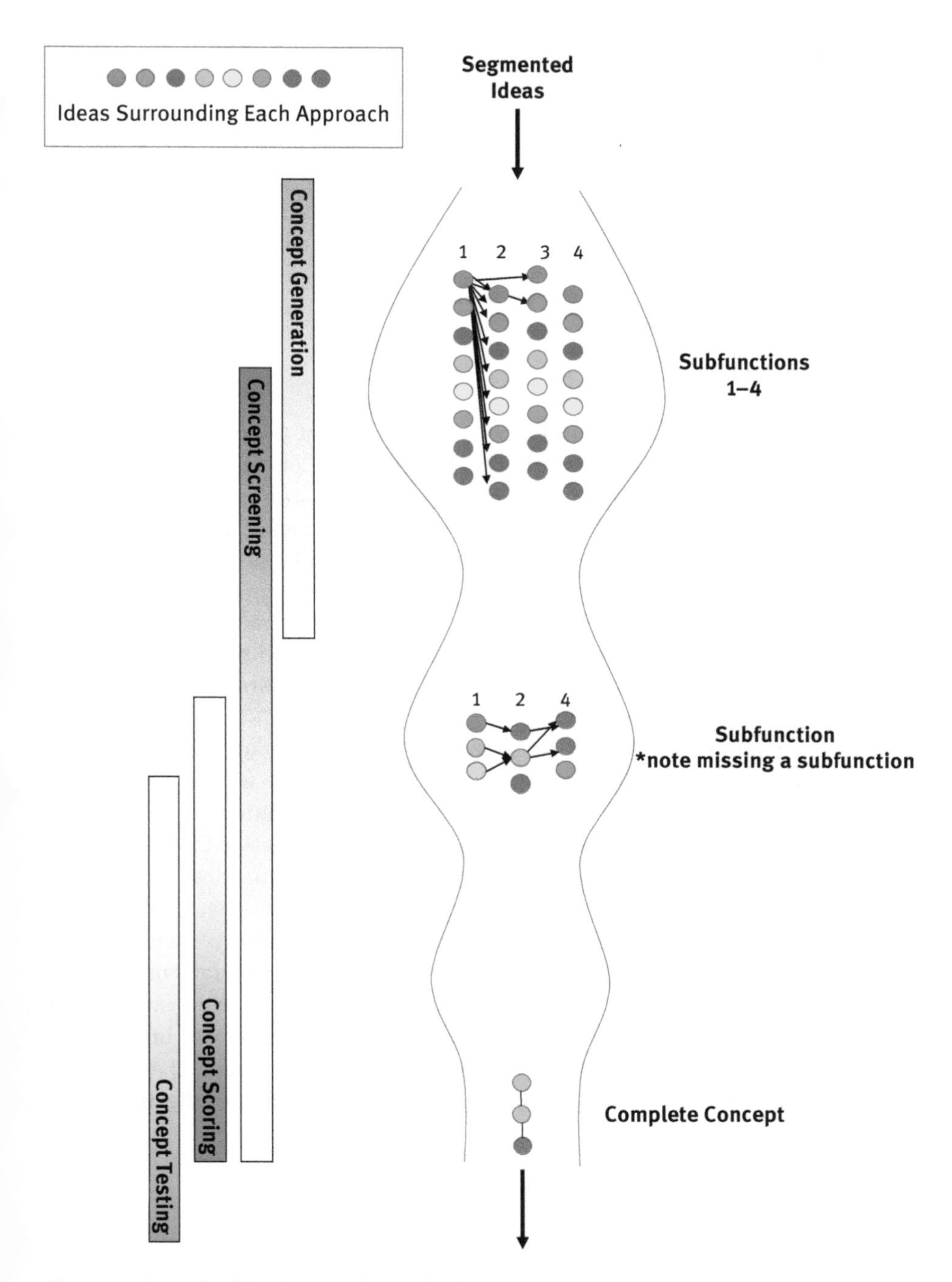

Figure 9.7: Example of the Concept Generation Process.

9.3 Preparing for Needs Finding Process and Concept Generation Process

The Needs Finding and Concept Generation processes take time and can require some forethought, in order to be most effective. Decision-making tools and preparedness are consistent for both Needs Finding and Concept Generation. Below is an overview of how to approach the Needs Finding Process and the Concept Generation Process.

9.3.1 Preparedness

When starting the Needs Finding Process or the Concept Generation Process, it is best to be prepared. The Needs Finding Process and Concept Generation Process take time to develop. It could take a couple weeks or a couple months. To start, set up a place that will not be disturbed, that is, create a "war room" for your team to work in. Find a dedicated space where nothing will be disturbed, that provides confidentiality, has multiple black/white boards, plenty of wall space for sticky notes, posters, charts, and so on. Whiteboard and wall space will be most important. At the end of each session, take pictures of the progress and upload them to a centralized location, such as a collaborative digital workspace, for team review.

The most important part to these systems is to come up with your own team's "process" for decision-making. This can involve the way to collect and categorize information from observations, the way to determine selection criteria, ranking and scoring of selection criteria, and more. Each team is different, so these processes will be different depending on team dynamics. The goal is to find a process that will allow the team to work well together, while being organized and efficient. Many decisions encompass interwoven factors (such as market viability and technical feasibility) that may be abstract or difficult to systematically assess. Individuals, and especially teams, often struggle to objectively assess multiple concepts, due to the level of complexity involved in the decision-making process. To avoid this, there are a number of tools that can be used by teams during brainstorming sessions that are effective decision-making models.

9.3.1.1 Decision-Making Tools

Decisions are often made by corporate executives, business owners, and managers through use of intuition. The results vary greatly and completely depend on the insight of a few key people in leadership. Systematic and data-driven decision-making processes have been able to demonstrate accurate outcomes. The development of decision-making tools allows leadership to use quantifiable analysis, such as cost–benefit analysis and predictive modeling, as a basis for making decisions. If implemented effectively, a decision-making model reduces individual bias and unforeseen circumstances that result in suboptimal decisions. That being said, some degree of bias will always exist. Data can be inaccurate or interpreted incorrectly. One common decision-making tool is the Pugh Matrix.

Pugh Matrix

A Pugh Matrix, also known as a criteria-based matrix, is a technique used to organize and direct a team-oriented project that requires sorting through multiple concepts or needs. It is a scoring system used to score and rank concepts or needs relative to acceptance (or "decision") criteria to help determine which items or potential solutions are more important than others. A Pugh Matrix can be a powerful tool to make objective decisions based on a combination of objective and subjective inputs. One of the key advantages of Pugh Matrices is that the matrix handles a large number of decision criteria. More importantly, it allows a team to systematically rank and agree upon top solutions in a logical fashion. Figure 9.8 outlines the process for developing a Pugh Matrix.

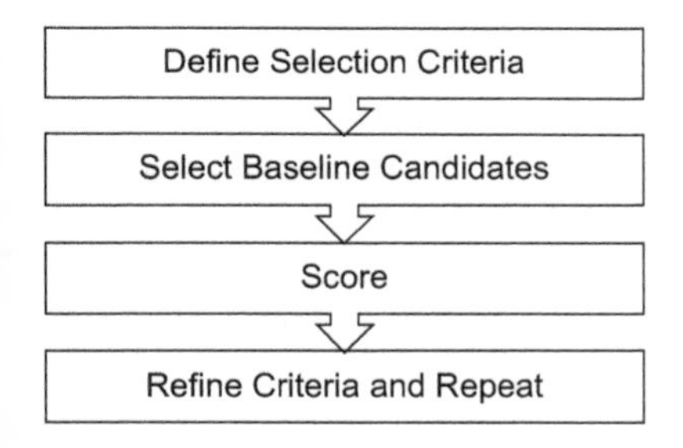

Figure 9.8: Process for developing a Pugh Matrix.

1 Selection Criteria (for Pugh Matrices and Other Decision-Making Models)

What are "criteria"? Criteria are the overarching requirements of a project. The ranking of the criteria can be subjective or objective, depending on the technicality of the product or the needs of the stakeholders. Criteria can be as vague or granular as needed, but *most importantly, criteria need to help differentiate the items that are being scored.* Often, during High-Level Validation that may cover a breadth of unmet needs, concepts, or other topics, the selection criteria are vaguer, then become more refined as In-Depth Validation occurs and iterations of the decision-making process occur. Developing Selection Criteria could consist of a couple of criteria or as many as the team sees fit. If there are going to be multiple iterations for filtering down the unmet needs, concepts, or other topics, then selecting a few key, well-defined criteria to filter out the outlying concepts will hone in on the most relevant criteria. Then, the most relevant criteria can be further refined by repeating the filtering process. Filtering at the early stages of criteria selection is more efficient than going through dozens of criteria from the outset. As shown in Table 9.2, the Pugh Matrix correlates multiple concepts to acceptance criteria for qualitative optimization.

Table 9.2: Example Pugh Matrix.

	Concept A	Concept B	Concept C	Concept D	Concept AC	Concept AD
Criterion 1						
Criterion 2						
Criterion 3						
Criterion 4						
Criterion 5						
Total +						
Total –						
TOTAL						

2 Baseline Candidate Selection and Comparison

The selection of a baseline candidate, or comparator, is needed to guide the decision-making process. Having a baseline candidate or norm to compare other options to streamline the process of comparing a long list of topics or concepts. Drawing conclusions based on comparisons made to the average is easier to comprehend than assigning scores or ratings according to varying, arbitrary scales, when making comparisons. Within a Pugh Matrix, the baseline candidate is often annotated by "S" (for same), "0" (for nonnegative or nonpositive), or "O." These annotations are typically utilized across industries. As shown in Table 9.3, the Pugh Matrix correlates multiple concepts to acceptance criteria for qualitative optimization to a baseline candidate. In this example, Concept D was scored highest, with Concepts A and B being scored equally. There can be different baseline candidates for each criterion.

Table 9.3: Example Pugh Matrix.

	Concept A	Concept B	Concept C	Concept D	Concept AC	Concept AD
Criterion 1	S	S	S	S	S	–
Criterion 2	S	–	–	+	+	–
Criterion 3	+	S	S	+	+	+
Criterion 4	S	+	S	S	-	+
Criterion 5	S	+	S	S	–	–
Total +	1	2	0	2	2	2
Total –	0	1	1	0	2	3
TOTAL	1	1	–1	2	0	–1

3 Scoring

The next step is to compare each candidate (i.e., concept) against the baseline candidate for each criterion. Once the selection criteria are determined, then a scoring scale needs to be established for all of the criteria. For most Pugh Matrices, the scores are noted by either (−1, 0, 1), (−, S, +), (−, −, S, +, ++), or (−2, −1, 0, 1, 2) depending on the granularity of the scoring. Other decision-making models use ranking scales (e.g., 1–3 or 1–5). The goal of the scoring (or ranking) scale is to be able to tally the scores for comparison. In the syntax of (−, S, +), S means "same," + means better, and – means worse. When scoring annotations −/S/+, these stand for −1/0/+1, respectively.

To better understand the scoring process, the following analogy shown in Table 9.4 is helpful: consider all of the goalies in a football (soccer) league, then arbitrarily choose a single goalie to represent the estimated "average," that is, this goalie represents the average of all goalies in the league in terms of characteristic "X," for example, a criterion such as saves, height, athletic ability, and reaction time. If one of the goalies (e.g., Robert) is selected as representing the average goalie throughout the league, then when scoring occurs, it is easier for managers to compare other goalies to Robert.

Table 9.4: Example football league Pugh Matrix.

	Weight	Robert	Renaldo	Boris	Paul
# Saves	−2 to 2	S	2	−1	S
Height	−1 to 1	S	S	S	+1
Athletic Ability	−1 to 1	S	−1	S	+1
Reaction Time	−1 to 1	S	S	+1	S
Interpersonal Skills	−1 to 1	S	−1	+1	S
TOTAL		S	0	1	2

To continue with the football example illustrated in Table 9.4, if a team is assessing the best goalie in the league out of Robert, Renaldo, Boris, and Paul, the criteria could be both quantitative and qualitative. The ranking associated with interpersonal skills (i.e., qualitative) would range from likable (a higher score, +1) to arrogant (a low score, −1). Similarly, the ranking for number of saves (i.e., quantitative) would be based on Robert's number of saves. In the case of number of saves, a low score of −1 would be assigned to any goalie that saved fewer goals than Robert and a high score of +1 would be assigned to any goalie with more saves than Robert. The league could also determine that the number of saves is more important than the new goalie's interpersonal skills, thus adding "weight" to each criterion as well. Weighting the criteria can be helpful so that the league does not need to re-rank each concept (e.g., each new goalie candidate), rather, the league would only need to recalculate the score based on weight. If a league collectively has 100 concepts (e.g., new goalie candidates), then filtering them against three to five criteria can make the decision-making process less cumbersome. An example of extra weight (demonstrated in Table 9.4) could be seen as a scale of −2, −1, S, +1, +2, when the other criteria are assessed on −1, S, +1.

For each candidate, the total score can be calculated by summing the number scores. Just because a candidate scores well on one criterion does not mean it is the best. The league should then discuss any anomalies, feelings, or candidates on which they would like to consider by conducting additional due diligence.

In following the football (soccer) goalie example, if too little information is known about a goalie, further due diligence could be conducted. This research could involve watching Paul play in a game, or interviewing Renaldo in a casual environment to get a better sense of his interpersonal skills. Additionally, the second and third ranked candidates are considered to be the alternatives, in case the top choice becomes a nonviable option (e.g., if the first pick goalie was signed by another league). Finally, record not only the results of the final Pugh Matrix, but the reasons behind why the team made key decisions. There are often "ties" between two good potential candidates, but upon further due diligence, the alternative choices may need to be referred to again.

4 Refine and Repeat

There could be many reasons to refine the criteria and repeat the decision-making process. For instance, the selection criteria can change, additional in-depth due diligence needs to be performed, validation could reveal new problems or issues with the concept(s), or a new concept could be added at a later stage.

9.3.2 Alternative Decision-Making Models

Not all decision-making models have to be as intricate, in-depth, or formal as a Pugh Matrix. As shown in Table 9.5, once the selection criteria are determined and defined, the criteria and how they are defined (i.e., scored or ranked) can be simplified to a scale system, with or without an added weight factor.

In Table 9.5, the ranking scale is kept the same, but the weights are different depending on the importance of the criteria to the team. The team decides the ranking scale. In the case of Technology Development Feasibility, a qualitative ranking approach is established based on how the team feels concerning their own strengths and the currently available technologies. Time to market may be based on how long it took another company to get a similar product to market. Market saturation and competition could be estimated from industry reports. Competition is a quantifiable number extrapolated from industry reports and market research.

Table 9.5: Example decision-making model with weighting (Wt.).

Criteria Example	Ranking Examples (Unfavorable) 1 to 5 (Favorable)		Score	Wt.	Final Score
Technology Development Feasibility	Definitely not = 1	We can do it! = 5		0.2	
Time to Market	10 years = 1	1 year = 5		0.2	
Market Saturation	100+ products = 1	No products on market = 5		0.2	
Competition	100+ competitors = 1	No competitors = 5		0.4	
Total				1	

9.4 Decision-Making Walkthrough

An example can help illustrate the decision-making process. A team was assigned to look for unmet needs by the Fictitious Company. The team consists of two general engineers and one business expert. In considering the strengths of the team, one of the engineers had a background in robotics, the other in programming, and the business expert managed a car wash while in college. Fictitious Company arranged for the team to observe colleagues in an assortment of fields. On the way to one of their observations, the business expert passed a car wash and asked to set up observations at the car wash, in addition to the other observation sites. Between the three colleagues, over 1 month of observations they collected over 200 unmet needs. In their war room, the team conducted online research to validate each of their potential unmet needs.

Iterative brainstorming was conducted, and the Pugh Matrix shown in Table 9.6 was used to filter down the unmet needs.

One need that showed potential was observed while a team member was at a partially automated car wash. The team member observed that (a) the car wash physically occupied a lot of land space, and (b) had no less than two employees operating the car wash: one employee serves as a money collector/operator and another employee wipes down the cars. The initial need observed and communicated was:

> *There is a need for car wash companies to reduce human resource expenses in order to be more profitable.*

The need was selected by the team as one of the stronger candidates. The team went back out to perform an In-Depth Validation that included interviews with the car wash attendants from multiple car wash companies, the multiple car wash owners, and multiple car wash accountants. The result of the interviews was the generation of a refined need statement of:

Table 9.6: Example Pugh Matrix.

There is a need for...	Technical Feasibility (1–5)	Team Alignment (1–5)	Cost Savings (1–5)	TOTAL
Car washes to occupy less space in order to be scalable	5	5	3	13
Car washes to require less human resources in order to increase profitability	5	4	5	14
Construction workers to have access to injury notification system in order to decrease emergency response time	3	4	1	8
Hospitals to be able to locate and track all of their mobile equipment (e. g., wheelchairs and beds) in order to prevent loss of equipment and reduce time for technicians to find equipment	5	1	3	9
Sports teams to be able to know when a player is healthy enough to go back on the field in order to reduce additional injury	3	1	2	6
Grocery stores to know when fruit is ripe enough to put out on the shelves in order to reduce warehouse space	3	1	3	7

There is a need for car wash companies to lower their overhead, while maintaining customer satisfaction, in order to sustain profitability.

From the need statement, an assessment of the process was conducted to perform a Functional Decomposition (illustrated in Figure 9.9). Each member of the team then independently thought about breaking down each step into as many substeps as possible.

To explore a broad range of concepts, the team decided to use the MECE list (refer to Table 9.1). Concepts ranged from utilizing dirt-eating bacteria for the cleaning of the car to using radioactive isotopes to identify residual dirt. Their ideas for addressing each individual process enabled abstract and complicated concepts to be brought together as a whole concept downstream.

Through Concept Generation, the team used the Pugh Matrix shown in Table 9.7 to determine the best concept. After multiple iterations, the final three concepts are scored as follows:

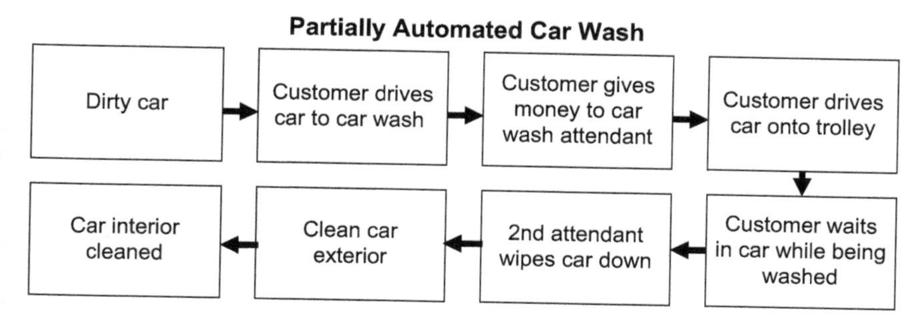

Figure 9.9: Functional Decomposition of partially automated car wash.

Table 9.7: Pugh Matrix for partially automated car wash.

	Existing Car Washes	Robots to Replace People	Fully Automated Car Wash	Bacteria Car Wash System
Technical Feasibility	S	−1	+1	+1
Time-to-Market	S	–	S	S
Profitability	S	−1	+1	−1
Land Requirements	S	+1	+1	S
Competition	S	+1	S	+1
TOTAL	S	0	3	2

The scoring showed that shifting from a partially automated car wash to a fully automated one would (a) be novel, (b) occupy less land, and (c) require less labor (e.g., profitability). The team's new Functional Decomposition (illustrated in Figure 9.10) was able to remove two functions from the partially automated car wash process.

The team took their findings and concepts and then worked with the Fictitious Company's patent attorney to develop a provisional patent application encompassing all of their ideations. For example, though the team discussed both a bacteria-based car cleaning method and a fully automated car wash, the team included both in their provisional patent application disclosure. By including both the bacteria-based cleaning reagent and the ideas surrounding the fully automated car wash in the provisional patent application disclosure, the team could later seek a nonprovisional patent for the fully automated car wash process and the bacteria-based cleaning solution separately (stemming from the same provisional patent application), even though the bacteria-based solution is not the solution that they intend to pursue. After extensive validation, the team agreed that bacteria-based cleaning reagents were not a viable direction (due to costs, bacteria disposal requirements, etc.) and decided to focus solely on the fully automated car wash, without needing to submit another provisional patent application.

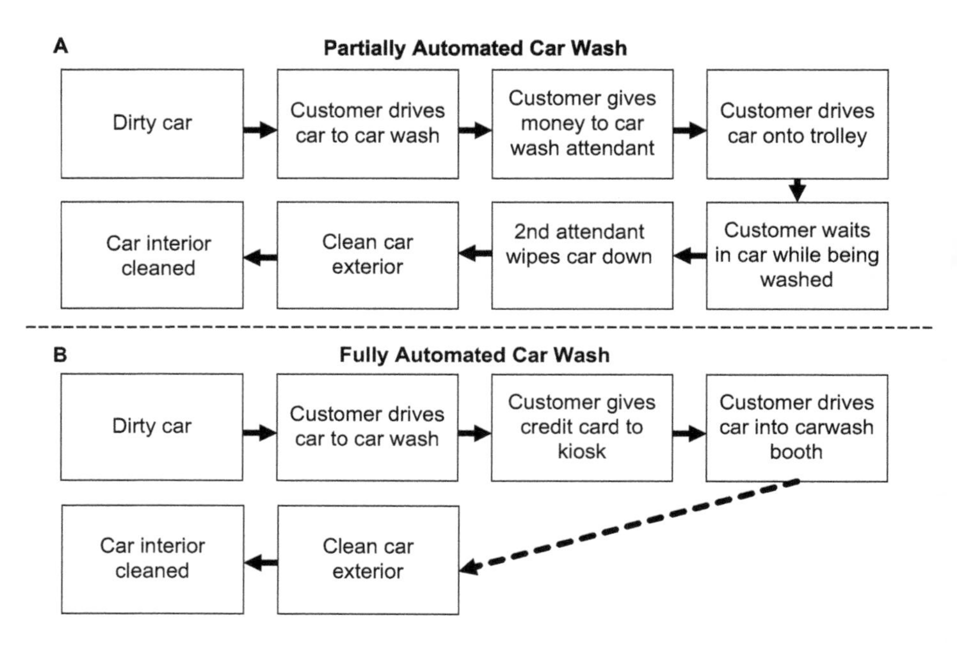

Figure 9.10: Comparison of Functional Decomposition of (A) a partially automated car wash to (B) a fully automated car wash.

9.5 Needs Finding and Concept Generation for Startups

The Needs Finding Process is often conducted in an academic setting or within a highly structured corporation that has the resources to categorically assess observations. An adaptation of the Needs Finding/Concept Generation Process can benefit startups. The Startup Need Statement Process illustrated in Figure 9.11 is not quite as resource intensive and can be tailored to the startup company's vision.

Most startup companies are created around a preconceived product and must work backward to arrive at an unmet need. All products have an unmet need that they address. In this case, it is recommended to look at the startup's preconceived product and reverse engineer the unmet need. Reverse engineering the unmet need can be difficult, due to team and founder bias and passions. Begin with a team brainstorming session to "take a step back" and assess the product. Once the unmet need has been clarified, it is easier for a startup to validate the Need Statement and move on to the Concept Generation Process. Even if a prototype for a singular concept has already been started, this process can not only help develop IP assets, but can possibly help with developing alternative product specifications of a truly innovative product.

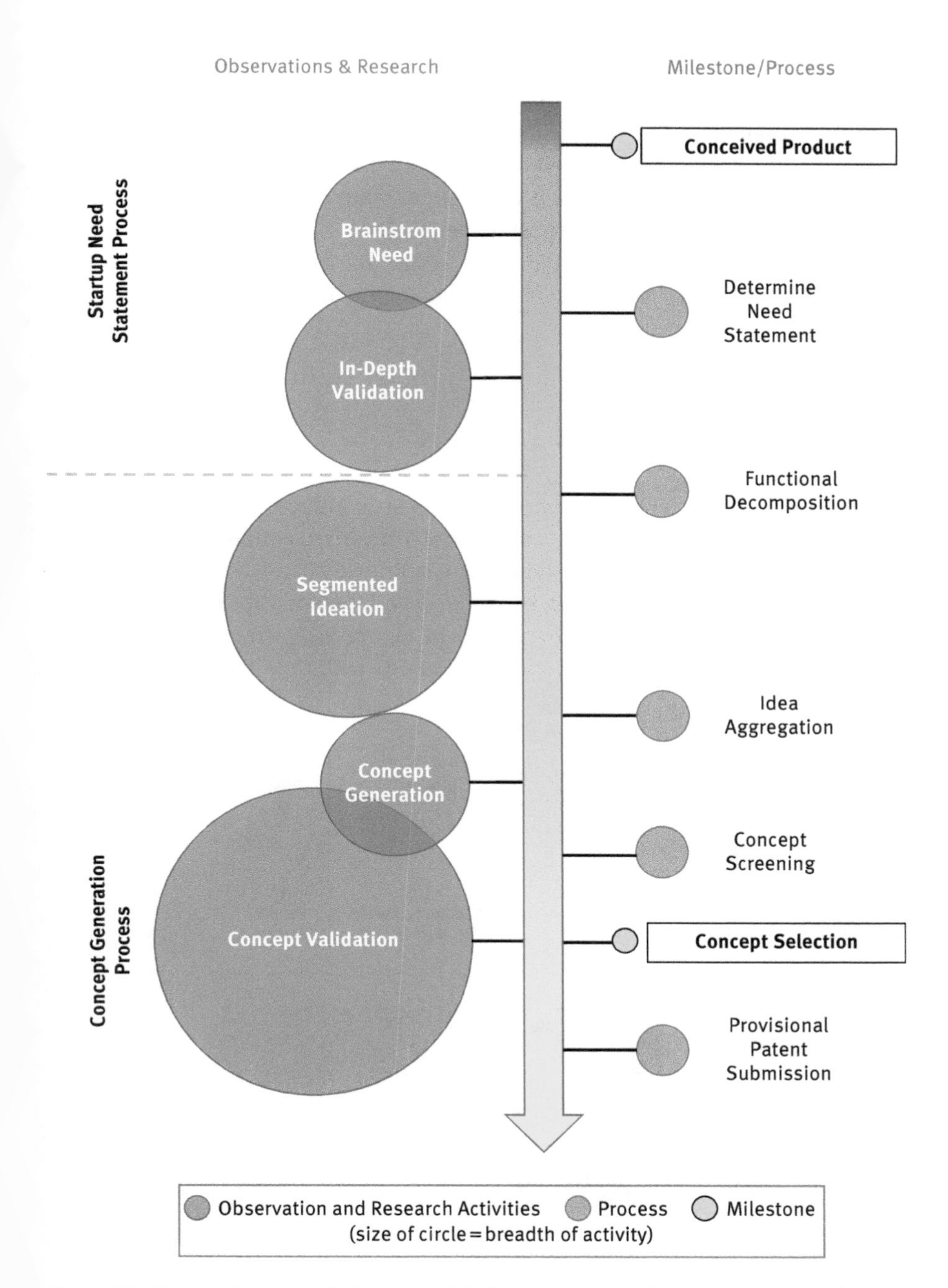

Figure 9.11: Process for extrapolating and validating an unmet need from a preconceived product used by startups.

Before moving to the next section of this chapter, which describes prototype development, this chapter so far has provided guidance on how to identify an unmet need and comprehensively and categorically develop concepts around the unmet need. The workflow process is broken down using Functional Decomposition. Concepts are generated, the top concept is selected, and any viable alternatives are included in an invention disclosure statement. Then the invention disclosure statement is given to an IP attorney, in order to start the process of securing patent protection. The next step in the Product/Technology Development process is to build a prototype of the innovation.

9.6 How a Concept Becomes a Product

Once Ideation and Concept Generation have been conducted, the next actions along the Product/Technology Development timeline are to conduct Prototype Development, followed by Product Development, and lastly Transfer to Manufacturing (Chapter 10). Throughout these stages of development, the technology will start transforming into a product. As shown in Figure 9.12, the entire process of developing a product starts with a concept and ends with the final product line or

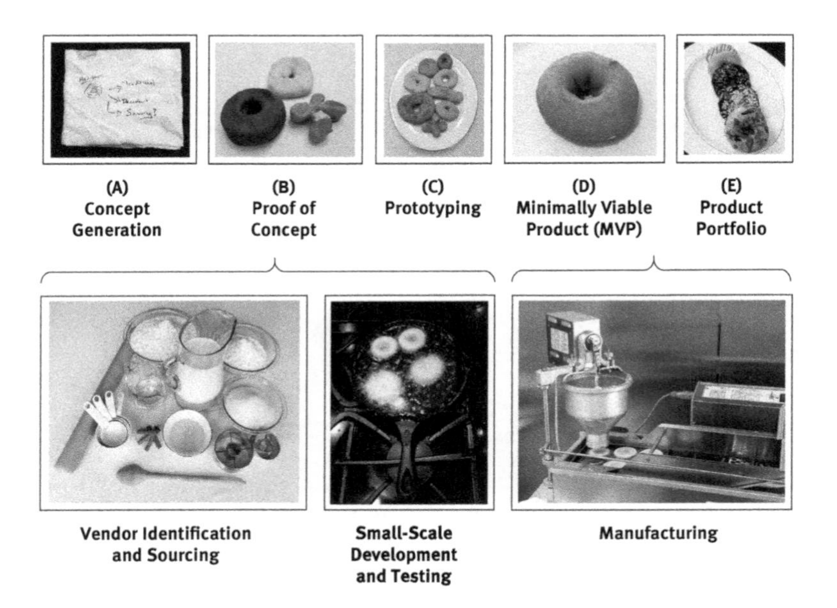

Figure 9.12: Progression of the Development Processes for donuts depicting the steps of (A) Concept Generation, (B) Proof-of-Concept, (C) Prototype/Product Development, (D) Minimum Viable Product, and (E) a Product Portfolio.

product portfolio. Figure 9.12 uses the example of the development of a donut to show that: (A) ideas must be translated into a concept, (B) the POC versions help to determine the best concepts to move forward with and their production feasibility, (C) early-stage ("alpha" or "α-") prototypes are developed to *determine and define* the user requirements and acceptance criteria, then the α-prototype is translated into a product intended for testing by the end user to refine the user criteria, user requirements, and acceptance criteria (i.e., "beta" or "ß"-prototype), (D) the minimum viable product (also known as the minimally viable product or MVP) is the product that has been transferred to manufacturing and is ready to be sold, and lastly, (E) next-generation products are expanded upon in a product line or portfolio.

When looking at the overall processes involved with Technology and Product Development, a big picture approach encourages the use of qualitative techniques, such as Design for Manufacturing (DFM), Design for Assembly (DFA), and Sourcing, to be implemented concurrently with Research and Development. The use of such tools during the Prototype Development and Product Development stages significantly reduces the amount of work being performed during Transfer to Manufacturing. DFM, DFA, and Sourcing are the tool sets that allow for the completion of quality work earlier in the developmental process, so that the work does not need to be repeated later.

9.6.1 Prototype Development

Prototype Development, also known as prototyping, begins by transforming your top concept(s) into POCs – the inexpensive, bare-minimum proof needed to vet your concept. The development of the POCs will help determine the most promising concept to move forward toward building the first iteration of a full system, or an Alpha (α)-Prototype. A prototype is a preliminary, or initial, "test" version of the product or technology that you are trying to develop. The purpose of prototyping is to create a working version of the product or technology, to test the concept, and better assess any assumptions you made in arriving at the design of the prototype. In the context of this book, Prototype Development encompasses the development of the POC and Alpha (α)-Prototypes.

A **POC** is the bare-minimum evidence needed to show that the concept is viable. POC demonstrates the "works-like" model of your prototype.[30] The POC does not need to function efficiently or be aesthetically pleasing; the POC simply needs to show that the concept is heading in a good direction. The POC is often made

30 The POC, α-prototype, and β-prototype (discussed more in Chapter 10) are analogous to the "works-like, looks-like, built-like" model, where POCs are "works like," α-prototypes are "looks like, works like, but not made like," and β-prototypes are "looks like, works like, and made like."

from off-the-shelf components and readily available materials. For example, if you want to develop a "Smart Home" system that would include a wireless computer-controlled sump pump in a home basement, you may begin by assembling a very rough prototype of your POC. For a POC prototype, it is not unheard of to develop a POC that comprises a circuit board and wireless receiver attached to a cardboard base using duct tape, which is then attached to a sump pump. The system could also be attached to a personal laptop, and you can use your programming skills to demonstrate that you can control the sump pump. The POC may not look professional, but it proves that the concept is technically feasible. As shown in Figure 9.12, it can take multiple attempts to prove out or demonstrate your conceptual donut.

An **Alpha (α)-Prototype** is typically the second iteration of a prototype, or the "works-like *and* looks-like" model of the prototype. This iteration of the product functions and visually looks more like the intended final product. To be consistent with the above Smart Home example, developing the α-prototype would involve preparing a Computer-Aided Design (CAD) drawing (Chapter 10) of the sump pump chassis that encloses the "off-the-shelf" components (e.g., sump pump, wireless receiver, and circuit board). Using the CAD drawing as a template for the α-prototype chassis that needs to be produced, the chassis can then be 3D printed. The α-prototype is usually of sufficient quality and functionality that it can be presented to early-stage investors.

Referring back to Figure 9.12, an early iteration of a donut can determine the user requirements and acceptance criteria. In the donut example, user requirements can include that the donuts must be able to be held in one hand, must be circular in shape with a hole in the center, and have a moist interior. On the other hand, acceptance criteria for the donuts might include that the donuts must have the dimensions of 15 cm diameter by 5 cm in height, the donut must maintain external shape for a minimum of 48 h, and the donut must maintain 40% moisture retention in the interior of the donut for 24 h after preparation.

9.6.2 Design for Manufacturing (DFM) and Design for Assembly (DFA)

While trying to develop the POC, many inventors singularly focus on trying to demonstrate that their idea works, rather than setting the stage for a smooth transition into manufacturing. In a "works-like, looks-like, built-like" model, the POC is only the "works-like" component. This discussion is an introduction to terminology and concepts which can dramatically reduce the amount of work throughout the Product Development timeline that will occur after you develop your idea. This can be applied to both highly technical and nontechnical industries.

What is often not thought about during Prototype Development is scalability, that is, what it will take to manufacture/produce enough units of the product at low cost to sell to the specific market and demographic you are targeting. Planning for scalability can be difficult, since you are at the beginning of the Development process. However, being able to produce enough units of your product for a Commercial Launch is the downstream goal.

Example

An example demonstrating the difficulty of anticipating the scalability of a product can be found in the sector of nanotechnology. Researchers spent decades conducting work to fabricate and understand materials and structures that are so small that we cannot see them with the naked eye. However, the excitement surrounding the research was overshadowed by the difficulties in producing enough individual nanosized objects to be useful on a larger scale. It took decades before the precision needed for most nanomanufacturing techniques caught up with the theoretical and fundamental research.

Currently, smartphones, computers, electronics, many plastics, and a lot of other household items utilize nanotechnology. It took a long time for these items to go from benchtop research to being developed for mass production. Furthermore, a lot of time and money went into basic research and development, before it was realized that the scalability of nanotechnology was an issue. As products and technologies keep getting smaller, faster, or more accurate, you will find that scalability will be a challenge.

9.6.2.1 Design for Manufacturing (DFM)

DFM is the practice of designing a product while proactively considering what is needed for the end goal of manufacturing. When brainstorming your initial prototype, DFM includes thinking about:

- Reducing the number of parts needed to make the product.
- The use of standardized parts.
- Modular components.
- How parts can be made to be multifunctional (e.g., a power supply being used as structural component as well).
- The use of multifunctional parts across products (for instance, one type of screw that is used in the three different types of computers you are producing).

The overlap and integration of these concepts (including the use of software and hardware) can make future strategies (such as how to approach scaling up) and transitions (such as transition to manufacturing) easier downstream. Relative to the donut example above (shown in Figure 9.12), DFM might refer to looking at the end donut, determining what dimensions you want, then prototyping dough extruding nozzles that would fit on a standard industrial batter dispenser during the prototyping stage. It may also include seeking advice from a vendor that is known for manufacturing other brands of baked goods, so as to ensure that the processes you are developing will work with the industrial baking equipment that your company plans to use.

9.6.2.2 Design for Assembly (DFA)

DFM generally focuses on individual parts and components of your product, where DFA focuses on reducing the cost and time of assembling the final product and simplifying the assembly process. This is approached by trying to reduce the number of steps that are required to physically assemble the product, and by reducing the number of parts that are necessary to make the product. To stay consistent with the donut example introduced in Figure 9.12, DFA might mean laying out the kitchen to maximize workflow. This may mean ensuring the pantry is close to the walk-in freezer, which would be easily accessible by the receiving department. At the same time, the mixer needs to be close to the pantry, freezer, a water source, dishwasher, open workspace, and fryers. The fryers should be close to the batter dispenser and cooling racks. The cooling racks should be close to the systems for applying glaze and toppings on the donuts. Lastly, the area where the donuts are packaged and labeled should be close to the shipping area.

Additionally, there is the process of Concurrent Engineering, which is the act of developing multiple products simultaneously with a standardized procedure, in order to save time to market during the development of a product. For example, this might be the development of plain, frosted and sprinkled, and glazed donuts, all at the same time.

Note: It is often recommended that startups do not practice Concurrent Engineering, because there are often not enough resources available at the startup stage to develop multiple product lines, and it is easy to get distracted from the beachhead with ancillary features or the next model.

DFA ties into the identification of the parts, or Sourcing. Sourcing is the process of assessing vendors that involves an analysis of cost per unit volume compared to the "guaranteed" quality of the component coming from the vendor.

When developing a product, each component will come from somewhere, whether you buy it from a vendor or decide to manufacture it in-house. For example, if you decide to start a bookshelf building business, you would get the unfinished wood, which you would then cut, sand, stain, and seal, and then assemble into a new bookshelf. Now you want to make more to sell. The disposable items you will need to source for your bookshelf will be the wood, the wood glue, the nails/screws, the stain, the sealer, the rags, Computer Numerical Control (CNC) mill, drill press, the miter saw, band saw, lathe, drills, drill bits, pens/pencils, drafting paper, CAD software, and so on. All of these items need to be accounted for.

When making the initial purchase of the components, you can ask the vendor/ supplier how they will guarantee the specifications that are important to you. You can even ask them to put their guarantee in writing. Risk management of vendors is important.

If you have a single supplier for your key component, what happens if that supplier goes out of business, cannot meet your specifications, or is consistently late? Any supplier that provides a key component that contributes to the performance of your product should be deemed "critical," and it is recommended to regularly check in and/or inspect the supplier. This includes regularly evaluating the supplier's financial stability.

Each supplier's quality should be assessed based on your needs. What happens when you rely heavily on wood glue provided by a supplier, but the last few 10-gallon buckets of wood glue that you have ordered from your supplier produces variance in bonding strength when you use it? Or you might have concerns about the consistency of the wood you source? Or the drill bits you have been getting keep breaking? Or the stain you order varies in color? Or the screws keep being shipped to you with a 4-month lead time? Or they keep sending you 6.4 mm nuts that don't quite fit the 6.4 mm bolt?

Ask whether the supplier has a quality manual, whether the supplier's employees are formally trained to their procedures, find out what their in-process and final inspections are, and do they provide certificates of conformance for each lot or batch produced for their customers, and so on. Implementing a sourcing policy, whether formal or informal, during the prototype phase, at least for the major components, can prevent the need to repeat a lot of work later.

Note: If you are developing a prototype that involves software integrated with hardware, then you will most likely need to redevelop the software or driver from scratch for each component, if you don't source during the development stage.

This leads to the Bill of Materials (BOM) (refer to Chapter 10), or the final approved list of all of the parts and components that are used to build your product. The BOM can be a simple checklist formed in a spreadsheet, or it can reside in a sophisticated resource planning system. In the donut example discussed earlier, the BOM is analogous to the ingredients list paired with quantity per unit and associated costs. A BOM for the donut example can be found in Chapter 10 (see Table 10.8). A resource planning system is an organizational system that identifies what each part is, where it is sourced, how it is traced or uniquely identified throughout production, and how many of each part are needed for each product that is produced. Data from the resource planning system should be tied back to the Accounting Department for tracking production costs, managing inventory, and placing orders.

Everything mentioned earlier can be overwhelming, especially when you are just starting a business. Most well-established companies follow ISO 9001 or ISO 13,485/9001 to implement a structure to guide these processes. More importantly, just thinking of these concepts as you are developing your next best idea can make building your new company much less stressful.

9.6.3 Sourcing

Sourcing is the act of finding reliable vendors or manufacturers for each component of your product. Important aspects of sourcing include Vendor Management (Chapter 7) and maintaining Quality (Chapter 11). Ideally, during α-prototype development, the selection of a contract manufacturer, vendor, or Original Equipment Manufacturer (OEM) that consistently produces/supplies components or parts that are of sufficient quality can be beneficial for a company, because that third party can be used throughout Product Development and into Manufacturing. This ensures that Vendor Quality Management does not need to be conducted from scratch through each stage of Prototype Development, Product Development, and Transfer to Manufacturing.

However, it may not always be possible to secure the "right" contract manufacturer, vendor, or OEM, early on in Prototype Development, especially when time and money is scarce. Often, startups will delay finding the best option for when they have time, but this just delays development and increases overall costs.

Additionally, there are often so many components in a product that it is best to find a contract manufacturer early on for sourcing key components. For example, the Smart Home company mentioned earlier may opt to source an ISO 9001-certified circuit board manufacturer for their wireless sump pump system during α-prototype development, so that they do not need to be as concerned later about finding a circuit board manufacturer for the MVP. Instead, the company can focus on the other details of Product Development.

9.7 Finding Resources

Accounting for DFM and DFA allows for a more seamless transition from Prototype Development to Product Development to Manufacturing. Knowing where (or at least what direction) you want to end up is key to this. So where to start? All of the due diligence you have performed will have identified the types of parts, components, hardware, software, and so on that you will need in order to produce a successful product. There are resources readily available for startups and professional manufacturers alike.

9.7.1 Hardware and Physical Components

The local hardware store will have basic parts. Electronics stores will have soldering irons, circuit boards, 3D printers, and so on. You can usually find tutorials of anything you need online. There are also vendors, such as McMaster-Carr and Grainger, that provide specialty industrial components. Contract manufacturers can often make or arrange to outsource parts that you may need. Alternatively, there are always machine shops that have access to full workshops, which operate lathes, CNC systems, and welders.

Other options include use of "shared work spaces," "innovation laboratories," and "maker-spaces." These are communal workspaces or workspaces accessible by the public, often in conjunction with an academic institution, nonprofit, or economic development organization with the intent of attracting and supporting startup companies. They may charge a monthly or annual fee, fees associated with individual equipment, or a combination of the latter.

9.7.2 Computer-Aided Design (CAD) and Computer-Aided Manufacturing (CAM)

CAD is the development of an electronic drawing for each of your parts. Computer-Aided Manufacturing (CAM) is the use of the electronic drawings by a specific machine to produce a part represented by the electronic drawings. To clarify, each part of your product that is novel and needs to be developed from scratch will need a drawing. If you use CAD software to draw each new part, then you can send the drawings to a contract manufacturer who can manufacture the part according to the specifications drawn. CAD software allows multiple parts developed together or separately to be compared to one another prior to construction to ensure everything fits together.

Once you have CAD drawings that are satisfactory, you may want to get the part(s) manufactured. You may need different types of processing, or manufacturing systems, for each part within a component. For example, for a simple pin that is used as a hinge for a cabinet door, you could purchase a block of metal and lathe part, or use molten liquid and an extruder, or use a polymer and an injection molding system.

Note: There are a lot more ways to make a hinge pin!

Since there are so many techniques that can be used to produce a part, a CAD drawing can be sent to multiple manufacturers to assess the quality of the part and the costs associated with each processing or manufacturing technique. When it comes time for the production of the part, the electronic file of the CAD drawings can be given to a manufacturer that owns the specialized equipment that will be used to make the part. Each manufacturing or processing system (e.g., CNC mill, 3D printer, and lathe) will have its own CAM software that translates the CAD drawing into a format that the specific equipment can understand. It is important to understand that each system has its pros, cons, and of course, limitations.

9.7.3 3D Printing

3D printing is a great way to test a design before manufacture. After developing a CAD of your technology, part, or component, you can easily export the CAD drawing to a file format that any 3D printer can read. The way a 3D printer reads files is analogous to a ream of paper. If you have a 3D object, imagine it being sliced into a thousand slices that are all the same thickness. When the 3D printer reads the file, it reads one piece of paper at a time, then moves up to the next piece of paper, and so on. 3D printing takes a lot of time, and the way different materials bind to one another like this is still being studied, so 3D printed parts are not always good for large-scale production and/or practical for heavy use or use in extreme environments. What 3D printing is very good for is iterative designing of prototypes or projects where you need to make small adjustments to prototypes before sending out final designs to be manufactured.

For example, some common processing techniques (such as injection molding and thermoforming) require molds to be made. A mold is a large block of metal that has an inverse etching of the part imbedded into it. Systems that require molds often have very high manufacturing throughput resulting in inexpensive production of parts. That being said, the molds themselves often have an expensive upfront cost. For example, a mold can cost $25,000 to make, but once in use, the mold can produce 4,000,000 units a year for 2 years before a new one is needed. Not including the cost of manufacturing technicians, electricity to run the mold line, or the raw material for the part, each part produced by the mold costs $0.003125 to manufacture. The use of a 3D printed mold to verify that the parts will be the correct size and dimension will cost a fraction of that amount to verify and can be done in a couple of days, whereas repeating the etch of a metal mold can take a month or two and will cost tens of thousands of dollars.

Example

Evelynn built an α-prototype robot for servicing the elderly with the goal to interest investors. The robot can move, speak, listen, ensure it does not run over obstacles, take pictures, open the refrigerator to get food (or beverages), provide alarms, and most importantly, dispense medication. All of Evelynn's hardware components for the robot were purchased online. She had a modest budget, so she wanted to prove her concept at low cost. At the same time, she wanted to make sure she would not have to start from scratch with the next design iteration of the robot. Evelynn's step-by-step process included:

1. Mapping the functionality of the robot system
2. Outlining the software architecture
3. Developing database architecture
4. Sourcing the hardware
5. Reverse engineering the drivers for the hardware
6. Developing CAD drawings
7. Developing a BOM

Once she had mapped the functionality of the system, Evelynn was able to see the software architecture that she should build for (a) stability, (b) the addition of additional functionality, and (c) the ability to not change the database when upgrading to industrial-grade hardware components. By developing the system this way, she would only need to reverse engineer the drivers for any new hardware, without needing to change the database or the software controlling any other functionality.

When sourcing the hardware, she did her best to ensure that M4 and M8 screws with hex heads were used throughout the system. While keeping in mind Design for Manufacture (DFM), she found that she could use the uninterrupted power supply as a power regulator for the rest of the system and as a structural support for the circuit boards. Evelynn also conducted extra research into manufacturers of the servo motors, since they are a key component of the robot. As part of her Sourcing efforts, she called each of the servo motor manufacturers to see if she could:

1. get free sample servo motors,
2. convince the manufacturer to put sales contracts in place for bulk purchases at lower prices,
3. request the manufacturer's Data Universal Numbering System to conduct background credit checks on the manufacturers, and
4. receive drawings and specifications for the parts of interest.

Lastly, Evelynn realized that if all went well, she would need to hire people to help her assemble the robots. When developing the CAD drawings, she was mindful about DFM, when she considered that the plastic components would most likely be thermoformed if mass produced, so she made sure the CAD drawings did not have any structures that would need to be redesigned, when it was time to transfer to manufacturing. Additionally, with a mind toward DFA, she did her best to stage assemble the robot system in easy-to-follow steps that required a single drill bit and a single hex key for all of the components. When putting all of these components together, Evelynn created a BOM that included a list of parts, their identifiers, what the part was used for, costs, supplier, picture (if available), and so on.

In the above example, Evelynn thought about what the MVP would be and how the likely manufacturing processes would be implemented, when designing her early stage α-prototype. She knew that there would be design changes to her robot system, but she identified key parts and components early on and conducted the research necessary to reduce her future work.

9.8 Moving on to Product Development

Now you have an α-prototype that works. It will not necessarily be an aesthetically pleasing prototype, but it will demonstrate to investors that your concept is technically feasible and that, with concurrent Business Development activities, there is potential to commercialize your product. The next stage of development is to take the α-prototype and iterate it into the version that is stable enough for consumer testing, that is, Product Development. Product Development provides the means to formally develop your product design into an MVP that can be manufactured for commercial launch.

10 Product Development and Manufacturing

Abstract: This chapter continues to explore the processes that follow making an early-stage, physical prototype (α-prototype) (as discussed in Chapter 9). The next step in the Product Development Process, and the heart of this chapter, involves iterating the α-prototype into a stable beta (β)-prototype intended for user testing, then translating user feedback from β-prototype testing into necessary design changes. The end result of this portion of the Product Development process is the genesis of a Minimum Viable Product (MVP) that is ready for commercialization. Product Development, Design Freeze, and Transfer to Manufacturing are key steps leading up to Commercialization of a technology-based product. Importantly, the development of the physical product is concurrent with the development and organization of design history and manufacturing documentation, the continued sourcing of key components, and the Verification and Validation (V&V) testing that proves that your product does what it is intended to do and that it meets user and customer requirements. The first half of this chapter will be oriented toward the development of the physical product, and the second half of this chapter will focus on the documentation that coincides with Product Development and Manufacturing.

Keywords: Product development, minimum viable product (MVP), design freeze, market and product requirements, project plan, design reviews, verification and validation (V&V), transfer to manufacturing, design transfer review, technology development Strategic Implementation Plan (SIP), β-prototype, documentation, design controls, design inputs, design outputs, design changes, design history file (DHF), device master record (DMR), device history record (DHR), installation qualification, operational qualification, and performance qualification (IQ/OQ/PQ), revision history.

10.1 Product Development through Manufacturing

Developing a prototype is exciting. Making a product come to life is a true accomplishment. Once your concept has proved to be viable and your prototype shows promise (Chapter 9), the next step is to conduct Product Development. Product Development is the process of developing, testing, and transforming a prototype into a manufactured product ready for commercialization, also known as a minimum viable product (MVP). The overarching themes involved with Product Development are:

- **Iterative Product Changes.** Iterative product changes are physical changes (structural, software program, aesthetics, etc.) to the product itself.
- **Testing.** Testing protocols and experiments are used to validate and verify that the product works.

https://doi.org/10.1515/9783110521900-011

- **Design Controls.** Design Controls are structured processes and documentation used to identify and track changes or problems with the product, testing procedures, and manufacturing processes.

10.1.1 Minimum Viable Product (MVP)

An MVP is a bare-bones version of the device that (1) has passed every test to meet your user needs, (2) is ready to be manufactured, and (3) is ready to be introduced into the marketplace. The MVP is the final manufactured product and is the end goal of Product Development.

Product Development requires expertise in many fields. During the Product Development process, *every* strategic partner, vendor/supplier, or contract manufacturer can impact the success of a product launch. Due to the complexity of any Product Development process, there will always be a part, component, or project that will need to be outsourced. The more parties involved, the more likely it is that small details can be overlooked.

Example

As an example, a startup company contracted a manufacturer to develop the key component (an electrode) of their new medical device. The sourced electrode was incorporated into the startup's β-prototype medical device, which the startup had planned to use to conduct animal testing to satisfy validation requirements for regulatory submission. However, the contract manufacturer did not clean the leads of the electrode appropriately before sending the electrode part to the startup company, and the dirty electrode was incorporated into the startup's β-prototype medical device.

The animal testing results were very important to the startup. The data from the animal trials would determine if the startup could implement a product Design Freeze and begin the US Food and Drug Administration (FDA) clearance process for their medical device product. Unfortunately, since the contract manufacturer did not clean the electrode leads, the startup company got varied and unreliable results during the animal tests.

Since the varied and unreliable data was unexpected, the startup company looked carefully for the source or cause of the unexpected data. The startup was ultimately able to identify that the source of the varied and unreliable data was from dirty electrode leads on the medical device; however, identifying the source of the problem cost the startup tens of thousands of dollars and significant amounts of time. If the problem had not been caught, the startup company could have failed or would have wasted precious time, resources, and funds to find another contract manufacturer to make the key component, which could set the company back years in terms of getting their medical device to market.

The above example can be applicable for products and technologies developed in any industry. Every process, every component, and every vendor that is involved plays a part in the successful development and launch of new products and technologies.

10.1.2 Planning How to Develop a Product/Technology with a Technology Development Plan

A Technology Development Plan outlines the timeline for prototype development, design review, and verification and validation (V&V). It is more technically specific than the technology plan discussed in Chapter 4. The Technology Development Plan also defines the Market Requirements (i.e., what the market research determined is needed) and Product Requirements (i.e., specifications of the product). As Product Development and Sourcing (i.e., the sourcing of premanufactured parts/components, services, and/or contract manufacturing) begins, (a) acceptance criteria for each component and the product, and (b) the risk assessments of the product, manufacturing processes, and vendors will be defined in the Technology Development Plan. The discovery involved in each stage of the Prototype Development and Product Development stages (see Figure 10.1) will become more refined and sophisticated as the product becomes more developed. For example, Sourcing during Prototype Development may involve obtaining parts from the local hardware store, whereas Sourcing during Product Development may involve obtaining parts from a contract manufacturer.

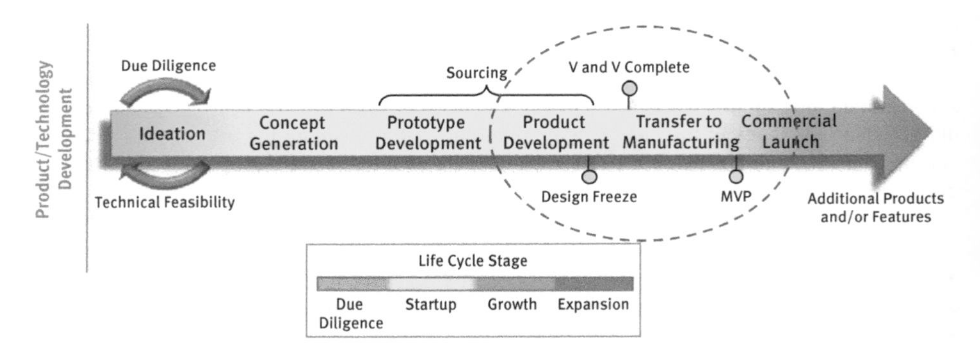

Figure 10.1: Product Development and Manufacturing stages of the Product/Technology Development timeline.

It is important to note that many actions and techniques in the early stages of Product/Technology Development overlap. In many of the early stages there will be gray areas, as the prototype is evolving into a product. The Technology Development Plan should outline:

- **Requirements** – What are the Market Requirements that need to be defined in order to understand the scale of production that will be required to meet demand? What Product Requirements need to be defined in order to determine what the product and its components are supposed to do on a whole system and subsystem level?

- **Acceptance Criteria** – What are the minimum performance standards for your product when the product is used as intended?
- **Costs** – What are the estimated COGS (e.g., sourcing costs for each part and component) and any other related costs of development (e.g., rates and estimated time requirements for a software development firm to develop a software driver for a piece of hardware)?
- **Risks** – What are the risks associated with:
  - Manufacturing Processes
    - Waste materials from the production process, or are there any resulting manufacturing byproducts from your product assembly that pose health and safety or environmental waste concerns?
    - Risks associated with manufacturing operator-intensive processes that cannot easily be automated, which leads to variability in the quality of the final product?
  - Product interactions with humans and/or the environment?
  - Vendors?
  - Sourced materials
    - Availability of existing components and/or technologies?
- **Utility** – Would potential prospective users and important stakeholders need this product? Specifically, make sure that the intended benefit(s) for the user are achieved (preferably in a fewer number of steps that require less time for the user than current solutions that are available on the market).
- **Usability** – How do the *usability and related human factors* of the concept affect the end user? Namely, make sure that the concept provides all the required functionality and form that is needed by the user.

Figure 10.2 depicts the steps, processes, and milestones for Product Development through to a design's Transfer to Manufacturing. The product development processes

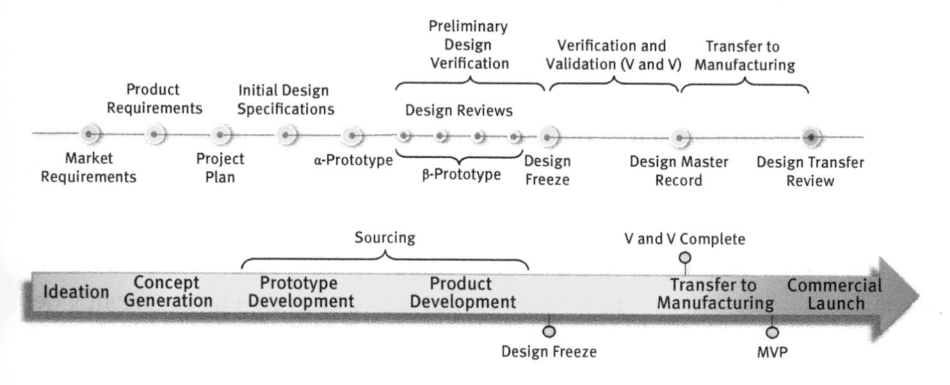

Figure 10.2: Product Development through Transfer to Manufacturing timeline.

outlined in Figure 10.2 are applicable to *any product or technology being developed in any industry*.

Market and Product Requirements – Market and Product Requirements are determined from conducting market research, when discovering your unmet need (see Chapter 9).

Project Plan – The Project Plan is the plan outlining what resources over time are needed to execute a project. There will be multiple Project Plans throughout Prototype and Product Development.

Initial Design Specifications – The Initial Design Specifications are developed after the best concept has been selected (Chapter 9). The selected concept(s) will determine the physical specifications of the system that fulfill the user requirements.

Preliminary Design Verification – Preliminary Design Verification is the process of evaluating your prototype to determine whether, and how, to proceed to late stages of Product Development. Once Preliminary Design Verification results in a finalized design, then the milestone of Design Freeze (Section 10.1.3.3) ensures that no additional changes to the design will occur prior to testing.

Design Reviews – Design Reviews (Section 10.2.1.3) are iterative review meetings that serve as milestones in the Design Control process, which formally document cross-functional team approval of various stages of Product Development and beyond.

Verification and Validation – V&V (Section 10.1.3.4) is the ultimate and final testing of a design before transferring to manufacturing. Completion of the V&V milestone is required, prior to proceeding to Transfer to Manufacturing (Section 10.1.4).

Transfer to Manufacturing – Transfer to Manufacturing is the process of translating a product design to the manufacturing processes that will produce the MVP. The development of the Device Master Record (DMR) (Section 10.2.1.10), the documentation for how to manufacture your product, ensures the ability to track, record, and introduce quality to your manufactured product. Once the product completes Prototype/Product Development and nears Design Freeze, then a Manufacturing Plan must be developed. A Manufacturing Plan outlines the tasks and timeline for transferring the design to manufacturing and the Design Transfer Review.

Design Transfer Review – The Design Transfer Review is the process of verifying that the design that was verified and validated remains so, after it has been translated to scalable manufacturing systems.

Once the Design Transfer Review has been completed, the MVP is ready for commercialization. The Business Development team should have been concurrently developing marketing and sales plans and implementation strategies to promote and sell your MVP. Transforming a concept to an MVP is an achievement that takes an incredible amount of time, resources, and planning.

10.1.2.1 Strategic Implementation and Project Plans

In Chapter 7, we discussed the purpose of having a Strategic Implementation Plan (SIP). Similarly, your technology development efforts require an SIP, namely a Technology Development SIP. A Technology Development SIP is a detailed, step-by-step plan for how to execute the development of your technology. The Technology Development SIP should outline each stage of Prototype Development [proof-of-concept (POC) and α-prototype, discussed in Chapter 9] and Product Development (β-prototype through MVP, which are discussed more in Sections 10.1.3.1 and 10.1.4). Identify who is involved at each stage, identify what resources are needed, and explain how you are going to verify that your product has been developed to satisfy your user requirements.

Between the earlier development of the technology plan (discussed above and in Chapter 4) and the multiple iterations of due diligence, most of the information you will need in order to be able to write your Technology Development SIP should be readily available to you.

From the Technology Development SIP, your team can create a Project Plan. A Project Plan is a component of the Technology Development SIP that provides the step-by-step details needed to execute the Technology Development SIP. The Project Plan will outline:

- What tasks/milestones need to be accomplished.
- Who is responsible for accomplishing each task/milestone.
- Timeframes for each task/milestone.
- What tasks/milestones are dependent upon one another.
- Costs associated with each task/milestone.

The Technology Development SIP will be specific during the early stages of Product Development and vaguer during later stages, due to unexpected changes that will undoubtedly be encountered during product development. In most cases, the team will simply not know all the Product Development details until later on during the project. This is referred to as the “rolling wave” principle in project management, because you can only see the details that are directly in front of you, due to the ambiguity of information over time. As each stage progresses, the Technology Development SIP can be updated (e.g., monthly and quarterly).

The Technology Development SIP should contain information about facilities, materials, equipment, personnel, storage, logistics, and any other processes necessary to develop the prototype, the product, and for facilitating the manufacture of the MVP. Additionally, the Technology Development SIP will outline what systems will be required for manufacturing. There can often be long lead times for capital equipment purchases (e.g., purchasing a $100,000 injection molding system), the equipment installation and equipment qualification (IQ/OQ/PQ is discussed in Section 10.2.1.12), and training of personnel. Alternatively, if you are a startup utilizing a contract manufacturer, you may not need to purchase your own capital equipment for the manufacturing assembly line until after process validation or prior to commercial launch.

Developing a detailed Technology Development SIP reduces the risk of getting sidetracked. The Technology Development SIP will help Product Development stay on task by highlighting where resources are being allocated. Questions like, "What is the core feature of the technology?," "What are the 'must-have' requirements that have to be met for commercial launch?," and "What are the ancillary, or 'nice to have,' features of the product that cannot delay commercial launch?" can allow a technology project manager and/or leadership to identify what is of importance.

10.1.2.2 Risk Management and Risk Analysis

Risk Management is the forecasting and evaluation of potential risks surrounding the use of the product, the design of the product, and the manufacturing of the product. Risk Management includes the identification of industry-recognized standards and compliance with internal procedures to avoid or minimize the impact of potential risks.

Example
For general manufacturing, ISO 9001 incorporates Risk Management into the ISO 9001 standard itself, whereas medical devices have their own standard, ISO 14971, "Medical Devices – Application of Risk Management to Medical Devices."

Risk Analysis involves coordinated communication between the R&D, Manufacturing, Quality, and, if necessary, Regulatory departments to identify critical concerns throughout the design stages. The identification of potential risks leads to the identification of the specific tests that can test the design or process against the potential risk, thus "qualifying" the design or process. Risk Analysis investigates potential failure modes, the effects of the failure modes, and the causes of the failure modes. Risk Analysis asks:

1. What are the different ways the product or process can fail?
2. What happens when the product or process fails?
3. Why does the failure occur?

Failure modes and effects analysis (FMEA) is a common risk analysis tool used to assess product risks. FMEA breaks down a product or process into a set of failure modes, that is, physical descriptions of how a product or process may not function as expected. An FMEA should identify the reliability and safety of critical components and materials that form the product. An FMEA provides quantitative ranking of potential failure modes (which can be tracked using Pareto Analysis, Chapter 11) and methodologies to track changes or improvements to a product and/or process. Table 10.1 outlines the content often found in an FMEA report.

Table 10.1: Example outline of an FMEA report.

Title	Description
Item Function (or Manufacturing Process if you are describing a Process FMEA)	The name and number of the part/component being analyzed and the function of the design being analyzed (e.g., motorized sample tray extension/retraction).
Potential Failure Mode	The description of which part, subsystem, or system could fail its design specification and/or purpose (e.g., electric motor could fail).
Potential Effects of Failure Mode	Description of how the failure would affect the customer (e.g., a broken electric motor would lead to the inability to load samples in the sample tray).
Severity	Severity (shown in a scale that is company specific, e.g., 1–5) indicates the degree of seriousness of the potential failure mode on the component, subsystem, or customer.
Potential Causes/ Sources of Failure	Potential Causes/Sources of Failure are the identified flaws or weaknesses of the design that would facilitate the failure (e.g., electric current to motor could not be regulated, thus burning out the electric motor over time).
Likelihood/Occurrence	Likelihood/Occurrence (shown in a scale that is company specific, e.g., 1–5) indicates the estimated number of failures that could occur per individual Potential Failure Mode over the design life.
Design Controls (or Process Controls if you are describing a Process FMEA)	List of applicable design controls relative to the Potential Failure Mode. There are three types of controls to take into consideration: (1) controls that prevent the failure from occurring, (2) controls that detect the failure, and (3) controls that detect the failure and lead to corrective actions.
Detection	Detection (shown in a scale that is company specific, like a scale from 1 to 5) is the ability to track and/or detect the design weakness as early as possible.
Risk Priority Number (RPN)	The RPN (shown in a scale that is company specific, like a scale from 1 to 125) is a compounded score of design risk based on the Severity, Occurrence, and Detection. The RPN is used to rank the importance of Potential Failure Modes. The higher the RPN, the more likely a corrective action will be needed.
Recommended Action	The recommended action(s) to address the Potential Failure Mode(s).
Responsible Party	The individual, department, or organization responsible for the risk mitigation and/or corrective action.
Action Taken	Description of completed action and the effective date of when action is taken.
Revised RPN	After the Action Taken has been completed, the RPN is revised/ recalculated.

10.1.3 Product Development

Product Development is the process of transforming an α-prototype (refer to Chapter 9) into a system that can be tested by potential consumers, or a Beta (β)-Prototype.[31] A β-prototype undergoes testing to verify and validate (Section 10.1.3.4) the refined system to ensure the system meets user requirements, product requirements, and acceptance criteria. β-Prototypes are used to conduct initial informal V&V testing and ultimately get to Design Freeze. In keeping with the donut example introduced in Chapter 9 and shown in Fig. 9.12, Figure 10.3 depicts the stages of Product Development. Figure 10.3 shows the transition from (A) to (B) as the process of selecting the most promising -prototypes for development into β-prototypes to test with consumers.

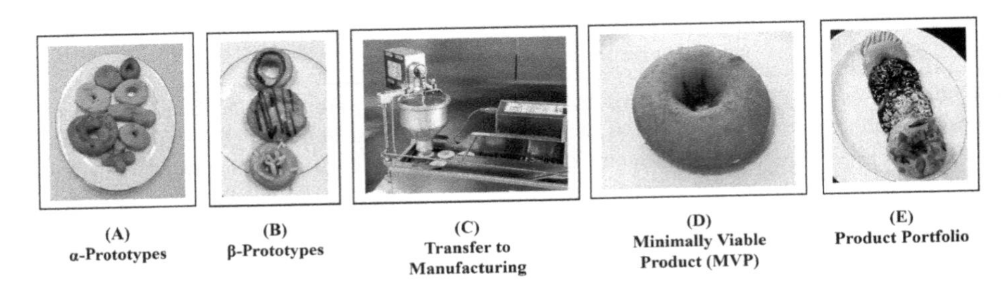

Figure 10.3: Progression of the Product Development of a donut depicting (A) Early-Stage (α)-prototypes, (B) Beta (β)-prototypes, (C) the Transfer to Manufacturing Process for donuts, (D) a Minimum Viable Product, and (E) Product Portfolio.

Product Development involves taking an early-stage prototype that has demonstrated its technical feasibility, and then transforming it into a version that is stable enough (e.g., the design is sturdy enough for handling, software will not crash and does not need to be reset between uses, and/or the system can be used by multiple users without failure) for field testing (i.e., the use of the product/system in its intended environment with prospective users). Product Development encompasses the development of the β-prototype, which is then subject to Design Freeze and V&V.

10.1.3.1 Beta (β)-Prototype

The β-prototype is the third iteration of the prototype (the POC is the first iteration, and the α-prototype is the second iteration). The β-prototype serves as the "works-like

31 The POC, α-, and β-prototypes are analogous to the "works-like, looks-like, built-like" model, where POCs are "works like," α-prototypes are "looks like, works like, but not made like," and β-prototypes are "looks like, works like, and made like." For more information, you can refer to Chapter 9.

and looks-like *and* built-like" model of the prototype. During this third iteration, the β-prototype is aesthetically refined once again, the functionality of the β-prototype is refined to a point that is ready for prospective user feedback, and the manufacturing process is improved and solidified (i.e., the determination of which manufacturing processes will be the best candidates for making the final product). The parts/components for the β-prototype are made and sourced with the intent to optimize and finalize the manufacturing processes, and the parts/components are provided by prospective suppliers (i.e., vendors).

β-Prototypes are often tested with prospective customers in the intended use environment (i.e., the β-prototype is tested by customers who use the prototype in the way that the product is meant, or intended, to be used) to confirm reliability and to uncover additional problems (such as minor issues with the ease of usability of the design of the prototype). Depending on the type of product, alternative β-prototype designs may be tested with customers to conclude which version of the β-prototype design will proceed to the next Product Development phase.

Lastly, the β-prototype iteration is where the development of the individual components with the intention of manufacturing is conducted. The Bill of Materials, or BOM (Section 10.2.3), is established and each component is itemized and assessed to determine cost effectiveness, likeliness to meet quality specifications, and quantity of parts/components needed to produce each product/device.

To add to the complexity of Product Development, the requirements for your β-prototype – as compared to the requirements for your MVP – will most likely change during the Product Development process. This is because the process of developing a product is, in a sense, still a discovery phase – you will learn so much from actually building the prototype, using it, and allowing others to provide feedback on the prototype. For instance, as voice of the customer surveys and interviews are conducted, the product requirements, risks, acceptance criteria, and your understanding of important human factors that will influence the design of your product will unfold before you. This process will continue and become more refined throughout the Prototype Development, Product Development, and Transfer to Manufacturing phases of the Product/Technology Development process.

Planning for Manufacturing During Product Development

Part of the β-prototype development process is looking forward to, and planning for, the Transfer to Manufacturing conducted just after the Product Development process. Product Development is not only the physical creation of the product or device, the development and organization of documentation, and where formal testing begins, but it is also where an engineering team plans for the later manufacturing of the product or device. The engineering team must formally identify the parts or components of the design that will be manufactured in-house by your company and which parts or components will be outsourced early in the Product Development process.

"Manufacturing a product" can mean that (a) every part, component, and the final product, device, or system is made in-house, (b) some parts or components are made in-house, while others are sourced, but the final product, device, or system is assembled in-house, or (c) every part, component, and assembled system is outsourced, but the final inspection and approval is conducted by in-house. The approach will be company dependent.

There are often multiple manufacturing techniques that can be used to produce the same part. There are hundreds of companies that can produce the same component for you. This is why the process of Sourcing and Vendor Selection is so important. When considering each part of a design, the decision will need to be made whether to manufacture in-house or outsource. As with vendor selection, deciding whether your company or a vendor will be able to address your needs will depend on:

1. Cost – What is the cost per unit of each part or component?
2. Volume – Can the vendor produce the necessary quantity of parts/ components?
3. Quality – Can the vendor maintain the quality of parts you require?
4. Reliability – Can the vendor get your parts to you when you need them?

When assessing *costs*, factors such as space, employees (e.g., cost of wages and benefits), and administrative costs (e.g., administrative support, human resources, and utilities) all factor into the calculation. Though many processes are specialized or developed internally by a company, there are many common manufacturing techniques, such as injection molding, thermoforming, extrusion, casting, and milling, that can be outsourced. Having a vendor manufacture a part can often be less expensive due to space requirements, personnel, administrative costs, or a combination thereof. More importantly, a vendor's expertise may enable the vendor to produce quality units more quickly than you may be able to in-house. Lastly, the cost of capital equipment adds up. For a startup company, it is common to use a contract manufacturer to produce parts until a certain amount of revenue has been reached, so that the process can be brought in-house. Manufacturing financial models can be used to assess and compare the cost of production. Financial models for estimating manufacturing costs should take into account:

- Labor – Wages and benefits and administrative personnel costs.
- Cost of Space – Cost of the rent, lease, purchase of property.
- Utilities – Gas, electric, water, sewage, waste disposal, and so on.
- Administrative Costs – Printer paper, staples, pens, software, and so on.
- Capital Equipment – Cost of purchase and/or lease and cost of maintenance.
- Raw Materials – Cost of goods sold (COGS).
- Time – Time needed to produce a specific quantity.

Case Study

A company produces a blood analyzer system that includes a disposable component. The company sells approximately 10,000 blood analyzer systems and 4,000,000 units of the disposable component annually. The blood analyzer system is a technological feat and is very complex. To add to its complexity, the disposable component consists of a specialized capillary tube designed to hold a blood sample. Once the blood sample is placed in the disposable tube, a plastic float and cap are added to the disposable tube. The cap seals the disposable tube, while the plastic float separates out a layer of the blood sample so that the system can analyze the sample to generate data about the sample. The plastic float needs to have very specific dimensions, shape, density, color, brittleness, and transparency (i.e., the design requirements); otherwise, the blood analyzer system cannot collect accurate data. Over time, multiple iterations, or models, of the blood analyzer system have been produced, while the design of the disposable tube has stayed the same. Currently, the company sells four different blood analyzer system models that all utilize the same disposable capillary tube component.

The company produces the float component in-house using an injection molding manufacturing process. The company relies on a sole-source supplier (Vendor 1) of a raw material used for making the plastic float of the disposable tube component. Vendor 1 informed the blood analyzer company that it would no longer be able to supply the essential polymer component (polymer A) that is needed to make the float. Proactively, the blood analyzer company purchased 5 years' worth of the supplier's remaining stock of the raw material, so that the company would have time to find a replacement vendor.

The blood analyzer company is faced with a problem: it now has a limited supply of a key polymer that is essential to the production of the float component. To solve this problem, the company looks to Design Control documents to understand how the float component was developed and is currently manufactured. The Design History File (DHF) (Section 10.2.1.9) mapped out the thought processes behind the development of every component of the blood analysis system, including the float component of the disposable tube. The information contained in the DHF was especially important with regard to how the float component was manufactured, since the disposable tube product line was purchased from another company 5 years prior (the blood analyzer company adopted the currently employed injection molding system and processes from the other company). The DHF also noted that the manufacturing process for making the polymer mixture used in the injection molding process was performed by a vendor located over 1,600 km away.

Further review of the DHF revealed that in order to make the plastic float component of the disposable tube, pellets of polymer A are purchased from a first supplier (Vendor 1), along with pellets of a second type of plastic (polymer B) having a specific concentration of red dye from a second supplier (Vendor 2). Polymers A and B are then sent to a third vendor (Vendor 3) to be ground and mixed into a fine powder. The fine powder mixture of the two polymers is then shipped back to the blood analyzer company for use in its injection molding system.

Once the processed mixture of ground polymer plastics arrives in-house, the powder is put into the company's injection molding system. The injection molding system melts the plastic powder into a liquid and then injects the molten liquid plastic into a stainless steel mold. The molds are expensive, costing approximately $25,000 per set.

The DHF noted that each of polymers A and B has its own material characteristics, or behaviors, such as cooling rate and how quickly the polymer shrinks during cooling. Ideally, the company would likely to be able to use the same molds and same (or similar) manufacturing process. Since every polymer is different, the company is concerned that by replacing Polymer A with a new polymer that additional, unexpected problems may be introduced into the manufacturing process. For instance, a replacement polymer may not shrink at the same rate as polymer A, thus needing to develop new molds. As costly as molds are, transitioning to a replacement plastic for polymer A may require that the molds be redeveloped to ensure that the size of any plastic floats produced by the injection molding process using a replacement polymer will be consistent with the size of plastic floats made from polymer A. Another potential problem would be the introduction of static charge, which would affect the degating (the removal of the individual floats from the injection molded rack).

The company decided to try injection molding the float component with a replacement polymer. The first two V&V attempts at injection molding the float component with the replacement polymer failed. The first attempt at a replacement float resulted in large deviations in results between all of the analyzer systems. In the second attempt to meet all of the requirements, they found the replacement floats to be too brittle, so clinical staff had difficulties using them. The engineers used the techniques outlined in the DHF to replicate the manufacturing process without success. When the company had only a year's worth of polymer A supply remaining, the company brought in a consultant for a final attempt to design a solution to the polymer A shortage problem.

The consultant assessed the requirements, manufacturing processes, and manufacturing equipment needed to make the float component. The consultant started by conducting interviews with the Management, Quality department, Shipping/Receiving, Purchasing, and most importantly, the technicians who produced the float component.

As a result of these interviews, management was able to provide a breakdown of the COGS, that is, the cost of resources for production (equipment and personnel) of the float component. The costs affiliated with the in-house injection molding technique included:

- Polymer A, which was no longer available for purchase
- Polymer B
- Cost of transportation of polymers A and B to Vendor 3 for grinding and mixing
- Cost of grinding and mixing the polymers
- Cost of transportation from Vendor 3 back to the company

As part of assessing the problem, the consultant also evaluated the cost of using alternative manufacturing techniques for making the float component. Switching manufacturing techniques was another potential solution, since the company already had a thermoforming system onsite. If the blood analyzer company did not already have the thermoforming equipment, then it would have been too costly to implement due to the cost of the capital equipment, installation, training, and lost time that would accompany the switch to a thermoform system. Additionally, in this case, the cost estimate for new molds for the thermoforming system would be $50,000.

The process for making the float component by thermoforming would involve heating the polymer mixture to form a malleable plastic and then stamping malleable plastic into the desired shape. A thermoforming process would be able to produce a float component that is of the same size and shape that is necessary and would be practical for producing the designed quantity of product (e.g., 4,000,000/year). The only drawback in using a thermoforming process is that the resulting thermoformed float components would have ridges that would have to be manually removed, thus adding to the processing costs.

The consultant identified another alternative manufacturing technique that could be used to make the float component – an extrusion process. Plastic extrusion involves pushing molten plastic through a die to form a long piece that can be cut (like pushing toothpaste out of its tube). Looking into the extrusion process option further, the consultant determined that while an extrusion technique could be used to produce the float component, there is too much variation along the edges of each cut that is made to the extrusion.

One final option that the consultant considered was replacing polymer A with a suitable replacement polymer. The consultant researched different plastic manufacturers and importers. The consultant found four plastics that were potential replacement candidates for polymer A. After multiple small-batch verifications with Design Review meetings (Section 10.2.1.2) to ensure the replacement polymer candidates met the manufacturing requirements, the optimal replacement plastic was selected. Fortuitously, the selected replacement polymer was also able to reduce processing costs. Once the calculations were made comparing the original costs of manufacturing the float component (processing and materials) to the potential alternatives, it was determined that continuing with in-house injection molding was the most cost-effective technique for the blood analyzer company.

The final outcome was that the consultant found a replacement plastic that met all of the manufacturing requirements of the polymer mixture (i.e., the consultant found a single plastic material that could realistically replace the polymer A and polymer B mixture) and found a supplier that agreed to be a contract manufacturer of the required red dye needed to make the float component the appropriate color for use in the company's blood analyzer systems. The dye supplied by the contract manufacturer was mixed directly into the replacement polymer, which removed the need for grinding and mixing. Once the replacement plastic was agreed upon as a replacement material, V&V studies (Section 10.1.3.4) were conducted to ensure that the replacement plastic met the requirements.

These changes to the manufacturing process associated with the production of the float component removed the need to purchase multiple polymer materials and send out the polymers to the grinder, which dramatically lowered the overall cost per unit of the raw plastic material.

The blood analyzer company put into place a contract with the new supplier/contract manufacturer guaranteeing a certain price and quantity of orders annually, in exchange for meeting the detailed specifications provided by the blood analyzer company. Specific Quality (Chapter 11) measures were included in the contract. Inspections and batch testing would occur with each new lot coming from the supplier to ensure consistency of the blood analyzer company's end product.

In addition to any technical challenges being addressed during the process of translating a β-prototype to an MVP, there will always be implementation obstacles that arise during Product Development through Transfer to Manufacturing (such as a shortage of a key material, as demonstrated by the above case study). Challenges often arise when integrating individual components of a β-prototype into a completed, scalable system (the completed, scalable system will be used to produce the final product), while ensuring quality and low COGS. The above case study demonstrates the intricacies of manufacturing. The analysis and approaches in the above manufacturing case study will be similar, whether utilized during Product Development (i.e., development of the β-prototype) or during Transfer to Manufacturing.

To add to the sourcing and Quality Management (Chapter 11) complexity of Product Development and Transfer to Manufacturing, most final products are comprised of at least some preexisting parts. Parts can be purchased from an original equipment manufacturer (OEM) (i.e., in bulk from the manufacturer) or from suppliers (e.g., companies that work with a lot of OEMs to resell parts, components, and equipment). By working directly with an OEM, a company can be provided parts/components in large quantities, usually at a lower price, and a contract between the OEM and your company can be established, guaranteeing individual part/component specifications.

Example

Consider the production of a new car. While the car, as a final system, is assembled by a final car manufacturer, all of the various parts and components that make up the final car are sourced from suppliers that are part of an integrated supply chain network, or OEMs. The metal for the frame of the vehicle must first be mined and then processed into usable metal sheets by a metal producer, and then those metal sheets must be provided to the car manufacturer. The nuts and bolts used to assemble the vehicle frame are most likely standardized and sourced from a separate vendor. The battery for the car could come from a company that specializes in supplying vehicle batteries. The windshield for the vehicle might be developed by an overseas supplier. And so on.

Getting everything to fit together and work properly in a final product is a huge undertaking, especially when you are creating a product that is made up of hundreds, thousands, or hundreds of thousands of parts that are sourced from different factories, vendors, or manufacturing processes. Even though all of these parts are supposed to fit together and work perfectly with one another, this is not always the reality of producing a final product from sourced parts. That is why technical due diligence during Prototype Development and Product Development confirms that (a) components can be incorporated into a complete system solution, and/or (b) system performance and operation are met under the anticipated operating scenarios.

10.1.3.2 Preliminary Design Verification & Validation (V&V)

Preliminary Design V&V is the process of developing, building small batches of product, and conducting small, structured studies to collect data verifying and validating the β-prototype design. The data collected from these studies is used to refine the β-prototype in preparation for going to Design Freeze. The Preliminary Design V&V also provides an opportunity to revise the studies themselves and should be used as a basis for the formal/final V&V, which occurs after Design Freeze. Preliminary Design V&Vs are conducted for the benefit of your company to determine the best path forward for Product Development, whereas the "formal/final" V&V is the set of final testing documentation.

The purpose of conducting Preliminary Design V&V is to ensure that all of the potential problems with the β-prototype design are addressed before the design is finalized for Design Freeze.[32] During the Preliminary Design V&V and final V&V processes, V&V Strategic Plans or V&V Qualification Plans are used to organize and

32 Propagating a faulty, flawed, or unsuccessful design to Design Freeze can result in failure of the product at its formal V&V testing stage, which translates to misappropriated investment, and the possibility that requests for additional investor funding may need to be made. To avoid this scenario, the Technology Development Strategic Implementation Plan (SIP) needs to include these Preliminary Design V&V to safeguard the investment that is required to execute the V&V.

record testing performed on a product. Table 10.2 is an example of the contents that go into a V&V Qualification Plan during the V&V processes.

Note: The structure of the Qualification Plan for the Preliminary Design V&V is very similar to the structure of the Qualification Plan for the final Design V&V; the significant difference is the scope and extensiveness of the final V&V.

Table 10.2: Example V&V Qualification Plan outline.

Title	Description
Test Number	The identifying number of the test that needs to be performed.
Test Identifier	An identifying number correlated to the test name for the purpose of reference.
Test Method	A description of the test that needs to be performed.
Acceptance Criteria	The identified criteria that must be satisfied for a product to meet specifications.
Requirements	The requirements that must be satisfied for a product to meet specifications.
Specifications	The specification that must be met in order to pass the test.
Test Location (Planned)	The location that has been identified where testing will be performed.
Test Phase (Planned)	The phase of development in which the test will be performed.
Sample Size (Planned)	The sample size that is planned for the test.
Sample Type	The type, or model, of sample to be tested.
Test Duration (Planned)	The estimated/determined duration of the test.
Start (Planned)	The planned start date for the test.
End (Planned)	The planned end date for the test.
Assigned To	The person, division, or department responsible for performing the testing.
Notes (Planned)	Any notes identified during the planning or writing stage of the Qualification Plan.
Test Report #	The identifying number of the report for the completed test. This could be the number of times the test was performed, version of the test, etc.
Test Report Identifier	A short descriptive name.

Table 10.2 (continued)

Title	Description
Status	The status of the test report. As tests get revised, older versions are often still kept as reference. This field would state whether a test is "active" or was "replaced by."
Start (Actual)	The start date for the performed test.
End (Actual)	The end date for the performed test.
Sample Size (Actual)	The sample size that was used during testing.
Test Results	The quantitative and qualitative results obtained from the testing.
Completed By	The person, division, or department who performed the testing.
Notes	Any notes related to the performed testing.

10.1.3.3 Design Freeze

Design Freeze is the point in Product Development, where a binding decision needs to be made concerning the final design of the product.[33] Effectively, Design Freeze is the process of stating "no more changes, this is what we are doing!" The Technology Development Strategic Implementation Plan needs to include this milestone. Once the Design Freeze is approved by a cross-functional team of technical, Manufacturing, Quality, and Regulatory leadership, the final product design will undergo V&V testing. Design Freeze is an important milestone in Product Development, because if changes to the design need to be made while conducting the V&V, then the V&V process will need to be repeated, which can be very costly and inefficient. Once the product has passed the V&V, then the design is transferred to manufacturing so that the MVP, which is mass-produced, can be commercialized. When the V&V is completed and is successful, then Product Development can shift to Transfer to Manufacturing, where the MVP is completed and prepared for commercial launch.

10.1.3.4 Verification and Validation (V&V)

V&V are the processes of *verifying* that your product or service meets the design specifications and *validating* that your product or service satisfies the needs of your intended users. Though referred together as the V&V, Verification and Validation are independent studies intended to give a "stamp of approval" to your product design. Verification ensures that all of the design and functional requirements and specifications are being met by the β-prototype design, whereas Validation ensures that

33 Quality standards such as ISO 9000 or ISO 13485 require a "freeze point" to distinguish between the design phase and change implementation.

the β-prototype design satisfies all of the user's needs. In regulated industries (e.g., industries that require ISO 9001 or ISO 13485:9001 conformance), validation and verification are conducted through a V&V Qualification Plan (refer to Table 10.2).

The V&V adapts the procedures and protocols from the Preliminary Design V&V tests into a final set of comprehensive studies to qualify your product for production. Verification Plan and Validation Plan are developed to determine why V&V studies are being performed, what procedures and protocols need to be referenced, and what resources are needed to execute the studies.

The Verification Report and Validation Report use the framework of the respective V&V Plans to document (in past tense) the execution of the Plans and include the results from actual testing. Figure 10.4 provides an example outline of the content that would be included in a Validation Report. The structure of the Verification Report would be similar to the structure of the Validation Report, though the testing associated with the Verification Report would be focused on design requirements and specifications rather than user needs. The V&V final reports (a Verification Report and a Validation Report) comprehensively assemble formal testing into qualifying documentation, showing that your product performs as intended.

Validation of [INSERT PROJECT NAME HERE] –
Protocol No. [INSERT PROTOCOL INDENTIFIER HERE], Original

Table of Contents

		Page
1.	Purpose	3
2.	Scope	3
3.	Background	3
4.	Responsibility	3
5.	References	3
6.	Acceptance Criteria	4
7.	Drawings	4
8.	Material	4
9.	Procedure	5
10.	Data Sheet(s)	6
15.	Results	7
16.	Records	8
17.	Revision History	9

Figure 10.4: Example Validation Report table of contents.

10.1.4 Transfer to Manufacturing

Transfer to Manufacturing consists of (a) the process of taking a developed product and adapting that product to be manufactured on a scale that is appropriate for your market, (b) testing the product designs on the manufacturing systems, and (c) the execution and organization of Design Controls (Section 10.2.1) documentation. Once management approves the Design Freeze, the drawings and documentation from the finalized β-prototype are then transferred to the systems that will manufacture each component. The final product that is produced using manufacturing techniques is the MVP.

Transfer to Manufacturing is a large undertaking. The parts, components, sourcing, and production methods used to develop the β-prototype will change significantly, when developing the manufactured MVP. The transfer of the design specifications requires that every individual component and the whole system needs to be verified and validated and compared to the product that is specified in the Design Freeze.

Vendor (or Supplier) selection (discussed more in Chapters 7 and 11) is of utmost importance. The developed product design needs to be adapted into scalable manufacturing processes, so that hundreds if not hundreds of thousands of units can be produced. The selection of a vendor for the purpose of adapting the product design for manufacturing will occur based on the importance of a vendor to your supply chain and the capabilities, cost, and reliability of your vendors. During Transfer to Manufacturing, there are two teams that need to communicate effectively with one another: The Product Development team and the Manufacturing Team. The two teams need to work together to seamlessly implement the mass production of a design, even though they may not be part of the same company (in the case of a contract manufacturer). This is not easy to accomplish. The well-planned design transfer process will be ineffectual, if it does not receive the necessary support from both teams. Common problems that arise during design transfer include:

- The Product Development Team does not finalize the design, even though Design Freeze was implemented.
- Business managers and technical experts underestimate the amount of time, effort, and resources required for design transfer.
- Manufacturing processes are not developed, documented, or validated.
- Capital equipment needed for production is not purchased, installed, or qualified in a timely manner.
- Manufacturing personnel do not have enough training on the manufacturing processes and/or on how to operate the required capital equipment.
- The Product Development Team has not developed approved vendor/supplier list(s).

- The Purchasing (or Quality) Department does not verify and approve new vendors.
- The Product Development Team does not address all costs, reliability issues, or safety issues before transferring to the manufacturer.
- Lack of coordination between the Product Development team and/or Manufacturing team with the Purchasing Department results in delays due to receipt, inspection, and release of production components.

Transfer to Manufacturing includes a process known as Design Transfer Review, which ensures that your mass-produced product meets the requirements and specification stated within the documentation associated with the product's Design Freeze. As you progress through the Product Development process toward Manufacturing, the collection and organization of the information surrounding your product will need to become more formalized and controlled.

10.2 Documentation

Research testing and documentation are critical to the Product Development and Manufacturing processes. The end goal of documentation is to ensure that each individual component of the product or service you are developing and the final product or service you ultimately offer for sale will function effectively (Verification) and will meet the needs of users and customers (Validation). Ensuring consistency relies on being able to identify, track, and remedy any problems, changes, discrepancies, or deviations that may occur during a manufacturing process. The processes and documentation that are developed during Product Development are the foundation to the documentation and processes used during the Transfer to Manufacturing and Manufacturing. Documentation is organized and monitored by a company's Quality Department (Chapter 10).

It is important to note that the processes and structures discussed in this chapter align with the development of a physical medical device regulated by the US FDA (unique aspects of bringing medical devices to market are discussed in Chapter 12). Medical device regulatory requirements are used as a "gold standard," but any industry should be able to benefit from the overarching concepts discussed in this chapter.

Medical device manufacturers operating in the United States need to follow Good Manufacturing Practices, according to 21 Code of Federal Regulation Parts 808, 812, and 820. Medical device manufacturing environments need to adhere to specific conditions that are carefully monitored and documented with regard to temperature, humidity, and the presence of particulates. Required safety measures for manufacturing operators would be documented within the manufacturing procedures and work instructions. Safety measures include antielectrical discharge mats, ventilators, chemical fume hoods, chemical storage units with materials safety data sheets, sharps bins, and personal protective equipment (such as gloves, eyewear, gowns, headgear, and lead aprons to prevent exposure to ionizing radiation). Also, medical device manufacturing facilities, manufacturing processes, manufacturing equipment and fixtures, manufacturing test methods, and manufacturing inspection methods (for instance, what constitutes a pass or fail product and how precise and accurate is failure detectability) must all be qualified, and these qualifications are documented.

Medical devices that are built for verification testing, validation testing, and for human use need to be monitored and documented during manufacturing. Manufacturing device history records (DHRs), including device batch/lot records, sterilization exposure records (if required), and shipping records, need to be stored and retained for a designated time period that is longer than the device's shelf life, according to medical device manufacturer's quality management procedures.

Manufacturing of medical devices is very precise and critical to device performance. Refer to the following examples.

- During the molding extrusion process, there may be inconsistency with surgical tube wall thicknesses, for instance, when introducer sheaths, catheters, and dilators are in contact with guidewires at the right angle, burrs may form at the most distal edges of these tubes, which could break off, leading to foreign particulates being introduced into the bloodstream.
- In-process quality can be very detailed. For instance, operators must exchange gloves between subassembly procedures to minimize polymer residue transfer. When silicone is involved in the manufacturing process, silicone can transfer from an upstream subassembly (for instance, one that utilizes silicone-based mold release) and contaminate a downstream subassembly (where polymer residue ends up on an exposed implanted surface).
- Curing processes (processes that utilize heat or ultraviolet energy to seal bonds) need to be optimized, so that results are reproducible and not operator dependent.
- Some processes are so delicate and require operator movement too small in scale to automate, so these processes are performed by hand, for example, precise stitching of nonwoven graft material for stents. These processes require transparent, detailed inspection criteria. The processes are so operator dependent that tribal knowledge needs to be captured in the work instructions; the more pictures and colorful examples, the better.
- Additional inspection steps may be required at first to execute a new manufacturing process, until operators are trained to know which details are critical. For instance, wires or connectors need to be placed in specific crevices/nooks in device enclosures to ensure that in the preceding final assembly there are no issues when powering up devices.

10.2.1 Design Controls

Design Controls are part of a company's Quality Management System (QMS) (Chapter 11). Design Controls are the collection of documents that explain the life (in terms of the Prototype and Product Development) and life cycle (in terms of Manufacturing) of a device. Design Controls apply to all changes, improvements, problems, to the device design itself, or any of the manufacturing processes related to the development of the device. The information needed to manufacture a product is typically comprised of documents like test plans, CAD drawings, inspections, test specifications, training materials, test reports, and manufacturing instructions. Since the Transfer to Manufacturing stage is often conducted by different departments and/or organizations, there is a need for effective knowledge transfer. This is why Design Controls have been developed. Design Controls are the transfer of knowledge in the form of structured, organized documentation. Design Controls contain all documentation, from the prototype development to manufacturing records. Table 10.3 lists Design Control documents specific to medical devices.

Though the documentation is cumbersome, this documentation provides the means to find the back-story when a problem occurs and provides the opportunity to assess cost-effectiveness and process efficiencies. Variations of Design Control documentation exist in other industries (such as microchip and automotive manufacturers) besides the medical device and pharmaceutical industries. Additionally, there are Design Control Plans and Reports. Often, within an organization there are multiple people, departments, and companies involved in Design Control. A Plan is developed by your company to outline what is needed and why (in future tense). The Report uses the framework of the Plan to outline its purpose, then documents (in past tense) the execution of the Plan, and includes the results from actual testing. Both Plans and Reports are kept together.

10.2.1.1 Design and Development Planning

Design and Development Planning consists of the following:
- Technology Plan (discussed in Chapter 4)
- Technology Development Strategic Implementation Plan
- Project Plans

10.2.1.2 Design Inputs

Design Inputs are the physical and performance requirements that are used as the basis for device design. Specifically, Design Inputs are all of the external considerations that need to be taken into consideration when building a device (or product), such as the customer product requirements, special characteristics needed, and standards and/or regulations that need to be observed. A general example checklist

Table 10.3: List of Design Control documents for a medical device.

	Design Control Document	**Description**
1.	Design and Development Planning	Design and Development Planning includes the creation of Technology Plans, Technology Development Strategic Implementation Plans, Project Plans, Verification and Validation Plans, Manufacturing Plans, etc.
2.	Design Input	The physical and/or performance requirements of a system that are used as the basis for a product's design.
3.	Design Reviews	Design Reviews are iterative review meetings that serve as milestones in the Design Control process, which formally document cross-functional team approval of various stages of Product Development and beyond.
4.	Design Output	The Design Outputs are the major output deliverables achieved from the Product Development process. These are the final specifications for the device, which are documented in models, drawings, testing/assessments, and other documents. The Design Outputs are directly correlated to the Design Input requirements.
5.	Design Verification	Design verification confirms design output that meets the design input requirements. (*Note: Will be documented in the Design History File (DHF), refer to Section 10.2.1.9.*)
6.	Design Validation	Design validation provides objective evidence that device specifications conform with user needs and intended use(s). (*Note: Will be documented in the DHF, refer to Section 10.2.1.9.*)
7.	Design Transfers	The procedures, protocols, Installation Qualification/Operational Qualification/Performance Qualification (IQ/OQ/PQ) documentation (refer to Section 10.2.1.12), and Design Transfer Review report(s) ensuring the device design is correctly translated into production.
8.	Design Changes	When a feature of a device needs to change, this is the collection of procedures identifying the change, justifying the need for the change and any associated documentation (e.g., validation, verification, review, and approval).
9.	Design History File (DHF)	A compilation of records that describes the design history (prototype development and all associated documentation) of a finished device.
10.	Device Master Record (DMR)	The DMR is the organization of all instructions, drawings, and other records for what it requires to *manufacture your device or product.* This includes the DHF, which includes V&V protocols and reports, Design Reviews, and other documents, including IQ/OQ/PQ of all equipment, etc.

Table 10.3 (continued)

	Design Control Document	Description
11.	Device History Record (DHR)	The DHR is the documentation recording all of the parameters, protocols, and specifications of the devices you have manufactured. The DHR will contain or reference where the following information can be found: (a) The date(s) of manufacture; (b) quantity manufactured; (c) quantity released for distribution; (d) the acceptance records showing that the device was manufactured following the DMR; (e) a copy of the identification label and any other labeling used for each production unit; and (f) device identification(s) and control number(s).
12a.	Installation Qualification (IQ)	IQ is the evaluation of installation and setup of new equipment in its performance environment (e.g., did all of screws and bolts come with the system for assembly, was it put together properly, can it live outside or work effectively on an incline?).
12b.	Operational Qualification (OQ)	OQ is the verification that new equipment operates the way it is intended (e.g., does the system turn on and do all of the knobs work?).
12c.	Performance Qualification (PQ)	PQ is the verification that new equipment performs the way it is intended (e.g., if you buy a car that says it has 350 hp, does it actually have 350 hp?).

outlining the Design Inputs is shown in Table 10.4. Checklists like the one shown in Table 10.4 are industry specific. For example, medical devices would also include toxicity, biocompatibility, and sterility.

10.2.1.3 Design Reviews

Design Reviews are multiple review milestones within Product Development. The Design Review process is intended to evaluate changes and/or updates to a design, as it is being developed. Though recommended for any Product Development initiative, Design Reviews are compulsory, as part of the design controls for medical devices. An example outline for a Design Review is shown in Table 10.5.

10.2.1.4 Design Output

A Design Output is the result of the effort put forth by the design team at (a) each design phase, and (b) the end of the total Product Development effort. The completed design output consists of the design and all of the affiliated drawings of the

Table 10.4: Example Design Inputs checklist.

Design Input Document Checklist

GENERAL: The following items have been reviewed by the **[Insert Management Position]** of **[XYZ Company, Inc.]** and representatives from the Engineering, Marketing, Manufacturing, Purchasing, and Quality Assurance Developments of **[XYZ Company, Inc.]**. This review involved the determination of the various design considerations for inclusion into the design input document for the **[XYZ]** Product Line.

Item Number	Description of Design Consideration	Required	Not Required
1	Intended Use	☐	☐
2	Application	☐	☐
3	User Interface	☐	☐
4	General Product Description	☐	☐
5	Product Source	☐	☐
6	Safety	☐	☐
7	Performance Criteria	☐	☐
8	Limits and Tolerances	☐	☐
9	Risk Analysis	☐	☐
10	Compatibility with Accessories/Auxiliary Devices	☐	☐
11	Human Factors	☐	☐
12	Customer Input	☐	☐
13	Physical/Chemical Specifications	☐	☐
14	Reusability	☐	☐
15	Labeling/Packaging	☐	☐
16	Shelf Life	☐	☐
17	Storage Conditions	☐	☐
18	Directions for Use	☐	☐
19	Reliability	☐	☐
20	Statutory and Regulatory Requirements	☐	☐
21	Voluntary Standards	☐	☐
22	Manufacturing Processes	☐	☐

Form No. XXX Rev. X

device, special considerations, reliability results, BOM (Section 10.2.3), Design Review results, its packaging and labeling, and any additional information pertaining to the development of a product and/or medical device.

10.2.1.5 Design Verification

Design Verification is the process that tests, documents, and ensures that a product, service, or system meet its design specifications. Design verification confirms that design outputs meet the design input requirements. Refer to Section 10.1.3.4.

Table 10.5: Example outline for Design Review.

Project Information	Outline the Purpose and Scope of the project that the Design Review is assessing.
References	Reference documents (such as laws, regulations, and standard operating procedures).
Action Items from Previous Meetings	Actions items that need to be discussed resulting from previous Design Review meetings.
Design Reviews	Updates from each department or contract manufacturer (such as the engineering department, manufacturing department, quality department, regulatory consultants, contract software vendor, design firm, and marketing department).
Milestones	Document accomplished milestones and the status of existing milestones.
Action Items	Note action items determined from Design Review process.
Notes/Discussion	Document notes, recommendations, and pertinent discussion relating to Design Reviews.

10.2.1.6 Design Validation

Design Validation is the process that tests, documents, and ensures that a product, service, or system meet the operational needs of the user. Design validation provides objective evidence that device specifications conform with user needs and intended use(s). Refer to Section 10.1.3.4.

10.2.1.7 Design Transfers

Design Transfer is the process of converting a design to manufacturing. It does not matter if the manufacturing is being conducted in-house or outsourced to an external manufacturer. Design controls require procedures to ensure that a device or product design is correctly translated to the manufacturing processes, while meeting the production specifications. A Design Plan consisting of phases is often developed to organize the transfer process. Production specifications are obtained from the Design Freeze that has successfully completed V&V.

10.2.1.8 Design Changes

No product or process goes unchanged, while the finished product is on the market. Even after the transfer of a design to manufacturing, there will be replaced parts, updated manufacturing systems, changes to vendors and suppliers, and so on. Product Revision and Product Revision Control are the processes for changing a product's design and the associated documentation that updates the records, conducts verification testing, tracks approvals, and logs the implementation of updated procedures. Design Change often falls under (a) document control (e.g., ensuring that a change is formally authorized and the documented change is updated in the appropriate locations throughout a company) and/or (b) corrective actions (e.g., a change is needed due to a detrimental customer or user complaint, which then requires formal authorization of the change, document updates, and documentation that the corrective action has been executed and verified for effectiveness). Product Revision and Product Revision Control documentation are maintained in the organization's QMS (e.g., in the DMR for medical devices/products). Product Revision and Product Revision Control documentation also follow similar outlines to Design Control documents.

10.2.1.9 Design History File (DHF)

The DHF consists of all of documents (such as drawings, prototype and product development reports, product revisions, design reviews, and V&V protocols and reports) that describes the design history of how a finished device was developed. Simply put, everything from Prototype (POC and α-prototype) and Product Development (β-prototype through Transfer to Manufacturing) of a finished device is documented, organized, and stored in the DHF. The DHF tells the history of all design activities conducted during design and development of the device, including any accessories, interactions with vendors, sourcing, labeling and packaging, and production processes. The DHF contains:

- Detailed design and development plans specifying design tasks and deliverables.
- Copies of approved design input documents – information on intended use, performance, labeling, and environment.
- Copies of design output documents, including specifications, top-level drawings, major subassemblies, and development of the DMR.
- Documentation of design reviews, including design review meeting minutes, assignments, and tracking issues.
- Validation documentation, including protocols, results of all tests, and testing methodologies.
- Design Transfer information, such as plans, protocols, and authority documents that show process ownership.
- When applicable, copies of controlled design documents and change control records.

10.2.1.10 Device Master Record (DMR)

Medical devices require a DMR, which is part of a company's QMS. The DMR contains all of the blueprints, instructions, drawings, training, and so on, that are needed to *manufacture your medical device or product*. The DMR can be electronically stored on your company's server or cloud storage, or can exist as a physical hard copy. An example outline of the contents of a DMR is shown in Table 10.6. For industries outside Medical Technology (MedTech), ISO 9001 requires structured documentation to be

Table 10.6: Example outline of a DMR.

Item	Description
Product Number	Identifying product number.
Document Revision History	Refer to Section 10.2.2.
Product Description	Short product description.
Design History File (DHF)	Refer to Section 10.2.1.9.
Raw materials specifications	Reference to the document containing the starting (raw) material specifications (e.g., type and properties of the plastic used for your injection-molded part).
Bill of Material (BOM)	List of parts and components for a single product along with vendor information, costs, quantities, etc. Refer to Section 10.2.3.
Drawing number	Identifiers for each related/affiliated drawing.
Manufacturing procedures	List of document numbers for every applicable Manufacturing procedure.
Quality assurance procedures	List of document numbers of affiliated Quality Standard Operating Procedures (SOPs) and forms.
Software	Software needed to build the product. List of software packages, number of licenses or seats, vendors, costs, etc.
Packaging procedure(s)	Instructions for how to package the product and number of units per package.
Ancillary packaging information	Any additional information regarding packaging (e.g., shelf life testing/reports, storage studies/reports, and transportation studies/reports).
Labeling	Procedure for labeling the product.
Label Printing Specification	Reference to the document with the Label Printing Specifications.
Packaging Validation	Reference to the document with the Packaging Validation protocols and reports.
Sterilization (*if applicable)	Reference to the document with the Sterilization procedure(s).
Post Sterilization Inspection and Testing (*if applicable)	Reference to the document with the Post Sterilization Inspection and Testing procedure(s).
Inspection Procedures	Reference to the document with the Inspection Procedures.
Shipping	Reference to the document with the Shipping Procedure.
Revision History	The revision history is an identifying iteration of the product's revision which includes the revision number, effective date, and short description of change.

developed, stored, and organized in a company's QMS (Chapter 11), even though this documentation is not formally referred to as the DMR.

10.2.1.11 Device History Record (DHR)

The DHR is the aggregation of the individual records of the production history of a finished device (such as the date, lot, batch number, materials, personnel involved, protocol, and variances/deviations).

10.2.1.12 Installation Qualification, Operational Qualification, and Performance Qualification (IQ/OQ/PQ)

The installation qualification, operational qualification, and performance qualification (IQ/OQ/PQ) are performed by R&D, Quality, and Manufacturing Departments. Documentation is required to verify that every piece of equipment or instrumentation has been installed properly and works correctly in its intended use environment. In some cases, only IQ and OQ are necessary. For example, a piece of furniture, such as a cabinet, may only need to be shown to have all its parts and be level to the ground (IQ) and that the doors open correctly (OQ). On the other hand, an analytical balance needs to be assembled correctly and leveled to the table or bench upon which it is placed (IQ), it needs to turn on, off, tare, and must be calibrated using certified weights (OQ), and, lastly, it must give reproducible results using those calibrated weights (PQ).

10.2.2 Revision Histories

A Revision History is a table at the beginning or end of every document showing the current and previous version number of the document, author, dates, approvals, and reasons for any changes to documents. Table 10.7 is a standard template for a document's Revision History.

10.2.3 Bill of Materials

The BOM is a list of all of the parts, and respective quantities, required to assemble the product. The BOM is the "ingredients list." From the BOM, metrics for determining product COGS, cost of labor, time of production, and waste can be determined. Every department within Operations uses the BOM. The BOM is developed and updated by a company's Research & Development or Engineering Department, is used and updated by the Manufacturing Department for product assembly, and is referred to by the Shipping and Receiving Departments for inventory control of incoming assembly components and to verify the assembled products are ready to be

able 10.7: Standard Revision History template.

evision umber	Author	Revision Date	Approved By	Approval Date	Reason
.0	Author Name	mm/dd/yy	Signature & Printed Name	mm/dd/yy	Initial Document
.1	Author Name	mm/dd/yy	Signature & Printed Name	mm/dd/yy	Update procedure

hipped to customers or distributors. Inventory control is the management of in-ound and outbound items in a company's warehouse. Inventory control involves he ordering of each component at the correct time so that not too much space is aken up in storage, while ensuring enough components are available at the time of nanufacturing. It is like managing the pantry in your kitchen. Though it sounds imple, the task becomes daunting, when there are hundreds of parts. Table 10.8 hows an example BOM using the donut example discussed throughout Chapters 9 nd 10.

able 10.8: Example Bill of Materials for manufactured donuts.

Donut Bill of Materials			
erves: 18 donuts			
otal Material Cost: $15.93			
tem	**Quantity**	**Unit**	**Cost ($)**
ilk, whole	210	Grams	0.94
ctive Dry Yeast	15	Grams	3.38
ggs	2	Egg	0.50
utter	40	Grams	1.96
ranulated (White) Sugar	100	Grams	28.88
onfectioners (Powdered) Sugar	420	Grams	0.53
alt	3	Grams	0.01
ll-Purpose Flour	700	Grams	1.00
egetable Oil	940	Grams	1.70
anilla Extract	5	Grams	1.10
prinkles (Colorful)	35	Grams	0.18
ark Chocolate Chips	170	Grams	1.99
mall Marshmallows	100	Grams	0.31
oney	20	Grams	0.23
rosting	140	Grams	1.55
hortening	40	Grams	0.42
ater	35	Grams	0.00

The development of a product is no simple matter. Developing a product is not just having a good idea. It involves a lot of ingenuity, sweat, engineering, and perseverance. No one person can do everything. Planning, documentation, organization, and teamwork are what make a simple concept become a successful, final product. The steps discussed during Product Development and Transfer to Manufacturing are only a part of what is needed to consistently produce a quality product. Chapter 11 discusses the intricacies of Quality and QMS in more detail.

11 An Introduction to Quality

Abstract: Ensuring consistent quality of a product involves developing and maintaining a quality management system (QMS) to identify, track, and monitor every component of a product at every stage of manufacturing. The purpose of a QMS is to provide a framework for executing processes, storing and archiving information, and executing a project according to the company's quality policy and quality controls through a system that is suitable, adequate, and effective, and enables technology and innovation safety and performance. This chapter reviews the types of QMSs, the design of QMSs and what process areas are included, the purpose of QMS audits, and the checks and balances that exist within an effective QMS. This chapter concludes with a discussion focused on Total Quality Management (TQM), which is the methodology that uses statistical tools to analyze the data that is collected and stored within QMSs. TQM tools such as Lean, Six Sigma, and the combination of Lean and Six Sigma are discussed with examples.

Keywords: Quality management system (QMS), International Organization for Standardization (ISO) 9001, CE marking, quality assurance (QA), quality control (QC), internal audit, external audit, quality certification, corrective and preventive action (CAPA), total quality management (TQM) tools, Lean, Pareto charts, Six Sigma, Lean Six Sigma.

11.1 A World of Quality

At some point in your life you have said the words, "Wow, what a piece of junk! Why did I pay money for this?" Clearly, quality was lacking in this product. "Quality" broadly refers to ensuring that products produced meet the design specifications and customer requirements (refer to the cell phone example later). Ensuring consistent quality of product involves developing and maintaining a system to identify, track, and monitor every component of a product at every stage of manufacturing to ensure accountability, known as a quality management system (QMS).

The international "gold standard" for QMS for general manufacturing is guided by International Organization for Standardization (ISO) 9001.

Note: ISO 9000 refers to the family of ISO quality management standards for QMS, whereas ISO 9001 is the standard that companies need to follow to be compliant with ISO 9000.

Additionally, some products have more specific guidelines, for example, medical devices are guided by ISO 13485. Manufacturing requirements for QMS under ISO 13485 are more rigorous to ensure patient safety, but align with ISO 9001.

https://doi.org/10.1515/9783110521900-012

Individual countries provide their own compliance standards. For example, within the United States, medical device manufacturers must comply with 21 Code of Federal Regulations (CFR) Part 820 – Quality System Regulation (referred to as 21 CFR 820). Similarly, the European Union (EU) under Directive 93/68/EEC requires a Conformité Européene, or CE Marking, to be allowed to commercially distribute medical devices in the EU. As the individual economies of various countries are becoming more intertwined due to globalization, the compliance standards of individual countries are becoming more and more aligned. Although 21 CFR 820 compliance is mandatory and FDA clearance/approval is required for medical device manufacturers to commercialize a medical device in the United States, ISO 13485 is an optional QMS certification that will allow a US-based medical device to be sold internationally where ISO 13485 conformance is recognized. The ISO 13485 certification is required for medical device manufacturers to obtain the CE Mark for their devices. As globalization continues, these quality and safety standards are being aligned and cross-referenced. The FDA has announced plans to modernize and harmonize 21 CFR 820 with the provisions of ISO 13485 in the near future.

Additionally, as the technological landscape broadens with artificial intelligence, nanomaterials, data science, and a better understanding of human conditions, it is necessary to update the processes surrounding the approval of medical devices for commercialization to ensure patient safety. In 2017, the EU adopted the Medical Device Regulation (MDR) to replace the existing Medical Devices Directive (93/42/EEC) that is required to obtain a CE Mark. By 2020, the MDR will be in full effect.

Even without regulations, quality is very important for customer safety and customer retention. QMSs are the structures put in place by companies to ensure the safe use of their products by end users.

An example of the need for a quality system would be if *Company X* manufactures the inner frame for *Company A*'s cell phone, then *Company A* will want to make sure the inner frame fits that model phone's components (e.g., phone chassis, volume control button locations, headphone jack location, and speakers) correctly. If *Company X* does not notice that the frame molds are wearing down, so that the frames no longer fit or align with the rest of the cell phone parts, then over time *Company A* will notice that their cell phones are no longer assembling well. *Company A* will either need to return or scrap the frames, which will result in delays for getting the completed product out to customers. If *Company A* tries to use the inner frames, then they will notice an increase in returns from customers, which will then result in *Company A* receiving more customer complaints, bad customer reviews, and decreased sales.

11.1.1 Making Consistent Quality Products: An Overview of Quality Management

Making sure everything is manufactured and assembled consistently is simple, right? Not so much. Machines wear down. People change jobs. Suppliers modify parts, change equipment, production techniques, get acquired, or just go out of business. This is where Quality[34] comes into the picture.

A medical device can be a product that is placed on or in the body of a patient, in order to care or treat a medical condition or disease (e.g., bandage, catheter, hip implant, and dental implant) or a piece of equipment that analyzes or diagnoses a health condition (e.g., blood analyzer, electrocardiogram, blood pressure cuff, and glucose meter). To give perspective as to why Quality is important; a device needs to not only perform its intended use(s), but it also needs to be proven to be safe, accurate, and effective. The parts of a medical device don't have to just "fit" together, they need to be repeatedly verified and validated (Chapter 10) in a way that demonstrates to regulatory agencies that the medical device manufacturer is responsible and accountable for the quality of its products. Any change to a process, part, material, and so on has to be documented and verified in a traceable and reproducible manner, to ensure that the change will not introduce new quality issues into the product.

This is accomplished with a QMS. A QMS is divided into a Quality Assurance (QA) Program and Quality Control (QC):
- **Quality Assurance** defines the overarching structure for how an organization ensures quality.
- **Quality Control** encompasses the organizational procedures and subsequent documentation used to show that the procedures have been executed within the QA framework.

When developing a QA program, it needs to be flexible enough to stay constant and provide structure, while allowing the details to be laid out in the QC documentation.

QA programs that are too detailed have the unfortunate consequence of bogging down an organization with frequent updates or changes, or they get the organization into trouble with regulatory or legal agencies during audits. In keeping with the above analogy, imagine the extent of revisions that would need to be made to all of the federal and state laws (i.e., the procedures and documents that make up QC) that would need to be changed if a single Constitutional Amendment was removed (i.e., changes were made to the QA).

A practical example of how developing a flexible QA system to prevent cumbersome administrative tasks downstream is assigning the annual internal audit to a specific position within the company, like the Director of Quality, rather than to assign the task to an individual, for example, Jane Smith. In most companies, the Director of Quality usually authorizes every quality document in the organization. By assigning the responsibility to the position, rather than an individual, a company eliminates the downstream effect of needing to update every document within the QC whenever there is a change in personnel. As a result, only a limited number

34 Here "Quality" is being used to refer to a Quality department of a company.

of documents in the QA system will need to be revised and approved, rather than every quality document throughout the company.

11.2 Quality Management Systems (and Examples)

Quality conceptually consists of QA and QC, but the actual implementation of Quality is the QMS. The purpose of a QMS is to provide a framework for executing processes, storing and archiving information, and executing a project according to the company's quality policy and QCs through a system that is suitable, adequate, and effective, and enables technology and innovation safety and performance. A QMS is deemed effective when process inputs yield or give way to consistent process outputs, or repeatable results. A QMS promotes safety by making it easier to detect and correct for problems in how the company processes are being followed. Think of a QMS as a roadmap and compass set for how to safely navigate and execute an implementation strategy according to the high-level mission, vision, and objectives of the business. This system is not meant to constrain innovation; instead it establishes a framework for innovation to flourish.

A QMS can be simple or complex in design, with various levels of checks and balances. The type of QMS used in any particular company should reflect the maturity of the company and how the company wants to operate. For example, a company that wants to move quickly, like a startup, will want processes that are more flexible and less stringent with fewer approval cycles. Startups need to be careful not to get bogged down in unnecessary or nonvalue added activities.

Company personnel who are involved in designing, creating, updating, and managing the QMS and executing QCs are typically referred to as Quality or QA. Quality professionals are typically very organized, attentive to detail, and are good communicators.

11.2.1 Types of Quality Management Systems

At a high level, there are four types of QMS:

1. 100% paper based
2. commercial off-the-shelf electronic QMS with little ability for user customization,
3. commercial off-the-shelf electronic QMS that is sector specific, with some ability for user customization (e.g., can be cloud based), and
4. fully customizable electronic QMS software that can integrate and manage multiple functions.

Hard copy or 100% paper-based QMSs are low cost to implement; quality documents are stored in a series of folders in protected, locked filing drawers. However, this low-tech QMS structure does not easily permit significant changes in the future, so it becomes more difficult to manage this system, as the business grows. Additionally, it can be difficult to access the information in the system, when one is not familiar with the organizational hierarchy. It can also be difficult and time consuming to access QMS data for analysis (Figure11.1).

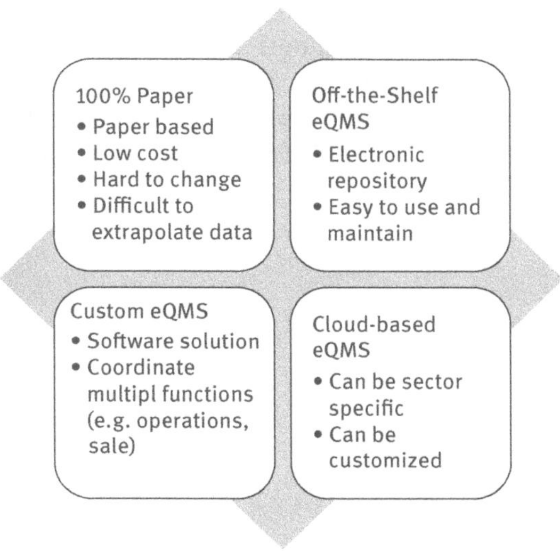

Figure 11.1: Four types of quality management systems.

The next option for implementing a QMS, which is more modern but scaled down in complexity, is a hybrid approach that utilizes electronic and paper versions of information. Most startups opt for this type of QMS. Commercial, off-the-shelf, cloud-based document repository solutions, such as Dropbox or Box, are easy to set up and use, but these digital solutions are not QMS systems in and of themselves – companies must develop their own QMS. User-specific, role-based access controls (e.g., Owner, Editor, and Viewer) can be assigned to the electronic folder structure. Users can review and sign hard copies of documents, and then upload and store them as master copies into the electronic folder structure. Users cannot easily change the functionality of these commercial solutions.

For businesses that would like to implement an electronic QMS that permits some user customization, there are more intricate cloud-based solutions that can be specific to a certain sector, for example, the Greenlight Guru QMS software solution can be used for medical device design and development. This approach is costlier than the previously discussed commercial, off-the-shelf solutions.

Finally, a fully customizable electronic QMS software solution can be designed and implemented according to your business-specific functions and workflows, integrating the entire supply chain, embedding QCs into procurement, manufacturing, distribution, and sales. This option is very costly but can be scalable to evolve and grow, as the company grows.

11.2.2 Design of Quality Management Systems

QMSs are designed to be in compliance with external inputs, including legal and regulatory standards, regulations, and industry guidelines, and are governed by a hierarchy of controls, beginning with a company quality policy (which defines the QA framework), then standard operating procedures (SOPs), work instructions (WIs), and forms (which define the QC) (Figure 11.2).

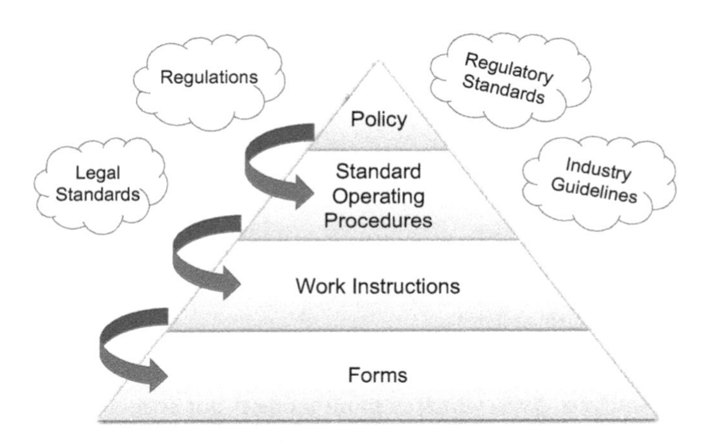

Figure 11.2: QMS governance framework.

A company's quality policy is a guide or manual that describes how the company will be organized and operate, including the responsibilities of management, how resources will be managed, and how products and services will be delivered to customers. A quality policy is high level and provides a summary for all the different areas where quality may apply across the organization. If a change is made in the quality policy of the company, there is a ripple effect across the company's QMS, and changes may be required in the SOPs, WIs, and forms that support the company's quality policy. Therefore, the company should not make changes to the quality policy too often, or even make too frequent changes to the SOPs for that matter, because changes carry the burden of ensuring that all processes involved in the QMS then agree with one another, that is, all processes involved in the QMS are in alignment with one another, after the changes have been made.

To use an analogy from athletics, SOPs make up a company's playbook. SOPs describe the scope, roles and responsibilities, requirements, and processes for different areas of quality, in a greater level of detail than the quality policy, but SOPs are not as detailed as WIs. Any change of content at the SOP level will need to be reflected in the appropriate WIs that are affected by the changes made to the (governing) SOP, that is, changes in the QMS "trickle down hill."

In regulated industries that require the auditing of companies' QMSs to verify compliance, the quality SOPs within a QMS must include:

- Design controls
- Risk management
- Supplier and purchasing controls
- Statistics and data analysis
- Verification and validation (design and process)
- Design history documentation
- Employee training
- Retention of records
- Equipment qualification, maintenance, and calibration
- Production and process controls
- Control of documents
- Corrective and preventive action (CAPA) planning
- Design reviews
- Management reviews
- Annual quality planning

To continue with the athletics analogy, WIs describe how a team would execute a play in the playbook. WIs are detailed, step-by-step instructions on how to perform a process that is outlined in an SOP. WIs describe processes at a finer level of granularity than SOPs. For example, the document control process and rules can be described in an SOP, but the step-by-step instructions for creating multiple copies of source documents, converting documents to PDFs, adding watermarks and headings to documents, and uploading released or approved documents into different areas of the QMS are included in a WI. Any change of content at the WI level will need to be reflected in the appropriate forms that are affected by the changes made to the (governing) WI. In an engineering project, technical drawings and specifications are also considered at the same hierarchical level as WIs.

In athletics, example forms would include your team roster or starting lineup. Forms are written records that are used to document adherence or compliance with a company's quality processes. The data that is documented in forms can be analyzed to confirm whether the form is being used correctly and whether the SOPs are being followed. Going back to athletics; the team roster can be checked against each play that is executed, to determine how well the play is executed throughout the game, and who needs additional training and coaching to execute the play better.

Forms are more detailed than SOPs and WIs, and are typically updated the most frequently. For example, document control WIs include step-by-step instructions for adding approved documents into the QMS, but there are change order forms to document the types of changes and the rationale for making these changes to the QMS. These forms also capture who needs to approve the changes, the impact of these changes (if any) to the QMS, whether new employee training is required, whether there is a change to a product or process, and whether these changes require follow-up or additional tracking.

If a QMS is required by regulatory authorities, for example, in order to design, develop, manufacture, deliver, and service your technology, then the QMS must be designed and adhere to a recognized quality standard. As stated earlier, medical devices approved or cleared for sale and distribution in the United States by the Food and Drug Administration (FDA) need to use QMSs that are compliant with 21 CFR Part 820: Quality System Regulation. Similarly, medical devices that are approved for sale and distribution in the EU need to use QMSs that are compliant with ISO 13485 Medical devices – QMSs – Requirements for regulatory purposes.

These standards, and others, such as internationally recognized ISO 9001 QMSs; software-engineering focused ISO 90003; petroleum, petrochemical, and natural gas-focused ISO 29001, guide companies when quality compliance is required. Companies can certify that their QMSs are in compliance with the standards and work with auditors and certifying organizations to review the QMSs. These certifying organizations will identify gaps, or opportunities for improvement, and issue certification, if they determine the QMS meets all requirements that are documented in the applicable quality standards.

Companies can ensure their QMS is robust and practical by creating QMS-specific objectives that are reviewed periodically at management reviews and audits. Management reviews and audits of the QMS, conducted on at least an annual basis, are helpful in confirming the robustness of the quality processes, verifying the company's adherence to these processes, and ensuring the effectiveness of the QMS at detecting issues and preventing errors.

Internal audits are conducted by a company representative, which could be an independent third-party consultant or contractor, whereas external audits are conducted by external regulatory agencies, for example, FDA, Occupational Safety and Health Administration, and Notified Bodies (Notified Bodies are discussed more in Chapter 12). External regulatory agency audits can be triggered by quality system certification requirements, for example, to achieve or maintain ISO 13485 certification, or these audits can be unannounced, for example, manufacturing or clinical monitoring inspections.

Quality can plan for and execute internal audits at a defined frequency per the company's SOP on internal audits, per a development project's achievement of key milestones or completion of development phases (such as the development team is preparing to conduct clinical trials or receive and ship product for the first time), or per a product's post-commercialization quality management plan. The results of these internal audits are reviewed by Quality to ensure project/product/process compliance with the company's quality policy and key quality system regulations, and to identify opportunities for improvement and deficiencies that must be corrected.

Internal audits may be scheduled more often due to findings from a previous internal or external audit, for example, an audit finding needs to be investigated further or a corrective action was identified that poses a high risk of impacting the product quality, and that corrective action needs to be implemented.

To conduct internal audits, an auditor should be experienced and hold the proper qualifications, for example, Certified Quality Auditor. If an independent, third party is not selected to perform the internal audit, then an employee may be selected to perform the audit, but this employee should never audit his or her own QMS elements and processes.

11.2.3 Checks and Balances within Quality Management Systems

QMS processes are typically executed across the company via multiple departments working together, each contributing to quality process inputs and/or outputs (refer to Figure 11.3). QMSs make it easy to collect and analyze data at each step to ensure process quality. Checks and balances are put in place, such as interim and final process approvals, audits, management reviews, to verify that QMS processes are completed correctly.

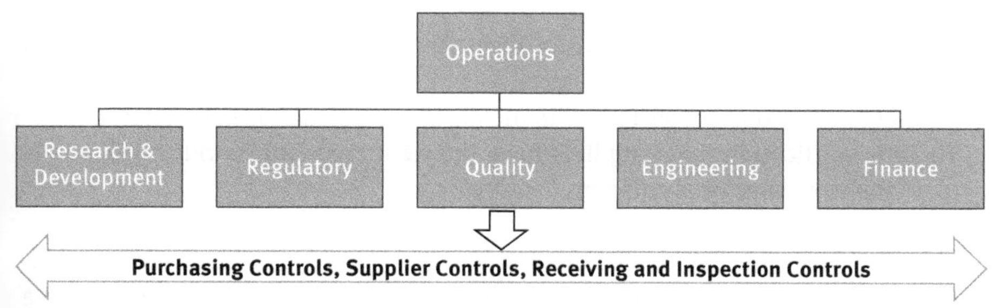

Figure 11.3: Interdepartmental execution of QMS processes.

For example, one of the processes typically included in a QMS is purchasing controls. In technology development, employees from research and development (R&D) or engineering may identify that a new product or service is needed, or a new supplier is needed, or a new contractor or consultant is needed to support a project (or set of processes). Quality, along with Engineering and Management, as needed, will review whether this new entity can and should be added to the company's approved vendor list (AVL).

Then, when a quote for services is received from a supplier, contractor, or vendor, but before creating a purchase order, Quality, or a delegate, should verify that the vendor is active or pending on the company's AVL. Quality should also verify whether a specification needs to be provided to the vendor, and ensure that the vendor is using the most current and approved specification. Then, Quality should think about the goods or services that are being provided by the vendor and determine whether additional documentation, such as lot release/batch records, First Article Inspection data, or Certificates of Conformance, is required and should accompany the finished product.

The QMS enables Quality to store signed and unsigned versions of Purchase Orders (accessed by accounting/purchasing, operations, and/or the appropriate departments) and to keep a log of such transactions, so that when an invoice is received from the vendor, Quality can check against the open Purchase Orders to verify that payment is due. Once proof of payment is collected and tied back to the appropriate invoice, Quality can officially close the Purchase Order in the QMS.

Another process in the QMS is supplier controls. Suppliers provide products or processes for your business that cannot be created or achieved internally. Suppliers may be asked to complete work temporarily (such as a project contractor or a consultant), or you may want to establish a continuing relationship with a supplier for a particular product or service, that is, the supplier is part of your regular supply chain, such as a contract manufacturer.

Quality performs a risk-based supplier review and approval. A risk-based supplier review and approval process ensures that:

1. Protections are in place with suppliers so that your confidential, valuable company information is not shared, for example, Non-Disclosure Agreements and Contracts are in place with suppliers.
2. Suppliers are vetted according to financial, technical, and quality criteria.
3. There is verification that suppliers have the right credentials and certifications to do the work that your company requires.

Additionally, there are risk-based Supplier Quality Agreements that can be established between your company and the supplier that discuss all the controls that the supplier has in place to consistently deliver the product or service that meets your company's requirements. Any company can require a supplier, for example, an original equipment manufacturer, to demonstrate its adherence to quality standards via certification. The quality certification only shows that a company is compliant to the standard(s), meaning that they have a QMS in place that follows the legal and regulatory guidance. The certification does not mean that the manufacturer's finished outputs are necessarily good or consistent, only that they have a system in place for monitoring and tracking their products and processes.

Case Study

In alignment with the case study discussed in Chapter 7 in Section 7.3.5, the same case study shows the value of QMS. As discussed before, the purpose of the stain kit is to detect the presence of malaria in a patient's saliva sample. If a positive detection result occurred, then the patient most likely had malaria and additional testing would be necessary. The manufacturer of the malaria stain kit was based within the United States, whereas the customer was located in Vietnam. The sales representative that interacted with the customer received a comment from one of their clients stating that they "liked the new color, but [it] did not seem as efficacious as previous batches."

Once the customer reported the discrepancy to the sales representative, a CAPA was submitted by the sales representative, in accordance with the company's SOP on CAPAs as dictated by the company's Quality Manual. The submitted CAPA form provided the Sales and Manufacturing Department managers, as identified in the company's Quality Manual and the CAPA SOP, with a notification that a problem was reported. The two managers spoke about the potential problems and began an investigation. The investigation included an assessment of the outbound shipping documents, the receiving documents of the components of the stain kit, and a review of the SOPs and the QC forms filled out during manufacturing and assembly of the kits. The shipping documentation was needed to determine which batch and lot of the outgoing malaria stain kit was under scrutiny. The Receiving department provided documentation on the receipt, inspection, and certifications of each of the ten components that went into the malaria stain kit. The SOPs and QC forms were pulled from the QMS. All of these documents and forms, how they are filled out, where they are stored, and who is responsible for them are dictated by the QMS.

The result of the investigation concluded that the primary-approved vendor for a component of the stain kit, located in India, was shipping inconsistent lots of the stain to the US-based manufacturer. This conclusion was drawn from a combination of observations.

First, when comparing the certificates of conformance, the dates were different, but the analytical values did not change from lot to lot over a period of 2 years. Once this was noticed, chemical analysis of multiple lots of the stain was conducted. It was observed that even within the same lot, the color of the stain varied. Additionally, the investigation revealed through the review of the QC documentation that the laboratory manager had left the company close to the same time of the production of the kit, which resulted in the technician not receiving proper training – instead the new technician had to rely on the SOPs as training tools. The SOPs were inadequate, so the technician adjusted the recipe of the stain to technically meet the acceptance criteria on the SOP. The US-based company then took action to revise the SOPs with clear instructions and more stringent acceptance criteria.

In summation, the first corrective action of the manufacturer's CAPA plan was to cease production of the stain until the investigation was completed. The second corrective action was to ensure that any "bad" product in inventory was quarantined and did not leave the building. The third corrective action was to remove the nonconformant vendor from the AVL and validate alternative stains and their vendors. The fourth corrective action was to revise the QC protocols and SOPs, to ensure the problem did not occur again.

This case study serves as a cautionary tale about how a single bad component produced by a third party caused quality problems in a manufacturer's product. The bad component caused the final malaria stain product to be nonconformant. It is also important to note how communication between the Sales and Manufacturing Departments was facilitated through the QMS. If the quality issue had not been caught and addressed, a potential malaria epidemic could have occurred in

Vietnam, due to the faulty malaria stain kits. This case study illustrates how QMS systems are useful for honing in on the root cause of quality issues in a manufacturing process.

11.3 Total Quality Management (TQM) Tools

The information within a QMS is continually changing. Startup companies often do not have all the resources needed to implement an extensive QMS, let alone improve upon it in the future. A QMS is not only a system to protect against lawsuits or to retain customers: If the system can successfully integrate manufacturing, shipping and receiving, accounting, and sales via robust analytical software, then precision budget forecasts, customer retention efforts, maintenance, vendor analysis, and so on, can be correlated to one another for predictive analytics. Analytics can be cumbersome for a startup company to undertake initially, but while a company is growing, management's allocation of funds for the expansion of the QMS and implementation of analytical tools will be imperative to maximize financial returns. In other words, the QMS and its analytical tools can be used to monitor a company's waste and productivity, thus enabling a return on investment (ROI) for the purchase of the QMS.

Whereas a QMS is a structured framework for executing processes, storing and archiving information, and executing a project according to the company's quality policy and QC through a system that is suitable, adequate, and effective, total quality management (TQM) is the methodology that uses statistical tools to analyze the data collected and stored within the QMS.

QMSs require a lot of resources to implement and maintain. Without analytics, it is difficult to quantify the benefits of not failing an audit, not getting sued, and retaining customers. If TQM tools are not used to assess the data within the QMS, then management all too often gets frustrated with the Quality Department, because Quality comes across as pointing out flaws in the finished product(s) without offering solutions. Quality systems are expensive and cumbersome, and unless TQM tools are used, valuable returns attributable to a QMS can be difficult to visualize. The truth is that there are always returns on investment in QMS, even if it is only in the form of legal protection and/or customer retention.

A Historical Background

The origins of TQM analytical tools begin with Eli Whitney, who in 1798 introduced interchangeable parts that were similar enough in fit and function to allow for random selection of parts in the assembly of muskets. The introduction of interchangeable parts led to the methodology for measuring parts and the establishment of dimensional tolerances, which were continually refined throughout the 1800s. In 1913, Henry Ford's moving assembly line shifted production inspection, and inspection and acceptance methods, toward a sampling approach rather than conducting an inspection of every single product. In 1924, Walter Shewhart at Western Electric introduced the first known process control chart with data collection and analysis, giving birth to the start of statistical QC. This Western Electric plant employed several well-known statisticians, including Joseph M. Juran, W. Edwards Deming, and Walter A. Shewhart, at some point in each of their careers.

After the devastation of Japan during World War II, in 1945 the United States shared US manufacturing management principles with the Japanese to help the Japanese economy rebuild. Deming was among one of the first American statisticians in Japan to teach statistics and US quality methods. He emphasized analyzing production and systems data against computed statistics to quantify variation in a system in order to predict future process performance, allowing for the identification of the source of the variation in the system and the extent of problems. This predictive identification process offered the opportunity for continual process improvement. As a result, Deming developed the Plan-Do-Study-Act (PDSA) cycle, which is an example of continuous process improvement. In other areas of industry, process improvement methods were gaining popularity. For instance, in 1950 the US government implemented statistically based levels of product quality, thus introducing the military standard MIL-STD-105A, which reinforced the implementation and use of the analytical tools Deming and others had been developing.

From this, Toyota developed the Toyota Production System, or Kaizen. Kaizen is a set of methods used to incrementally reduce waste and improve efficiency. In the 1970s, Kaizen evolved into Lean Manufacturing for global production systems. Later, Six Sigma, a system of process improvement tools and techniques, was created under the leadership of Bob Gavin at Motorola, in 1986. Motorola executives married the concept of process capability and product specifications, introducing the calculation for process capability through defects per million opportunities. In 1987, the ISO, based out of Geneva, introduced a series of quality standards that were widely adopted throughout the world. By 1988, IBM, General Electric and other companies had adopted TQM methodologies. In the mid-1990s, ISO 9000 was introduced.

TQM can be comprised of individual analytical tools (Table 11.1), depending on the organizational need, or can be comprised of multiple, integrated, and/or adaptive tools, as seen in Six Sigma. TQM offers a way for an organization (big and small) to:

- Have an organizational structure in place to keep track of all manufacturing details.
- Ensure that procedures are in place so that when someone leaves the organization, any trained individual can assume the former's responsibilities without a change in process outputs.
- Analyze suppliers.
- Analyze the quality of manufacturing production lots.
- Analyze wear and tear on R&D or manufacturing equipment.
- Analyze productivity of employees.
- Track customer feedback.

Table 11.1 describes analytical tools that systematically and quantitatively provide continuous process improvement within organizations. Comprehensive toolsets have been developed by many organizations to implement TQM, and these toolsets account for:

- Reduction in product defects.
- Increased manufacturing process yields.
- Reduction in variation between parts.

Table 11.1: Common TQM tools.

Acronym	Tool	Description
FMEA	**F**ailure **M**ode **E**ffects **A**nalysis	A granular and systematic approach for identifying all possible failures in a design, a manufacturing or assembly process, or a product or service.
DMAIC	– **D**efine the problem – **M**easure process performance – **A**nalyze the process to determine root causes of variation or poor performance – **I**mprove process performance by addressing the root causes – **C**ontrol the improved process and future process performance	Data-driven quality strategy used to improve processes.
DCOV	**D**efine **C**haracterize **O**ptimize **V**erify	An approach within the Six Sigma toolkit to prevent potential problems through methodology design.
Gage R&R	**G**age **R**epeatability and **R**eproducibility	A method used to measure variations between production measuring processes.
PDSA	**P**lan **D**o **S**tudy **A**ct	PDSA is a cyclical process tool for testing a change through the development of a structured test plan for that specific change so that informed decisions can be made on how to iterate or adjust the change.
SIPOC	**S**uppliers **I**nputs **P**rocess **O**utputs **C**ustomers	A tool to map high-level business processes from beginning to end.
DMADV	**D**esign **M**easure **A**nalyze **D**esign **V**erify	Part of Six Sigma that focuses on the development of a new service, product, or process.
SPC	**S**tatistical **P**rocess **C**ontrol	Methodology for real-time measurement and control during the manufacturing process.

he goal of implementing TQM is to improve customer satisfaction and thus share-older value.

1.4 Six Sigma and Lean Six Sigma

ix Sigma, developed by Motorola, and Lean Six Sigma are a combination of man-gement tools that help companies achieve a culture or environment of process uality, so that desired results are consistently achieved. Lean Six Sigma concepts re helpful to understand – as a problem solver, as a decision maker, as a manager f people and process, and as an innovator. Lean Six Sigma relies on data-driven nalyses to improve all aspects of a business or project – ranging from ROI, cost eduction, maximization of productivity, and quality improvement. Lean and Six igma will be discussed individually as concepts, and then combined as Lean Six igma, because industry has found that Lean and Six Sigma principles work syner-istically together.

1.4.1 Lean

o understand Lean Six Sigma, we need to talk about waste. Waste is all around us, ecause we never use our time and resources (e.g., financial, human capital, and aterial) 100% effectively. An example of waste in technology development is not eeting customer or user requirements, and without that your product cannot meet arket needs; therefore, your product is not a viable solution. In manufacturing, rocess yields, the ratio of manufactured product that meets final acceptance crite-ia versus the total number of all manufactured product, are never 100%, that is, esources will be utilized to build product that never makes it past interim or final nspection. Also in manufacturing, unnecessary in-process visual inspections and ther types of manufacturing QA tests add time and consume resources. To state it lainly, waste drives up the costs to produce product.

The more waste that is eliminated from processes results in more output being chieved with the resources that are available. Reduced waste translates to reduced osts, which in turn results in more satisfied customers who pay lower prices for the nal product, that is, both internal customers that rely on the outputs of the less-waste-ul processes and external customers that rely on the final product.

To appreciate the vast problem of waste, waste can show itself in many differ-nt forms. For instance, wasteful movement or motion is prevalent in industry and an be found in all sectors. For example, hospitals that do not have medical labora-ories on-site must send out samples and specimens for analysis at external labora-ories. The samples must be physically transported to the external medical aboratory. Transport increases the amount of time it takes to receive an analysis and reporting of patient results, impacting the clinical decision-making process and subsequent clinical workflows.

Another example can be found on any manufacturing floor, when tools and disposable supplies are not conveniently located near respective workstations. When tools and disposable supplies are not easily accessible and near workstations, manufacturing operators are required to start and stop their work whenever they need a specific tool or supply, which results in waste.

Similarly, in any laboratory setting, a scientist having suboptimal means of executing an experiment and collecting data is wasteful.

Additional forms of waste include the following:

- Wasteful time and effort – working harder but not smarter.
- Waste due to waiting – personnel are sick and unable to complete tasks, equipment is not available to be used, manufacturing lines are down for maintenance, and so on.
- Physical waste – oversupply of components that will not be used or products that will not be sold.
- Process inefficiency – multiple layers of process review and approval cycles or data entry that is repeated in multiple forms.
  - In other words, multiple sources of the same information are created for no reason, or paperwork is being completed at different steps in the process for the sake of having additional paper-based evidence that is nonvalue added to the final outcome.
- Waste in performance – product does not perform as intended.
- Waste in errors – erroneous calculations, low process yields, poor training of staff that leads to rework.
- Waste in implementation – a product is delivered which does not operate as intended, which leads to recalls, design fixes, budget overruns, and a plunge in customer trust.

The worst waste is the waste of human ingenuity. For example, teams work hard to design and deliver a product for which there is no value to be gained by users – that is, a similar second-generation product that cannot take or steal market share away from the first-generation; a "breakout" product; or a product is delivered that solves a problem that does not exist. Another example could be a financial modeling project that bases its models on faulty trading assumptions, which results in over promising on the model's ROI. Based on these promises, company leadership decides to form a new, internal organization and invest in new hires, only to learn later that expectations need to be realigned and the models need to be reworked.

So how do you apply Lean principles to reduce waste? How do you bring order to disorder? How do you add value instead of adding nonvalue activities or tasks? And once you have reduced waste, how do you know waste is reduced so that solutions can be implemented, monitored, and improved?

Recall that a product or service's value is determined by customers and end ers, not by inventors and manufacturers. If a customer is not willing to pay for a oduct or service attribute that a company deems important, then that attribute is nvalue adding and not worth the company's time and resources. To determine hat truly adds value to a product or service, a company's long-term vision for its oducts and/or services needs to reflect years of customer and industry research d fact-finding. And, once a commercially viable solution is launched into the arket, continuous improvement initiatives are required to ensure that the product service's value does not decline and continues to meet the needs of customers d end users.

Remember, Lean Six Sigma depends on data-driven analyses to make recom- endations and enact change. Therefore, the current state needs to be measured d effectively communicated, so that a baseline can be created. To create a base- e, observation of the current state is important, and multiple observers can in- ease the credibility of the findings. Observation can lead to the creation of Pareto arts, which are data visualization tools that can identify the sources or reasons r errors or waste and the frequency of occurrence.

For example, the Pareto chart in Figure 11.4 describes the reasons why customers a clothing store decided to return their merchandise during a particular month. Cus- mers provided the following reasons: the merchandise did not fit, the customer anged his/her mind, the merchandise was too expensive to keep, the customer und a defect in the merchandise, and additional reasons, which the store categorized "other." Count or frequency of occurrence and cumulative percentage for each type customer reason are depicted in the Pareto chart example.

gure 11.4: Pareto chart – reasons for customer-returned merchandise.

e categories of defects are arranged from the most frequent to the least frequent. Fre- ency of occurrence is one way to measure areas of waste or process errors. Leaders also need to understand the impact to business profitability for each category of waste and the cost to implement a solution. Leaders will tend to fix errors that have the greatest impact on business profitability. In the example above, store leadership may investigate why items do not fit and determine that it is because of defects in manufacturing. They would then insist that the product suppliers improve their products. If that does not work, then leadership may decide to change suppliers. They will want to fix merchandise defects, because the defects represent poor quality and the store does not want to lose its customers. Leadership, therefore, reevaluates the types of visual inspections that occur before merchandise is made available for display on the store floor and prior to customer purchase. Incremental improvements in process quality are implemented and monitored over time to verify a difference has been achieved.

In a different example, Pareto charts can be used in a manufacturing environment to understand and document the number of and the different types of defects that occur during a manufacturing process. Using the merchandise example, the Pareto chart in Figure 11.5 documents the frequency and types of errors that occur during a nonwoven manufacturing process that creates handbags. Stitching errors account for approximately 30% of the errors, while mistakes in gluing accounts for another 12%, and so on. Leadership may decide to fix processes that fail less frequently, for instance, because those fixes may require less time and less capital equipment changes.

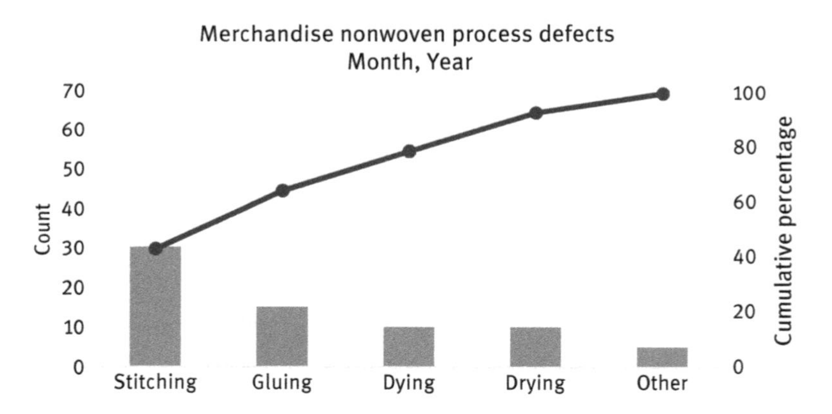

Figure 11.5: Pareto chart – merchandise nonwoven process defects.

In a manufacturing environment, Pareto charts can also compare the types of mistakes or errors made by different operators and/or different shifts of teams. Pareto analysis can also be performed to understand if more process variation occurs at certain times of the month or year. Reason(s) for variability may include:

- Operations equipment requiring preventative maintenance or calibration.
- Variation with newer or older raw materials that behave at different shelf life stages.
- Variation in humidity and temperature changes on the manufacturing floor.

egarding waste reduction and the application of Lean principles to fix process uality issues, success requires that evidence is measured, collected, and presented or discussion, and the "pathway" to achieve desired process quality is methodially planned, communicated, and implemented. These pathways, or roadmaps see Chapters 7 and 8), in general are so important for teams – by seeing that the athway to success, ownership, and accountability can be felt by all.

1.4.2 Six Sigma

igma (σ) is a letter of the Greek alphabet; sigma is also used as an indication of deviaon from the average or mean of a range of measurements. Six Sigma is three measres of deviation on either side of an average of measurements: -3σ and $+3\sigma$ from the ean.

Six Sigma is a philosophy for establishing and measuring process variation. he less variation there is in a process, the lower σ will be; there will be a narrower istribution of measurements and the process will be more controlled, with more epeatable process outputs and a lower probability of process failures. The greater ariation there is in a process, the higher σ will be; there will be a wider or broader istribution of measurements and the process will be less controlled, with less reeatable process outputs and a higher probability of process failures. The broader istribution of measurements indicates process quality issues.

Graphs of process variations are described as bell curves. Bell curves represent normal distribution of measurements above and below the mean. The bell curves n Figure 11.6 illustrate the variations between two processes that have the same ean but different measures of deviation. The top bell curve has a σ of 1, and the ottom bell curve has a σ of 2. Note that the area under the curve in the bottom bell urve is larger than the area under the curve in the top bell curve. This indicates here is a broader distribution of process measurements in the bottom bell curve, nd, therefore, the process that is highlighted by the bottom bell curve is not as ontrolled as the process that is highlighted by the top bell curve.

Bell curves can also be skewed to either the positive or negative side of ne mean. This indicates that a majority of the process measurements are either oing to measure above or below the mean of the distribution of the measurements.

Another great tool for tracking and visualizing process variation data over me is the process control chart. For example, over time manufacturers can rack product defects, number of workplace accidents in a given period, or a

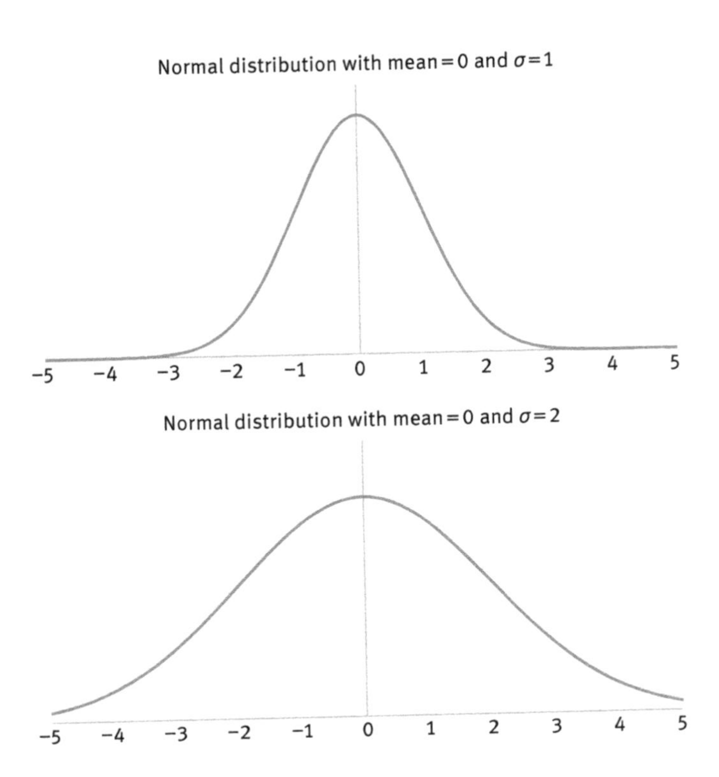

Figure 11.6: Example Bell curves.

piece of equipment's operating temperature or pressure. Process control charts can establish a baseline for a new pilot process so that manufacturing engineers understand normal process variation and unplanned of unintended process variation. Process control charts can indicate when manufacturing operators need to replace or maintain pieces of equipment, and process control charts can be used to confirm that process improvements are taking effect. Process control charts are used to track observed data (according to a predefined sampling plan) and compare it to the known average of the data set (centerline) and ±3 standard deviations from the average (the upper control and lower control limits that appear on the process control chart). In a manufacturing environment, engineers and operators may recognize that a process is out of control, when multiple data events are beyond 3 standard deviations above or below the average in a given period of time.

Process engineers need to understand the threshold for process failure (meaning establish upper and lower bounds of process operating limits) and design a process in which process data can be collected and analyzed, and in which failures happen as infrequently as desired. Failures should be targeted to be small in mpact and to occur outside the Six Sigma range of typical process variation. This vay, an engineer will know when an outlier process measurement is a process ailure (outside of Six Sigma) or when an outlier is acceptable (within Six Sigma nd at the accepted failure frequency rate).[35]

1.4.3 Lean Six Sigma

Iolistically, Lean is an optimization tool and Six Sigma applies a mathematical onstraint to maintain quality. Lean and Six Sigma, combined as Lean Six Sigma, orm a hybrid of waste reduction and process variation reduction tools, through the se of data collection and data analysis. Together, Lean Six Sigma assures product nd process quality. The following examples describe how Lean Six Sigma is implemented, and these quality principles can be applied to any engineering and technology innovation.

---

lanufacturing Example

s a production or manufacturing engineer, you are asked to optimize the floor plan of a production facility. The following questions need to be answered:

- How will the product flow from process to process?
- Will the product flow one piece at a time (one-piece flow), or in batches?
- Will the product flow be a pull or push?
  - A pull flow is when product is moved to the next part of the manufacturing process only when there is a demand.
  - A push flow is when product is moved to the next part of the manufacturing process, regardless of whether the next process is ready to accept the flow of product.

fter the product flow is designed, you need to optimize and shorten the distances for people and roduct to move. Think about the following:

- Where raw materials are stored, and where and how do they enter the manufacturing process;
- When and how product is manufactured, and how finished product exits the process;
- How equipment and individual process steps or manufacturing cells are spaced (typically there will be space constraints);
- The movement of product flow within a manufacturing assembly line;
- Where in-process test articles will be collected and where will they go (e.g., labs, test benches, or measurement stations);
- Where inspections will occur; and
- The degree of flexibility applied to the manufacturing floor (e.g., chairs and tables on wheels).

---

5 Company leaders will know whether a process is in control or not in control by measuring process data over time and determining whether a statistically significant number of process data oints exceed the process control limits, which in terms of Six Sigma methodology are ±3 standard eviations from the average. In other words, approximately 99.7% of the process data in a set of ormally distributed data will measure between the upper and lower process control limits.

Technology Development Example

As an R&D lead in the early stages of product development, and while multiple product concepts are being considered for further development and refinement, it is your job to ensure that prototypes get built and tested to collect the most information possible. This iterative process is typically executed with a limited number of resources (such as time, funds, and personnel), especially in startups that are operating with pre-seed money. As a leader, look to utilize local resources, rapid prototyping capabilities, machine shops, and test labs. Use simulations and modeling techniques, where possible.

Remember, the goal is not to overcomplicate the design of your product to meet a multitude of user needs, and there is no decree that the design needs to be the most elegant. Sometimes, the simplest design is more innovative. If this is your company's first product, you need to get your product to market as quickly as possible to attract the most market share and create a revenue stream. Think Minimally Acceptable Product. As a design team, you can always continue to innovate and release next-generation versions of your product.

Quality Management Example – Training

As a quality manager, one of your roles is to ensure that personnel receive the required training, so they can perform the expected roles and responsibilities for their job functions. Training can take many forms: lectures, self-taught reading or coursework, learning by doing, informal discussions, mentoring and coaching, and so on.

A QMS needs to track a master list of training documents (such as SOPs, policies, and WIs) that personnel need to be familiar with, the current version that is released in the QMS, and the version in which the appropriate personnel need to be trained. For instance, an operations role would need to be trained on supplier controls and procurement, but not necessarily post-marketing regulatory authority reporting. Personnel training needs to be assigned when substantial updates are made to these important documents, and when new content can help personnel in their job functions.

Quality managers need to monitor when personnel are hired and what training needs to be completed, to ensure personnel are prepared to complete the duties and tasks of their particular job role, and also to improve the likelihood that personnel will meet the objectives of the business. Quality managers also need to monitor when personnel are out of the office and when new training is assigned. If personnel miss key training, then their training gaps need to be addressed when they return. Quality managers also need to monitor when personnel change job roles, and as their individual training objectives change (e.g., after every annual performance review). To monitor effectively, a QMS needs to be organized, such as comprehensive training records with employee signatures and date(s) of completion.

One of the management objectives you can create for your organization is a timeframe for when new training assignments need to be completed. An objective for every employee can be to complete the training that is required for their role and to seek new training opportunities for professional development.

2 Medical Devices

bstract: This chapter of *Engineering Innovation* will reference prior content from arlier chapters and will explore the unique management, technical, clinical, regu-tory, and other aspects of bringing medical device products and services to mar-et. This chapter will mainly refer to regulations and examples from the United tates, because the authors' academic and industrial experiences are with medical evices primarily of US origin.

eywords: Regulatory strategy, US Food and Drug Administration (FDA), benefit-)-cost analysis, medical devices, Class I, Class II, Class III, pre-market approval ’MA), 510(k) clearance, de novo classification grant, substantially equivalent (SE), ot substantially equivalent (NSE), predicate device, presubmission meetings, clini-al trials, animal studies, cadaver studies, benchtop testing, marketing, medical evice reporting, Manufacturer and User Facility Device Experience (MAUDE), arking, reimbursement.

2.1 Introduction

he following business development, product/technology development, and opera-ons concepts in Figure 12.1 will be discussed at a high level, specifically address-g how they are applicable to medical devices.

At this point in *Engineering Innovation*, you should have an idea about how chieving development, precommercial, and commercial milestones determines the nancial value of your idea and your company. The primary medical device startup aluation milestones, some of which will be explored further in this chapter, are sted in Figure 12.2.

2.1.1 Impact of Medical Devices

generic terms, a medical device is a machine, mechanism, or apparatus .g., hardware, software, and/or material) that provides a biological or bio-edical function that enables health or wellness. The science and engineer-g that is utilized to design, build, and/or sustain the medical device can be escribed as medical technology. Medical devices and innovative medical chnologies are incredible inventions with the potential to improve, change, nd save lives. For instance, a hip or knee replacement can restore motion to part of the body that does not function anymore due to degenerative arthri-s and the wearing away of joints. The lives of individuals who suffer from continence have been greatly improved by advances in urinary catheter

tps://doi.org/10.1515/9783110521900-013

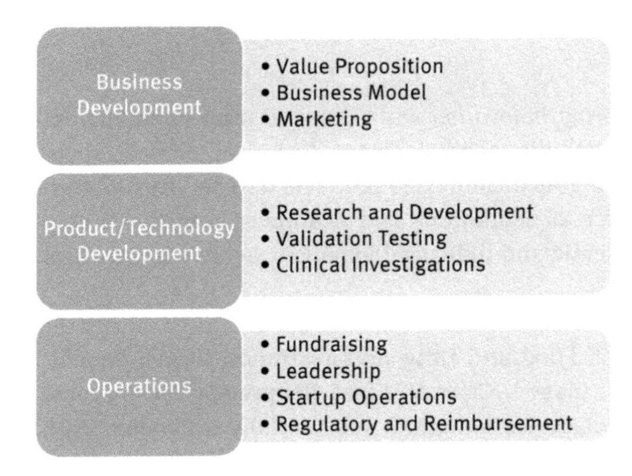

Figure 12.1: Summary of chapter content.

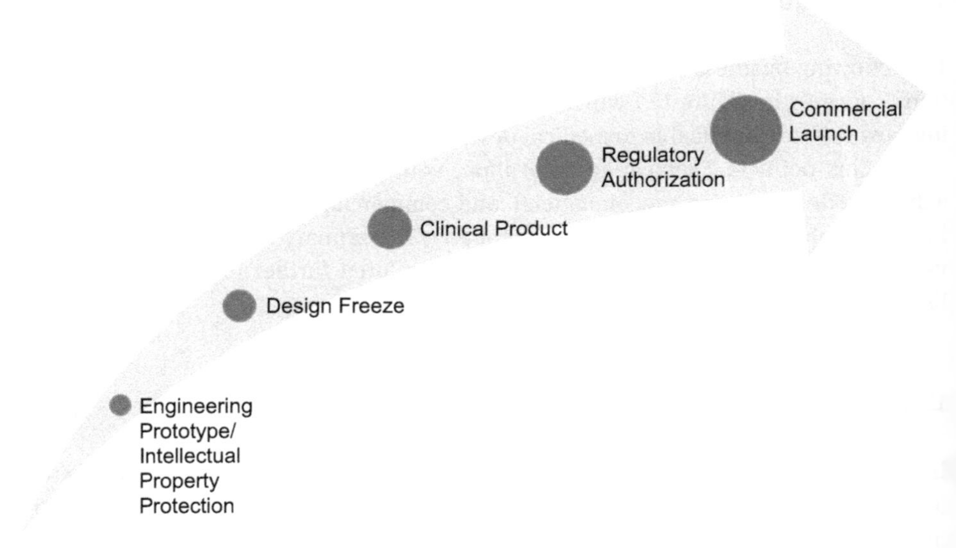

Figure 12.2: Primary medical device startup milestones that impact valuation.

design and urological devices. And an ultrasound image can reveal the miracle of new life thriving inside a mother's womb, or can reveal the unfortunate news that a breast tumor is now metastatic.

The United States Food and Drug Administration (FDA) defines medical devi-
as[36]: "an instrument, apparatus, implement, machine, contrivance, implant,
itro reagent, or other similar or related article, including a component part, or
ssory which is...

intended for use in the diagnosis of disease or other conditions, or in the cure, mitigation, treatment, or prevention of disease, in man or other animals, or

intended to affect the structure or any function of the body of man or other animals, and which does not achieve its primary intended purposes through chemical action within or on the body of man or other animals and which is not dependent upon being metabolized for the achievement of any of its primary intended purposes."

lical devices restore health, provide relief, and improve the quality of patients'
s via a broad range of applications. You do not need to look far to find evidence
medical devices fulfill a critical, and even life-sustaining, purpose. For exam-
heart disease is on the rise, but there are many different circulatory and cardio-
cular medical devices that help sustain a failing heart. Cardiovascular devices,
n as pacemakers, electrically stimulate the heart to help pump oxygenated
od throughout the heart, so that the heart muscle does not need to work as hard.
liovascular devices known as stents are metal structures that are implanted into
tient's vasculature and exert an outward force to keep vessels and arteries open
he midst of atherosclerotic plaques. Stents can also be coated in drug formula-
s, which degrade or elute over time. These types of stents are known as drug-
ing stents (a type of medical device known as a "combination device" because
le the stent itself is a medical device that is implanted into the vasculature of a
ent, the stent also includes either a biologic or pharmaceutical agent that elutes
n the stent over time) that help prevent scarring or restenosis.[37]

Many medical devices are implantable in the patient, but there are thousands of
lical devices that exist *outside* the body. Medical devices are widely used in hospi-
, especially outside of the operating room. For instance, laboratory technicians
in vitro diagnostic medical devices to analyze a patient's blood serum chemistry,
atology, and pathology. Similarly, medical devices are used to prevent hospital-
uired infections, for example, "smart" patches and wound dressings, and devices
educe the number of microbes in open wounds are used in many in-patient and
very areas of the hospital. Large-scale, noninvasive medical devices are also used

f a product is labeled, promoted, or used in a manner that meets the definition in section 201
of the Federal Food Drug & Cosmetic Act, it will be regulated by the FDA as a medical device
is subject to premarketing and postmarketing regulatory controls.

Restenosis is the recurrence of an abnormal narrowing of a blood vessel or heart valve that oc-
after surgery.

in the hospital's radiology department, for example, radiation-emitting magnetic resonance imaging and computed tomography (CT) scanners.

Additionally, medical devices are not limited to physical products. Even software can be considered a medical device – especially when the software provides clinical decision support, that is, a clinician will use information that is analyzed by software to make a decision about a patient's diagnosis or treatment.

Modern medical devices are also prevalent in your own home. To find medical devices in your home, look no further than the bedroom or bathroom: hearing aids, dentures, sleep apnea machines, nose strips, pregnancy tests, tampons, condoms, toothbrushes, contact lens solution, adhesive bandages, prescription glasses, at-home blood glucose meters, and so on are all home health and consumer medical devices.

Medical device technologies are advancing rapidly, thanks to a tight knit global community of physicians, medical device designers, manufacturers and distributors, service providers, lobbyists, patient advocacy groups, strategists, and regulators. This community, known as the Medical Technology or MedTech community, is recognized for sharing knowledge and insights, presenting new data, and forming partnerships in academia, industry, and government.

12.1.2 Overview of Medical Technologies

Devices are classified as either *therapeutics*, which is when a type of treatment or remedy is delivered to the patient, or *diagnostics*, which is when a patient's condition or physiological attribute is measured, monitored, and/or identified. Both have a primary means of device function (the medical device primarily functions via the material or structure of the device, electrical aspects of the device, mechanical properties of the device, thermal capabilities of the device, magnetic properties of the device, optical features of the device, acoustic capabilities, radioactive functionality, etc.). Device mechanisms may occur at the macroscale (such as a mechanical stent provides chronic outward force to ensure an open arterial lumen for healthy blood flow), microscale (for instance, temporal antimicrobial or anticoagulant coatings on a device), or even the nanoscale (e.g., polymer fibers that make up a tissue engineering scaffold used to facilitate cell growth and promote cell signaling).

This wide range of device function and mechanism leads to a diversity of medical device products that can be categorized in many ways. In the United States, the FDA, the regulatory agency that authorizes and ensures safe and effective medical devices for commercial use, organizes the medical device marketing applications it receives according to the following categories. There is a separate device review panel for each category (refer to Table 12.1).

Medical technology research has changed over the years. Advances in other areas of science and engineering such as microscopy and characterization

le 12.1: FDA device review panels (medical device categorization).

esthesiology	Orthopedic
neral Hospital Devices	Clinical Chemistry
, Nose, and Throat (ENT)	Neurology
neral and Plastic Surgery	Pathology
nunology	Toxicology
hthalmic	Dental
diology	Hematology
rdiovascular	Obstetrics/Gynecology
stroenterology/Urology	Physical Medicine
robiology	Molecular Genetics

chniques, smart materials, computational modeling and optimization of nonlin- r systems, and so on, are advancing medical technologies. For instance:

Consumer wearables are changing how patients monitor their health. The FDA recently approved the first Apple Watch medical device, which monitors a user's blood pressure, displays a user's heart rate signal (electrocardiogram), and reports whether a user's blood pressure is normal or abnormal. For an extra monthly fee, consumers can upload their data to the cloud and can customize and send reports to their physician.

Cloud-based technology allows for portability and accessibility of patient data, that is, patients can retrieve personal electronic medical records from the secure cloud. With this information, patients are better informed, patients can monitor their own health trends, and patients can ask their physicians more pointed questions about their condition(s).

Digital technology and robotics are changing the way surgeons are saving patients, even when surgeons and patients are physically located thousands of kilometers apart. Using minimally invasive surgical tools, surgical robotic innovations, and remote surgical care digital technologies, "virtual" surgeries are no longer science fiction. Medical experts from all over the world can converse and make decisions about a surgical procedure, and they can perform "live" surgical maneuvers remotely.

Who says surgical procedures need to be risky? CT images of a patient's diseased organ (e.g., a tumor-riddled kidney) can be taken, and 3D printers can create an exact replica of the organ that looks like and feels just like the diseased organ. Using this 3D-printed surgical tool, the patient's surgeon can practice the surgery in advance on the exact replica of the patient's organ and based on that practice surgery can customize the patient's surgical plan.

chnologies and learnings from other sectors are also being utilized to advance edical technologies. For example, reservoir and petroleum engineers are teaming up with cardiovascular surgeons and medical device engineers, because both of these professionals study and solve problems related to fluid flow, fluid viscosity, and fluid management. The difference is that petroleum engineers study crude oil and refined products, whereas cardiovascular surgeons and medical device engineers study blood flow in the circulatory system. Learnings from the oil industry can be leveraged by the medical industry and vice versa, because the fluid mechanics challenges and the challenges of scale in both industries are similar. For example, the oil industry requires pipelines to move its product across long distances, and blood travels across long distances throughout the circulatory system via arteries and veins.

Now that you have a better understanding about what medical devices are and the benefits they provide, let us discuss the unique aspects of medical device business planning, development, and commercialization preparation.

12.2 Unique Management Aspects of Bringing Medical Devices to Market

12.2.1 Medical Device Value Propositions

As we discussed in Chapters 1 and 4, the Value Proposition is a short summation of your company, your product, your position in the market, and your target market that explains why your product is attractive to customers. A company's value proposition is a promise of the value to be delivered by the company or the product that is purchased. In addition to what is discussed in Chapters 1 and 4, for a new medical device to be successful there needs to be an unmet clinical need (discussed in Chapter 9) and the device needs to be as good or better than the current paradigm of care. For instance:

- Does the new medical technology reduce healthcare costs?
- Does the new medical technology decrease patient morbidity?
- Does the new medical technology decrease healthcare complications?
- Does the new medical technology decrease patient length of stay in hospitals or rehab care centers?
- Does the new medical technology prevent hospital readmission?

To determine unmet clinical needs and come up with a stellar value proposition, medical device manufacturers need to understand the needs of payers (such as insurance companies), healthcare providers, and patients.

So, how can a medical device manufacturer go about determining unmet clinical needs? As outlined in Chapter 9, companies can do the following:

- observe actual surgical cases and procedures,
- interview clinicians, healthcare professionals and patients,

- listen to patient advocacy groups,
- study clinical literature,
- analyze medical databases,
- review statistics and data being reported by payers and provider systems,
- review medical device adverse event repositories, and
- brainstorm with people from diverse backgrounds to understand current processes and clinical methodologies, and to assess gaps present in healthcare.

A medical device's value proposition is in part dependent on the total addressable market (TAM) for the device/technology. To reiterate, the TAM is the size of the potential market or total market opportunity. Considerations to be taken into account for determining the TAM for a given medical device may be the number of people that are afflicted by a certain medical condition and the number of medical procedures that are performed to treat a medical condition. Answers to the following questions can be helpful:

- Is the technology intended for use by multiple types of patients, such as male/female, adult, juvenile, pediatric?
- Is the technology applicable to a variety of different medical/surgical procedures? (For example, a device initially intended to be used for cardiovascular procedures could also be used for interventional vascular access procedures in the lower extremities of the body. Both device uses are intended to promote healthy blood flow.)
- How many procedures are performed in the US annually that require use of the device, or how many procedures are performed in other leading markets annually that require the device (for instance, European Union (EU), Japan, India, China, Brazil, etc.)?
- In which markets will regulatory approval for the device be sought?

The TAM, and therefore the value proposition, can – and most likely will – change over time. For instance, more procedures may need to be performed because patient demographics and/or patient needs change over time. Or a change in a patient subpopulation (e.g., an increase in the elderly patient population, or an increase in the number of patients with diabetes or cardiovascular failure, etc.) can affect the TAM.

The value proposition has to weigh the costs to the healthcare system versus the benefits that will be provided to the healthcare system. Benefit-to-cost ratio is calculated by dividing the 1) monetary value of qualitative benefits that a medical device provides to the healthcare system by the 2) monetary value of qualitative costs that a medical device adds to the healthcare system. The benefit-to-cost ratio is calculated on a case-by-case basis for each medical device, its indicated medical procedures, and intended patient population. If a device has a poor benefit-to-cost ratio (a value that is less than 1.0), then the device introduces additional cost(s) to a procedure or treatment, and the device does not alter current clinical practice by adding benefits or improving healthcare outcomes. New medical device benefit-to-cost ratios are always compared against the benefit-to-cost ratios of commercially available medical devices and technologies. A benefit-to-cost example is described below.

Value Proposition Benefit-to-Cost Example

Your company is developing an orthopedic diagnostic system that will be used on patients in a physician's office, but your company does not plan on obtaining a reimbursement code before commercializing the device. Therefore, your company is betting that patients will pay out-of-pocket to use the device (utilizing high deductible healthcare plans or flexible spending accounts (FSAs)). The doctor would be the purchaser of the system, but would have to "sell" the use of the system to his or her patients to make a profit and offset the purchase cost. Think about this from a patient's perspective and a doctor's perspective. Your company needs to influence prospective consumers and prospective users of your device. What questions should you ask to describe your device's benefit-to-cost ratio?

Patients: Would you rather pay $15 and spend an extra 10 min at the doctor's office to avoid the burdens of managing conditions such as arthritis, osteoporosis, joint reconstruction, fracture repair, or soft tissue repair? These burdens pose extra visits to the doctor's office, extra trips to the hospital, extra days spent in the hospital, extra days spent out of work, extra hours spent in physical therapy, additional cost(s) of in-patient/hospital services, cost of outpatient services (for instance, rehabilitation centers), cost of prescribed medications, and other miscellaneous costs (such as for caregiving).

Doctors/Providers: How would you prefer to spend time with your patients? Would you prefer to spend more time physically examining your patient and discussing your patient's health condition in one-on-one conversation? Would you prefer to spend less time analyzing patient data to uncover key trends/insights? Would you prefer to spend less time transcribing data/information into your patient's electronic medical record? Would you prefer to spend more time postconsult with your patient? Can you increase the margins on the exam so that you can make a profit?

Ensure that the value proposition offers more than a "one-and-done" solution. This is especially important if your company is seeking acquisition or merger as an exit strategy. Medical device company leaders should be striving to develop products that can answer "Yes" to the following questions:

- Can a comprehensive solution be formed around your device?
- Are there service offerings that can be improved or new service offerings that can be developed with your device?
- Can a future portfolio of product offerings and complementary products be built from a single initial product?
- Are there extensions/next generation products that can fulfill a research and development pipeline?
- Can your device be integrated into existing products that are on the market?

If your company is targeting a particular strategic partner (which is a multinational medical device manufacturer) for acquisition, merger, or partnership (with, e.g., an entity like Medtronic, Johnson & Johnson, Boston Scientific, Stryker, etc.), then ckground research should be conducted on the needs and requirements of the ategic "customer." For example, how will your product complement their portfo- of existing products? Will your device be an easy add-on to their sales pipeline? n your device integrate easily into their sales and distribution activities? Will you ed to do additional clinical studies or development work to make the device more ractive to the market? Is there a regulatory approval pathway into the markets in ich their products are already authorized?

.2.2 Medical Device Business Models

discussed in Chapter 2, a business model needs to be determined and validated – is involves validating the business (commercial) assumptions and the conceptual chnical) assumptions regarding the new idea or technology. Business models can tested by poking holes in the assumptions that underlie your business model and scussing "what–if" scenarios with stakeholders, representative users, customers, vestors, and business partners.

In general, medical device business models are different from other types of mpanies for the following reasons:

- Medical device companies require significant financial resources to execute device design and development activities, gather preclinical (e.g., through animal studies, cadaver studies, benchtop testing) and clinical data to validate the safety and effectiveness of their device, apply and receive approval from regulatory authorities to market their product, and to adhere to strict manufacturing and process validation controls.
- The timeline from concept to commercialization of a medical device is on the order of years, not months, and this timeline depends on the complexity of the device and the regulatory pathway that is required. Medical device companies have been known to take more than 4 years, more than 7 years, and even more than 10 years to go to market.
- Quality is not something that should be taken lightly in the medical device industry. The cost of poor quality, for example, repeat of design verification or validations, repeated failure investigations, rejected regulatory submissions, medical device recalls, and so on can severely cut into a company's gross margins and tarnish a company's reputation with customers and with regulatory authorities.
- The medical device regulatory approval pathway is resource intensive and challenging, and a majority of companies do not "get it right" the first time that an application is submitted. Medical device companies should anticipate a high probability of additional information (AI) requests and required testing from regulatory authorities.
- In the United States, once a device is approved for market, it is not always clear that the medical device manufacturer's customers (namely providers) will be

reimbursed by public or private payers (insurance companies), to use the device. It can take up to several years for medical device products and services to be granted the approval to obtain reimbursement.
- Medical device companies typically do not sell directly to their customers, unless the medical device can be used by nonmedically trained consumers. Medical device companies need to work with procurement teams that represent provider healthcare systems, to get their products into hospitals, clinics, rehabilitation centers, and so on.
- Medical device investors will expect medical device companies to protect the intellectual property of their design(s). If a new idea is not patented, there is a lesser chance that a medical device company will be funded and lower likelihood that a medical startup will be acquired.

Medical device business models should reflect 1) the type of medical device technology being developed, 2) the medical device go-to market strategy, and 3) the company's exit strategy.

An important medical device go-to-market strategy needs to answer questions like:
- Will the device be licensed to another company so that it can be incorporated into the other company's solution, especially if the medical device is offered to the other company's customers as one piece of a more comprehensive solution? Remember, even software can be licensed (see Chapter 5). This go-to-market strategy (i.e., development of the device and then licensing the device) would utilize another company's marketing, sales, and distribution infrastructure.
- Conversely, does the medical device developer want to be the "frontrunner" of the product, which involves establishing a company and distinguishing the product brand, developing a sales force, building a network of distributors, and establishing a unique presence in the market?
- Should the emphasis be on leasing the medical device to customers while pushing for customers to purchase the disposables and ancillary services that are required to use the device?
- How will the device reach customers, that is, will the medical device manufacturer sell directly to customers or indirectly through distributors?
- What will be the Average Sales Price of the device?
  - How much will provider systems pay to use the device?
  - Will there be reimbursement from healthcare payers?
  - Can the device be included with the existing overhead costs of the procedure without a separate reimbursement?
- What will be the device Cost of Goods Sold, including labor and materials?
- How much profit can be made per device? Does profit increase with the number of devices produced?

;eneral, companies that are seeking an early exit via merger or acquisition may
x to grow to a limited extent and to develop the technology with fewer internal
ources. For example, company leadership needs to decide whether a majority of
company's functions should be outsourced, or whether company functions
uld be developed internally to create a vertically integrated company (where the
npany itself handles design, development, manufacture, and distribution of the
ice).

Keep in mind that if a company's functions are outsourced, the business model
ypical of a "virtual" company, where most of the work is performed electroni-
y by resources that are not colocated together. These companies, which are not
ically integrated, are forced to rely on partners, vendors, and consultants (hope-
y all included on the company's approved vendor list, refer to Chapter 7) to pro-
medical device design, development, manufacturing services, and so on. Very
ailed proposals and/or supplier quality agreements between a medical device
npany and its vendors are needed to define scope, deliverables, milestone dead-
s, preferred ways of team engagement, and team communication. But as a vir-
business, or one that is not vertically integrated, you relinquish control and
e to rely on vendors to devote experienced resources (according to standards
er than your company's own) that are fully committed to your project. Compa-
s do not want their external resources being stretched too thin by multitasking
overcommitting to multiple projects. With vendors, there is also resource turn-
r on projects that your company cannot control. There is a risk of losing valuable
ividual project knowledge, which can be difficult to transfer to new resources
ernal or external) that join your company's project.

On the positive side, the virtual business model enables startups to do more
n fewer full-time employees, less capital requirements, and lower operating ex-
ses (for instance, with the virtual business model, medical device startups do
need to develop a controlled environment/clean room, engineering labs,
nufacturing processes, and warehouse space). However, the progress being
de by vendors needs to be monitored closely (Chapters 7 and 11) because startup
npanies need to demonstrate tangible, visible progress to their investors.

In addition to medical device business models reflecting 1) the type of medical
ice technology being developed, 2) the medical device go-to market strategy,
3) the company's exit strategy, the business model needs to be flexible.

Business models need to be flexible to changing commercial strategies.
- For example, companies can change from engaging with an outside vendor to manage distribution and warehouse operations, shipping, handling, receipt of return product model, and so on, to instead performing those processes internally.

Business models also need to be flexible to changing financial and regulatory strategies.

  - For example, companies can withdraw from certain markets to conserve available cash and refocus efforts on regulatory submission/application and clinical studies in other markets. Employees need to be flexible to changing priorities, deliverables, and work projects.
- Business models also need to pivot, if there are more promising indications for use and larger TAM for the device. For example:
  - If you market your device as an endovascular device, can it be used in both arteries and veins?
  - Can your device be indicated for pediatrics as well as adults?
  - Can your device be incorporated into commercially available kits (which already include accessory devices)?
  - Can your software technology be licensed to another company that can incorporate the technology into its platform?

There are many other areas that may impact a medical device business model, and company leadership needs to be flexible. Other examples include Quality Management System (QMS) process needs, additional capital and laboratory equipment requirements, additional resource requirements, additional regulatory compliance and certification requirements, additional supplier contracts, and so on.

However, even if the company's business model is sound, fundraising challenges may halt a high-performing and high-potential company in its tracks.

12.2.3 Medical Device Fundraising

Your company may be developing a truly innovative device with enormous life-saving potential, but your story needs to be told. It is also not just about what you say about your device, but also how you say it. How can you invite an audience "in" with a powerful, emotional call to action to make an audience stop and really listen to you? Can you tell an engaging story about your device without going off on tangent thought processes and side stories? Prospective investors do not have a lot of time to listen, so you need to get straight to the point. What makes your device so unique? Why will it transform healthcare? Can you explain technical features or functions about your device to a nontechnical audience? What is the profit potential for your device?

Once your storytelling technique is mastered, be prepared to answer detailed questions from prospective medical device investors.

- Medical device investors want to see a timeline of your valuation milestones (for instance, IP filings, regulatory submission, preclinical (animal) and clinical (human) studies) and how much funding will be required to increase the value of your company and to get to market.

Medical device investors want to know if your device has attractive and multiple growth opportunities, for example, if you were developing an introducer sheath, there are opportunities in large bore catheter procedures and percutaneous coronary interventions and other endovascular procedures. At the same time, they want to see that your go-to-market strategy is focused on a beachhead market (refer to Chapters 4 and 7).
Medical device investors want clear answers about revenue sources, such as the reimbursement pathway, plans for the sales price and profit margins for your device, your sales and distribution strategy, how you will get your device into hospitals, clinics, or homes, how you will market your device domestically and internationally (if that is the strategy), plans for licensing IP assets, and so on.
Medical device investors want to know that your technology is patent protected appropriately (provisional applications, nonprovisional utility patent, design patent, and international filings under the Patent Cooperation Treaty as appropriate; see Chapter 5). Investors want to know the details of the intellectual property portfolio, such as filing and issue dates, the law firm(s) that filed the patents, details of any assignments made of IP rights to the company, and whether prior art searches were performed.
Medical device investors want clear answers about your regulatory pathway. For example, will clinical data be required prior to a regulatory submission, or will animal data suffice?
Finally, medical device investors want to know how their investment will prosper. Investors are going to look for favorable terms and opportunities to make a profit on the order of a 5× to 10× return. Will their investment be diluted in later rounds of funding, how many rounds of funding are anticipated, how will their investment equate to company share options or company shares, what are the convertible note terms, and what is the maximum return they will receive for their investment?

- For example, medical devices that require a 510(k) regulatory pathway to market generally are able to offer investors a quicker return on their investment than medical devices that require a PMA regulatory pathway to market. PMA medical devices have additional preclinical and clinical testing requirements to meet, typically over a longer period of time than 510(k) devices. Since additional funding is required to develop and launch PMA devices, investors can expect for their investments to be diluted, especially throughout later rounds of fundraising as more and more capital is needed to complete development and prepare for commercial launch. PMA devices typically have higher exits than 510(k) devices, however, and investors can expect a higher rate of return with PMA devices. This is because PMA devices generally do not have to compete with substantially equivalent (SE) devices that are already legally sold on the market, unlike most 510(k) devices (as many 510(k) device submissions are based on substantial equivalent predicate devices).

If the medical device involves software, wireless communication with the cloud or interfacing with the internet, investors may want to learn about your cybersecurity policies, especially if your solution has a digital health component. Questions asked by investors may include: Will your company need to protect patient data from cybersecurity breaches? What measures will your company put into place to protect against cyber vulnerabilities? Are these security measures in place presently? Has your company's firewall ever been penetrated, and how sensitive is the information that is stored on the company's network? Has Penetration Testing been performed?

If the software within your medical device relies on an extensive Information Technology (IT) network, investors may want to know about your IT policies. For example, if your solution comprises custom-built software that is based on a third-party product, are your contracts with the third parties still valid? If the third-party product is modified, can your device be upgraded? How are you establishing version control and regression analysis testing between different versions? What is the software version release process? How secure is the software? What are the backup processes and systems? Is antivirus software being used? How secure are company servers? How do you plan to maintain Health Insurance Portability and Accountability Act compliance to protect and manage protected health information, if necessary?

The more complex the technology, the more comprehensive due diligence you should expect from investors. As discussed in Chapter 6, investors want to know that you have thought of everything and that you have a plan and a back-up plan in place when issues arise. Investors will most likely ask for an executive summary, a prospectus, a business plan, and financial models to review. Investors are going to want to see how their money will be spent and how you have been controlling your spending so far – make sure your balance sheets and Profit and Loss statements are comprehensive and accurate. If need be, hire an accountant to help prepare your financial statements.

Now that your business model and fundraising supporting documents are in place, investors look to the leadership team to determine whether your company is a “safe bet” for their investment.

12.2.4 Medical Device Leadership

Medical device executive leadership (i.e., the CEO or President) is accountable to ensure there is enough funding available to operate the business, that enough funds are available to satisfy all accounts payables (e.g., lawyers, regulatory and quality consultants, accountants, designers, developers, vendors across the supply chain, etc.), and that there is enough in financial reserves to meet key development milestones.

The development of medical devices is risky, but many medical devices possess the potential to be highly lucrative. Any form of technology development is prone to setbacks; issues ranging from lack of data reliability to design failures to poor vendor management will occur. It is important for technical leadership to protect the company from unfavorable deals with partners and outsourced vendors, for example, designers, developers, and manufacturers. Unfavorable deals may encompass contracts that are built solely on time and materials spent (versus payment that is contingent on achieving project milestones), with vendors who are slow to hand over deliverables and are poor at managing a steady cash burn rate. In other more specific examples, good leadership is able to make difficult decisions when it comes to preserving financial resources and capping spending by:

- Urging the team to come to a root cause decision on Corrective and Preventive Action device failure investigations prior to device design freeze, especially if the results of the investigation hold up future verification and validation (V&V) testing.[38]
- Balancing regulatory versus commercial risk by finding ways to meet internal regulatory submission/application deadlines (which are dictated by the interests of current and future investors), with the intent of submitting information later to a regulatory agency in a way that will not negatively impact the submission/application:
  - For instance, submitting a sterilization validation protocol only (without the final data) and then submitting the sterilization validation data and final report to the regulatory agency when requested, or
  - For example, including unaged device data (nonclinical or bench testing data) with the submission and then submitting aged device data later, once accelerated or real-time aged devices have completed testing.

Medical device leadership teams should include Scientific or Clinical Advisory Board members with experience in the device's applicable therapeutic area. Scientific/Clinical Advisory Board members can be co-Principal Investigators on grants, can solidify the clinical strategy, can collect and publish device data in peer-reviewed journals, can participate in clinical trials, and can vet other clinical investigators. Clinical data bolsters your device's credibility – clinical data can establish that your solution is safe, effective, and (hopefully) meets an unmet clinical need.

38 Investors do not want their investments to pay for drawn-out failure investigations so that companies can chase down root causes (whether singular or aggregate) by conducting multiple scientific experiments with the support of expensive consultants with specific areas of expertise. Investors only care about outcomes and adhering to budgets and timelines. At some point, there is limited gain that can be made from a team's failure investigation efforts.

Note: In the United States, clinical data will also be needed for medical device manufacturers to apply to the American Medical Association (AMA), or other specialty provider association, for their device to receive a new Current Procedural Terminology (CPT) code or approval to utilize an existing CPT code. Obtaining the appropriate code(s) enables your device to receive monetary reimbursement from private insurance companies or publicly from the Center for Medicare and Medicaid Services (CMS). For more information about medical device reimbursement, refer to Section 12.4.6.

12.2.5 Medical Device Startup Operations

Early-stage medical device startups need to focus on how their device provides an ideal solution to an unmet clinical need, that is, what is the value that is being provided to patients and providers. Startups cannot get bogged down in adding technical features that do not warrant a lengthy development effort or get delayed by completing additional, unnecessary rounds of testing. A company's minimum viable product (MVP) needs to be designed, developed, tested, built, and launched to market as early as possible, in order to be the first to market or to claim the market share that was estimated in the company's business models. Future product iterations that address product improvements (i.e., MVP version 2.0, 3.0) can be launched to market later.

As discussed previously in Chapters 7, 10, and 11, startups rely heavily on the work produced by, and the capabilities offered by, third-party vendors, such as contract manufacturers, contract research and test labs, and consultants. As such, vendor management, with the emphasis on quality, needs to be a huge focus of a startup's Quality team. Startups typically outsource complex research and development activities (such as animal studies, device builds, and safety validation testing), because the cost to develop and maintain laboratory space and manufacturing facilities is cost prohibitive prior to commercial success.

To ensure that lack of investment funding does not impede operations, early-stage startups can apply for grants (Chapter 6). Applying for grants is risky, because it takes a long time to prepare the grant application and then wait for the reviewers to make their award decision. It is not guaranteed that startups will receive grants, but if they do, it can bolster a company's credibility. Early-stage startups should also apply to speak at technology competitions and business plan competitions – grants and competitions not only provide invaluable feedback to startups but also raise a startup's profile and probability of securing funding.

Early-stage startups also may want to consider medical technology accelerators, incubators, and/or fellowship programs to further develop a business model, value proposition, market research, clinical utility, intellectual property due diligence, and reimbursement and regulatory strategies. For example, in the United States, there are opportunities for startups to apply to accelerators and incubators in many cities.

12.3 Unique Technical Aspects of Bringing Medical Devices to Market

This section will discuss the unique technical aspects of bringing medical devices to market – topics related to research, development, and testing (medical device manufacturing is discussed in Chapter 10). Successful medical device projects eliminate and minimize technical risks and minimize the effects of risks to patient health. The benefits to patient health need to outweigh the risks to patient health. In other words, your device has to meet an unmet clinical need and provide enough of a health benefit to a sufficient number of patients that meet the indication for use criteria for a sufficient duration of time and to a sufficient extent. Think: Who benefits (which patients) from the device? What are the benefits of using the device? By how much do patients benefit? For how long do patients benefit?

12.3.1 Medical Device Research and Development: Transitioning from Bench to Bedside

In early benchtop research or translational research,[39] minimum viable prototypes should be utilized at a high frequency to verify proof of concept, simplify the design for manufacturability, and validate user needs via information and observations collected during focus groups or surveys. Research teams also need to select commercially available devices that are similar to the device being developed to conduct benchmarking tests. Benchmarking tests[40] need to be conducted early in the

---

39 When referring to research regarding the life sciences (such as medical devices, pharmaceuticals, biotechnologies), there are three categories of research: basic, translational, and clinical. Basic Research (also referred to as Fundamental Research) is research that is intended to discover a building block of nature for the knowledge of mankind. Basic Research is often funded by the federal government, foundations, academic institutions, or a combination of these. Translational Research (or Applied Research) is the research conducted to take Basic Research out of the laboratory and "translate" it to practical applications. Clinical Research then takes the Translated Research and looks at the effects on a large scale, with a controlled group of subjects over time. This terminology is important because it will determine what type of funding you will be able to obtain and the time frame for commercialization. Although not explicitly discussed in this book, the concepts of Basic Research, Translational Research, and Clinical Research are analogous to the Proof of Concept, α-prototype, and ß-prototype (discussed in Chapters 9 and 10), respectively.

40 Benchmarking tests are nonclinical tests that are performed to compare the performance attributes of a device that is currently in development to similar commercially available medical devices, in order to refine the device's technical specifications.

design and development stages, so that design teams know how the device compares in terms of performance and which device concepts to develop further.

For example, let us say your company is developing a new introducer sheath, that is, a tube or cannula that is inserted into a patient's vasculature as a means to insert other surgical device tools, for example, balloons, stent graft systems, or catheters. The sheath needs to behave and function similar to the existing sheaths in basic ways; otherwise users will refuse to switch from using commercially available sheaths to using your new introducer sheath. The sheath cannot be too stiff because introducer sheaths on the market will be more flexible and perhaps more attractive to end users, and the new sheath should not require higher forces (e.g., insertion and withdrawal forces) to use than the sheaths that are currently available on the market. During surgical cases, the new sheath should not add additional steps/time for physicians to prepare the sheath prior to insertion into the patient. Also, if your introducer sheath includes marker bands that are radiopaque, or if the polymers that compose your sheath cannula include radiopaque additives, then your device should be visible to physicians using fluoroscopic imaging, similar to other sheaths available on the market.

As your research team continues to refine your device concept, preclinical studies (such as animal, cadaveric, user validation studies) are great sources of representative user feedback, information to improve/simplify the design, and improve device efficacy and performance. Preclinical studies can be used to fix issues before going to V&V, that is, to expedite the path to design freeze. Producing small batches of medical devices to prepare for preclinical studies is a great opportunity to identify manufacturing issues that are critical to quality and can reveal where quality improvements can be designed into the manufacturing process, for instance, through the usage of tools and fixtures, and improved operator training.

Many great ideas do not make it further than the laboratory bench or workshop bench because the design of the device is too complex. If a device has a lot of moving parts or parts that require frequent replacement, if the device requires a lot of power to operate, or if the device relies on a lot of firmware and diagnostics, more effort and resources are needed to maintain the device. This increases operational costs for users and potentially increases usability issues, which are reasons that users will not want to purchase these complicated devices.

In another example related to device complexity, the operator–device interface needs to be intuitive for users. If the interface includes lights and sounds, the lights that exist on interfaces need to be visible from multiple angles and visible whether the device is facing down or up. Sounds need to be audible and loud enough to be heard through the noisiest of operating rooms and for the right duration in length to draw the attention of the clinician and surgical team. The colors that are used for the operator–device interface should also be consistent with the colors that are used in the device logo, on the device's marketing materials, and the device abeling and packaging, because the brand and image of the device product should e consistent and familiar to device users.

Many great ideas also do not make it further than the laboratory bench or workhop bench because late-stage development and manufacturing challenges are not mitgated. If the Cost of Goods Sold of a device is too high, then the business model for the evice will not be financially viable. For example, if a plastic device is more complex in esign than it needs to be, then it might be difficult to produce via a standard plastic nolding technique. Since a more complex molding process is required to make the deice, more trial-and-error manufacturing runs will be required and more material scrap vill be wasted before the molder can deliver a product that meets specifications.

The adage "if it's not broken, don't fix it" applies here. If part of your device utizes a nonpatented design that is typically seen in the industry, that is okay. Not every art or component of your device needs to be unique. Going back to the introducer heath example, when you select a stopcock for your device, consider selecting a hree-way stopcock (versus selecting a stopcock that can rotate freely at 360°). Cliniians typically utilize the three-way stopcock to prevent bleed back, that is, to prevent he stopcock from staying in the open position, which would allow for a patient's lood to bleed freely out of the body. With three-way stopcocks, there is tactile feedack when the stopcock is in the closed position, and the stopcock cannot move or wist any further. For introducer sheath material selection pick colors, dyes, additives, nd coatings that are already approved for use in human patients by regulatory agenies. There is existing biocompatibility data for these materials, so regulatory agencies lready know that the material selection is safe and will not produce a toxic effect, vhen in contact with the body.

Before getting too far along in the device research and development timeline, if our device is required to adhere to a regulatory standard, then make sure it does so nd that you have the evidence to prove it! If you do n'ot have the resources to test our device, then outsource this testing to a certified and recognized test facility.

2.3.2 Medical Device Validation Testing Requirements

As discussed in Chapter 10, validation testing is different from verification testing. Verification testing confirms that you have built a device that meets its product equirements at a system, subsystem, and component level. Validation testing onfirms that your device (when used as intended) meets actual user needs, such s the needs of the patients and their caregivers, providers, and other healthcare rofessionals. Validation is probably the most important part of any regulatory ubmission/application because it demonstrates device safety and effectiveness nder actual use conditions.

Consider three broad types of validation testing: safety validation, performance alidation, and user (usability/human factors) validation.

- Safety validation testing confirms your device is safe and effective for the intended use as described in the device Instructions For Use/service manuals that are provided with your device.
- Performance validation testing confirms your device performs and functions as intended.
- User validation testing confirms your device can be used by the intended user (whether the intended user is a clinician, healthcare professional, or patient).

Regulatory agencies require medical device manufacturers[41] to provide evidence of safety, performance, and user validation. This is especially true for Class II and Class III devices that are regulated by the US Food and Drug Administration. Class II devices require the evidence of general controls[42] (such as device operating instructions, device adverse event reporting, device registration and listing, Good Manufacturing Practices (GMP)), *and* special controls[43] (such as nonclinical mechanical performance data, sterility data, biocompatibility data, postmarket surveillance) to demonstrate safety and effectiveness. Class III devices, because they represent the highest risk to patients, require the evidence of general controls, special controls, and their own type of requirements known as Premarket Approval (PMA).

Preclinical or human clinical studies can be designed and performed to validate device safety *and* performance *and* usability by confirming that the device meets indications for use (i.e., the disease, condition, or pathology and the patient population for which a device is intended) and intended use statements that are made by the medical device manufacturer.[44]

41 According to ISO 14971, Medical Devices – Application of Risk Management to Medical Devices, a manufacturer is a "Natural or legal person with responsibility for the design, manufacture, packaging, or labeling of a medical device, assembling a system, or adapting a medical device before it is placed on the market and/or put into service, regardless of whether these operations are carried out by that person himself or on his behalf by a third party."

42 General Controls are the basic requirements of the May 28, 1976 Medical Device Amendments to the United States Food, Drug, and Cosmetic Act. The General Controls in the Amendments apply to all medical devices. Class I devices only need to adhere to General Controls.

43 According to the FDA, "Special controls are regulatory requirements for class II devices. FDA classifies into class II devices for which general controls alone are insufficient to provide reasonable assurance of the safety and effectiveness of the device, and for which there is sufficient information to establish special controls to provide such assurance." See https://www.fda.gov/MedicalDevices/DeviceRegulationandGuidance/Overview/GeneralandSpecialControls/.

44 According to 21 CFR Part 801, intended use is "The objective intent of the persons legally responsible for the labeling of devices. Objective intent may, for example, be shown by labeling claims, advertising matter, or oral or written statements by such persons or their representatives. It may be shown by the circumstances that the article is, with the knowledge of such persons or their representatives, offered and used for a purpose for which it is neither labeled nor advertised."

- **Safety.** Animal studies may provide histopathology evidence to prove that your device, when inserted and used in the animal, does not cause trauma.
- **Performance.** Testing demonstrates that your device meets confidence and reliability performance levels, for example, 38 out of 40 devices must perform as intended.
- **Usability.** Questionnaires can be included for the physician to document their answers about every procedure or test sample, thereby addressing usability concerns (e.g., catheter preparation, insertion, handling), with responses ranked according to a scale.

2.3.2.1 Safety Validation Testing

xamples of safety validation testing include biocompatibility, sterilization, and ging (accelerated device aging versus real-time device aging) testing.

iocompatibility

iocompatibility testing is performed per the International Organization for tandardization (ISO) 10993, Biological evaluation of medical devices family of andards, which include individual test method and animal model guidance. iocompatibility testing ensures that your device does not cause an adverse or armful reaction when in contact with biological fluids or tissues. Biocompatiility testing is performed according to the level of risk associated with the use f the device. For instance, long-term implantable devices require more biocomatibility tests compared to devices that are implanted only temporarily, and deices that come into contact with blood or cerebrospinal fluid require more iocompatibility tests than devices that do not come into contact with such flus. Biocompatibility tests include testing for irritation, sensitization (testing for ensitivity to chemicals used in the device), thrombogenicity (causing blood ots), genotoxicity (testing for genetic mutagens), and pyrogenicity (testing for e induction of fever), to name a few. Some tests can be completed in as little as veral days, and others may take up to several weeks.

terilization

terilization of medical devices is performed to prevent microbial contamination of nished medical devices. Validation of a sterilization process ensures that a steriliation method (such as using either gamma radiation or gaseous ethylene oxide) ill achieve a device sterility assurance level/microbial contamination probability f 10^{-6}. Medical device manufacturers select the "worst case" of their device family sterilize, that is, where it would be most difficult for the gas/radiation to permete. For example, if you are developing a range of introducer sheaths, you would elect the smallest inner diameter size to validate the sterilization process because the lumens are smaller, or there are tighter interference fittings between the introducer sheath and the dilator.

Sterilization validation ensures that a sufficient dose of gas or radiation is administered to sterilize the product. Once sterilization process parameters are finalized, three different lots of product are sterilized and tested to validate a sterilization protocol. If results pass, those three lots of product can be used for clinical or commercial purposes. But if sterilization validation fails, then the three lots of product cannot be used.

Sterilization validation is very costly and comes with risk; be sure to utilize a microbiological/sterilization expert to review your sterilization validation protocol. Sterilization validation also is used to validate the design of your packaging and to confirm whether your packaging materials create a sterile barrier for the medical device.

Aging

Aging testing, or shelf life testing, validates that the device and its packaging will remain effective over the lifetime or shelf life of device. For instance, if your device has a battery, will the battery be fully functional after the period of complete shelf life? If there is a coating on your device, will the coating perform as intended after the period of complete shelf life? Aging testing determines the shelf life of your device, that is, how long can the device be stored prior to use (e.g., 6 months, 1 year, 2 years, 3–5 years).

The longer a device's shelf life, the less often unused devices will be returned or disposed of by customers. Device returns and disposals pose additional manufacturing and operating costs for medical device manufacturers, pose an inconvenience for customers and medical device distributors, and pose lost revenue across the value chain.[45]

Regulatory agencies want to see evidence that the design performs as intended and passes bench testing requirements, after the complete period of shelf life. Shelf life testing offers controlled environments to simulate storage environments. Shelf life can pass in real time or be accelerated in order to save time. Medical device manufacturers can accelerate the aging process by exposing the device to periods of extreme temperature, pressure, and humidity to mimic the aging process.[46] This testing can be conducted by the medical device manufacturer using ISO-established

45 Packaging and product stability need to be validated before release, and a device cannot be used by a provider past its expiration date. Shelf life studies and establishing a device expiration date provides device manufacturers legal protection and the ability to control the product life span/life cycle. With this information, medical device manufacturers will know when to inform customers that they will need additional or replacement products.

46 See ASTM F1980, Standard Guide for Accelerated Aging of Sterile Barrier Systems for Medical Devices.

tocols specific to their device, or the packaged product can be sent to third-party t labs that specialize in shelf life testing of medical devices. If the latter is cho-, medical device manufacturers are responsible for verifying the third-party test is using the proper ISO protocol. In addition, medical device manufacturers will l need to conduct functional performance testing of the actual device. Functional formance test data that is collected from accelerated aged devices will be in-ded in a regulatory submission/application. Then, functional performance data m real-time aged devices, which need to be stored real-time under typical tem-ature and humidity conditions for the complete shelf life period (for instance, years), will be submitted later to the regulatory agency after FDA clearance/ proval has occurred.

3.2.2 Performance Validation Testing

example of performance validation testing includes packaging transportation, tribution, and storage simulation testing. This testing simulates transportation d handling conditions to test whether shipping, distribution, and storage affects performance of your device and its packaging system.[47] This testing validates t the device and its "sensitive" or "delicate" components, while encased within packaging, will stay intact even while undergoing extreme storage or travel con-ions, such as very frigid temperatures, high levels of humidity, extreme high or v levels of atmospheric pressure, or travel over bumpy terrain. People in the in-stry refer to some forms of packaging simulation testing as "shake, rattle, and l." Think about it like this: if your device is packaged in a shipper box that is ing on a tarmac or warehouse in the middle of the desert or on the top of snowy untains, will the packaging still hold up and will there be no evidence of wear d tear on the device?

3.2.3 User Validation Testing

er validation testing was discussed previously in Chapter 10. Medical device er validation testing may involve interactive discussions with representative ers. User validation testing can be performed in settings that mimic actual use, example, if the device is used in a catheter lab or an operating room, you can c prospective users to look at and handle the device in an environment with mparable noise and lighting levels. Users can be asked questions about the

See ASTM D4332, Standard Practice for Conditioning Containers, Packages, or Packaging Com-nents for Testing, and ASTM D4169, Standard Practice for Performance Testing of Shipping Con-ners and Systems.

look and feel of the device, how clear is the device Instructions For Use (IFU) wording, how legible and visible is the packaging, how complete is the labeling, and so on. Users can provide feedback on the device–user interfaces, whether audio or visual notifications are easily interpreted, and any other specific feedback on features you want the users to notice.

While validation testing is occurring, medical device manufacturing processes should also be finalized and qualified.

12.4 Other Unique Aspects of Bringing Medical Devices to Market

This section will discuss other unique aspects of bringing medical devices to market. Topics such as medical device regulatory strategy, medical device clinical trials, medical device marketing, and medical device reimbursement will be discussed.

12.4.1 Medical Device Regulatory Strategy

A medical device manufacturer's regulatory strategy needs to be based on the intended markets where the manufacturer will need approval or clearance to market the medical device, that is, there are regulatory agencies in the United States, Canada, Australia, Japan, and the EU, each with different regulatory requirements and pathways to market.

Regulatory strategy in the United States is dependent on the answers to several very important, seemingly straightforward questions. The complexity lies in the details. FDA guidance documents are helpful, but their recommendations can be confusing and answers to medical device manufacturers' questions are not always straightforward.[48]

---

48 FDA guidance documents can be referenced here: (https://www.fda.gov/medicaldevices/deviceregulationandguidance/guidancedocuments/default.htm). In every FDA guidance, you will read the following: "FDA's guidance documents, do not establish legally enforceable responsibilities. Instead, guidances describe the Agency's current thinking on a topic and should be viewed only as recommendations, unless specific regulatory or statutory requirements are cited. The use of the word should in Agency guidance means that something is suggested or recommended, but not required." FDA guidance documents are updated frequently. Recognized Consensus Standards (standards that are produced outside of the FDA but recognized by the FDA) can be referenced here: (https://www.accessdata.fda.gov/scripts/cdrh/cfdocs/cfStandards/search.cfm).

ich type of application is required for your device to achieve regulatory ap- val or clearance?
nary medical device applications to the FDA include de novo classification requests, (k) submissions,[49] and PMA applications.[50] 510(k) devices are either classified as ss I or Class II. PMA devices are classified as Class III. If you believe your device is ss I or Class II with no recognized single predicate,[51] there is a de novo classification hway. The de novo classification requires a lot of premarket data to be included in classification request, a comprehensive risk–benefit analysis to be performed, and dical device manufacturer implementation of risk controls, because a de novo appli- on will serve as the benchmark for future device applications.

ich device classification does your device fall under?
A Device Classification is dependent on the types of controls that are required to en- e patient and user/operator safety – as class number increases from I to II to III, the re controls are required, because there is greater risk of adverse effects or harm to patient. For example, a toothbrush will have fewer required controls and less regu- ory oversight than a dental implant.[52] A stethoscope will have fewer required con- s and less regulatory oversight than an implantable pacemaker.

ss I device. These devices are subject to general controls to provide reasonable urance of safety and effectiveness, for example, labeling, packaging, registra- n, records and reports, and GMP. Class I devices include devices such as tongue ressors, bandages, medical examination gloves, reagents that are used in clini- labs, and so on.

ss II device. These devices are subjected to general controls and special controls assure safety and effectiveness, for example, biocompatibility, sterilization, endo- in, transit and accelerated aging testing, and electrical safety. A Class II device is ically a short-term implantable device (the device may come into contact with

---

"A 510(K) is a premarket submission made to FDA to demonstrate that the device to be mar- ed is at least as safe and effective, that is, substantially equivalent, to a legally marketed device CFR §807.92(a)(3)) that is not subject to premarket approval."
"Premarket approval (PMA) is the FDA process of scientific and regulatory review to evaluate safety and effectiveness of Class III medical devices. Class III devices are those that support or tain human life, are of substantial importance in preventing impairment of human health, or ch present a potential, unreasonable risk of illness or injury."
A predicate device is a medical device that is already legally marketed, but there is more to it. CFR § 807.92(a)(3) states "A legally marketed device to which a new device may be compared a determination regarding substantial equivalence is a device that was legally marketed prior May 28, 1976, or a device which has been reclassified from class III to class II or I (the predi- e), or a device which has been found to be substantially equivalent through the 510(k) premar- notification process".
Medical device classifications are discussed in 21 CFR Part 860.

blood or cerebrospinal fluid for a specific period of time, such as during a medical procedure), and/or the device does not perform life-saving or life-sustaining functions. Class II devices include introducer sheaths and catheters, ultrasound devices, powered wheelchairs, infusion pumps, air purifiers, biopsy needles, and so on.

Class III device. These devices are subject to general controls, special controls, long-term performance data (e.g., stent corrosion testing, cyclic wear and tear testing, testing for the degradation of materials which leads to particulate generation), and clinical data. A Class III device is typically a permanent implantable device, or a device that performs life-saving or life-sustaining functions, or prevents impairment of human health, or can pose an unreasonable risk of illness or injury when used improperly. Class III devices include artificial hip joints, pacemakers, automatic defibrillators, heart valves, ventricular assist devices, HIV diagnostic tests, and so on.

Is your device SE or not substantially equivalent (NSE) to an existing, cleared, or approved (legally marketed) device?

In the United States, the answer to this question will determine whether a device manufacturer pursues a 510(k) submission or PMA application when the device in question can utilize an existing medical device product code,[53] or a de novo classification request when there are no known product codes for the device in question. The regulatory application pathway will determine the duration and resources needed to develop the device, how much money will need to be raised, and what data will be collected and submitted to the regulatory agency.

Note: A product code identifies the generic device category for the FDA. For example, the product code "DYB" refers to catheter introducer devices, which are overseen by the Cardiovascular Panel at the FDA. DYB devices are Class II devices that require a 510(k) submission type.

To determine whether a device in development is SE or NSE,[54] medical device manufacturers need to think about the type of device and its function(s), the device indications for use (i.e., how the device is used in practice and for how long, what types of patients need this device, and in what types of medical procedures will the device be utilized), the technology (e.g., design, materials, energy source, other features) that

53 FDA Guidance, Medical Device Classification Product Codes – Guidance for Industry and Food and Drug Administration Staff, 2013. The Product Classification Database can be found here: https://www.accessdata.fda.gov/scripts/cdrh/cfdocs/cfPCD/classification.cfm.

54 Refer to FDA Guidance, Benefit-Risk Factors to Consider When Determining Substantial Equivalence in Premarket Notifications (510(k)) with Different Technological Characteristics, 2018 and FDA Guidance, The 510(k) Program: Evaluating Substantial Equivalence in Premarket Notifications [510(k)], 2014.

ne device utilizes to achieve a clinical benefit, device risk and possible harm to pa-ents and users, and what device data is needed to prove safety and effectiveness.

Which preapproval studies will be required to validate the device user re-uirements and to ensure device safety and effectiveness?
o answer this question, medical device manufacturers need to understand how neir device will be used:

- How will clinicians use the device to diagnose or treat patients?
- Is it a life-saving, or life-sustaining device?
- Is it the primary device that will be used in a medical procedure, or is it complementary as a safety device (nontherapeutic)?
- What back-up options do clinicians have if there is a device failure?

o determine which preapproval studies will be required, medical device manufac-urers need to understand the level of risk associated with using the device:
- If the device contains electrical equipment, are any of the functions considered essential to performance, that is, would an absence or degradation of a function result in unacceptable risk?[55]
- If the device contains software, what level of concern should be applied to the software: major, moderate, or minor?[56] The software's level of concern estimates the severity of injury a device can cause to a patient or user because of the device's failure to function, inherent design flaws, or through normal, anticipated use.

o also determine which preapproval studies will be required, medical device man-facturers also need to understand clinical precedents:
- If the device will utilize the 510(k) submission pathway, does clinical data already exist for the identified predicate device(s)?
- Is the device similar to other devices that are complex in nature, where clinical data is typically requested?
- Is there an animal model that can provide similar, applicable data about device safety and effectiveness before testing the device in humans?
- Would it be unethical to perform human clinical studies with the device? If utilizing animal or cadaveric data, can this data correlate to human data?

5 See IEC 60601-1-2, Medical electrical equipment – Part 1-2: General requirements for basic afety and essential performance – Collateral Standard: Electromagnetic disturbances – Require-ents and tests.

6 See FDA Guidance, Guidance for the Content of Premarket Submissions for Software Contained n Medical Devices, 2005.

Should a medical device manufacturer hold presubmission meetings with regulatory agencies, such as the US FDA?
The FDA offers medical device manufacturers the opportunity to utilize the FDA presubmission meeting process (referred to as "presub") to receive feedback on: device classification and predicate device selection, what controls are required to demonstrate safety and effectiveness, what preclinical and clinical data would be required for the submission/application, and so on. Presubmission meetings are beneficial, because they foster an open dialog between the regulatory agency and medical device manufacturers. These meetings can also save time and money for the medical device manufacturer, if the FDA agrees to fewer test or data requirements prior to marketing authorization.

The discussions and decisions made during presubmission meetings are thoroughly documented and included as part of a medical device manufacturer's file with the FDA. Any changes that are made by the medical device manufacturer to the device design or V&V plan that differ from the feedback that is received during presubmission meeting(s) would be met with scrutiny from the FDA and would have to be defended at length by the medical device manufacturer. This is a great reason to solicit the expertise of at least one regulatory consultant to help prepare the presubmission meeting application and help facilitate the presubmission meeting. A good regulatory consultant will ensure that whatever the medical device manufacturer discloses in the application and during the meeting does not cause difficulties later in the regulatory process. If your company decides not to conduct a presub, then you will rely solely on the guidance of your regulatory consultants in finalizing your V&V plan and preparing your regulatory application or submission.

12.4.1.1 FDA Challenges

The FDA is a government agency entrusted to protect the American population from harmful new medical device technologies and products. The FDA is the medical device manufacturer's first customer, that is, if you do n'ot meet the FDA's needs, then your device would not be cleared or approved to be sold in the United States. Marketing authorization decisions from the FDA include PMA approval, 510 (k) clearance, or de novo classification grant.

For each type of regulatory application, the FDA has review calendar duration goals recently established by the US Department of Health and Human Service that the FDA must meet for a majority of the applications it receives. For example, the FDA has a goal to finish the reviews of at least 55% of its de novo classification requests within 150 calendar days by fiscal year (FY) 2019 and 70% by FY 2022.[57] For most types of regulatory applications, there is an initial checklist that is completed

57 FDA Guidance, FDA and Industry Actions on De Novo Classification Requests: Effect on FDA Review Clock and Goals, 2017.

by the FDA reviewer, and if the reviewer determines that something is missing in the application package, then the FDA will not begin the review of the application.

While the FDA is reviewing your submission/application, if the agency determines that AI is required (which is not already included in the submission/application), then a request for AI will be sent to the medical device manufacturer. The FDA review "clock" will stop until the agency receives a complete response from the medical manufacturer to that request. And, if more than 180 days go by and the medical device manufacturer has not submitted the additional info/data that was requested, then the FDA can withdraw the manufacturer's application. Upon the manufacturer's reapplication, the manufacturer will be required to pay the application's user fees again.[58]

12.4.1.2 CE Mark Challenges

As briefly discussed in Chapter 11, the EU under Directive 93/68/EEC requires a Conformité Europêene, or CE Marking, to allow medical device manufacturers to commercially market and distribute medical devices in the EU. To address changes in medical technology, the EU adopted the Medical Device Regulation (MDR) in 2017 to replace the existing Medical Devices Directive (93/42/EEC). By 2020, the MDR will be in full effect, and medical device manufacturers will need to update their QMS processes to comply with the new MDR.

To receive a CE Mark, one of the milestones that medical device manufacturers need to complete is the audit process with Notified Bodies. Notified Bodies are third-party organizations that are authorized by the EU to grant CE Mark to medical device manufacturers through a standard review process. Notified Bodies will conduct a series of audits with medical device companies and will assign a technical reviewer to review the CE Mark technical file submission. The CE Mark technical file is similar to the FDA's regulatory application that medical device manufacturers submit in the United States. However, there are some differences. For example, to receive CE Mark Approval, the technical file needs to include a clinical validation report, whereas clinical validation reports are not mandatory for all 510(k) submissions in the United States.

There are two audits that need to be completed – Stage 1 and Stage 2. Stage 1 audits ensure all of the pieces are in place to meet CE Mark requirements, and Stage 2 audits review the content and evidence that requirements are being met. The length of each audit, that is, the duration of time a Notified Body auditor spends with a company, depends on the size of the company (number of full-time employees).

58 The FDA collects user fees from medical device manufacturers who submit applications. These fees can be costly, but the FDA establishes lower user fees for small businesses, pediatric device manufacturers, and government or academic (nonprofit) entities. See FDA Guidance, Medical Device User Fee Small Business Qualification and Certification, 2018.

ISO 13485 certification of the QMS is required if medical device manufacturers wish to market their device in the EU. ISO 13485 certification follows a similar Stage 1 and Stage 2 audit format; the scope of the audits is dependent on the wording on the ISO 13485 certificate. In other words, does a company claim to provide design services, design and development services, or does a medical device company manufacture and distribute product, or does a medical device company distribute sterile product?

For example, if a medical device manufacturer wishes to include the words "sterile medical device" in the ISO 13485 certificate, then a Notified Body needs to perform a microbiological audit and involve the company's contract sterilizer. The Notified Body will want to review sterilization validation protocols and reports, bioburden data, audit plans, personnel training records, equipment V&V, equipment maintenance and calibration records, and sterilization release records to ensure that sterilization standards are being followed.

Once a medical device manufacturer receives a CE Mark, the Notified Body may follow up on an annual or biannual basis (depending upon how long it has been since the CE Mark was granted) to ensure that the company is still compliant and deserving of the CE Mark. The same is true for ISO 13485 certification. As of 2017, Notified Bodies have begun the Medical Device Single Audit Program. If a medical device manufacturer wishes to go to market in the EU, Canada, and other countries, one series of audits versus several can be conducted for the multiple markets. This is a cost savings for medical device manufacturers, but also places a burden on Notified Bodies.

If a medical device manufacturer does not wish to establish a commercial location in the EU, it will need an authorized EU Representative to act on its behalf and be the distributor of the product in the EU. Authorized EU representatives must be in place before receiving CE Mark. It is a key partnership with medical device manufacturers, because the Authorized Representative will be the "face" of the medical device company in the EU. The Authorized Representative will help companies navigate the unique marketing, regulatory, and commercialization requirements in each market.

12.4.2 Medical Device Labeling

There are other industries besides medical devices where labeling is heavily regulated, for example, chemicals, consumer electronics, and food. However, medical device labeling follows unique regulations[59] because medical device labeling protects users *and* patients, and the majority of medical device labeling is written for a

59 Medical device labeling requirements are discussed in US 21 CFR Part 801 Labeling.

healthcare audience, that is, physicians, with an expected level of medical training and expertise. Medical device labeling is defined as any information that is provided with your device, for example, in the form of symbols[60] or words on your device itself, information that appears on packaging, and product information that is included in accompanying device instructions (which could be provided via hard copy with the device or electronically).

Labeling strategy is based on the individual markets where the device will be sold. Labels need to be translated into languages that are most prevalent for the intended users and recipients of labeling information in each market. Label translations can be costly, and each language requires a different, individual label part number.[61]

Labeling strategy is also based on the capacity in which the device will be used, for example, there are different labeling requirements for research and development devices that are not yet approved for use in humans versus devices that are being used for clinical investigations versus devices that are cleared or approved for a given market.

Labeling is an important part of a company's brand. For example, are the correct colors (according to Pantone or hex number) used to represent the company's logo on the label? Does the font used on the label match the font that appears on the product itself and the company website and other marketing materials? Does the labeling appear smudged? Do the labels have a crisp, glossy finish? Are the labels securely fastened onto your product and packaging, or do they peel off easily? Are the labels wrinkled? The appearance of a device's labels can say a lot about the company's quality.

Labeling provides the following information about your medical device to users and customers:

- Labeling provides key information to users, for example, device size, accessory/compatible device size constraints.
- Labeling provides key information to manufacturers and distributors, such as label identifier, lot or batch number, device serial number, date of manufacture, manufacturer's address, and the number of devices that are contained in a package.
- Labeling provides elementary information, for instance, how to turn a device on/off, how to silence a device, and what the device looks like.

---

60 See ISO 15223–1, Medical devices – Symbols to be used with medical device labels, labelling and information to be supplied – Part 1: General Requirements.

61 In place of words on a label, try to use universal symbols. You do n'ot need to say "Use-By" if you are using the ISO 15223 recognized symbol for Use-By-Date, you do n'ot need to say "Manufactured By" if you are using the ISO 15233 recognized symbol for manufacturer, and you do n'ot have to say "Quantity" if you are using the recognized symbol of the box that contains a device quantify number.

- Labeling includes important claims about your product, for example, is your product single use or can it be used multiple times?
- Labeling includes important safety information, mainly in the form of symbols, for example,
  - Is your product sterile, and if so, how is it sterilized?
  - Instruction to not use your device if the packaging is compromised.[62]
  - Instruction to not use your device after an expiration or use by date has passed (FDA recognizes the expiration date format YYYY-MM-DD).
  - Is your product nonpyrogenic (i.e., bacterial endotoxin/pyrogen and/or material mediated pyrogenicity is tested on your product and is part of routine batch sampling/in-process testing)?
  - If your device contains electrical parts or components[63]:
    - Are your device's electrical components protected from the risk of exposure to particulates or water droplets?
    - Can a defibrillator be used on a patient that is also receiving the device?
    - Is there a possible risk of shock to the patient and/or user?
    - What are the electrostatic discharge and electromagnetic immunity protections?
- Labeling provides direction about storage and operating/use conditions for your device, for example, temperature and humidity limits or instructions to not store in direct sunlight.
- Labeling provides warnings to users. For instance:
  - Does your device emit radiation or does it contains lasers?
  - Must your device be prescribed by a physician?

Labeling also informs users of your device whether a device IFU needs to be reviewed before using your device, and whether the IFU is located online. Instructions for Use typically contain:
- A description and pictures of the device.
- Indications for use.
- Intended use.
- Device sizing information.
- Information about how the device is supplied and what is included in the packaging.

---

62 Labels can be applied to the different levels of packaging that protect medical devices during shipping, storage, and handling prior to use, for example, trays, cartons, and shipper boxes.
63 See IEC 60601–1, Medical electrical equipment – Part 1: General requirements for basic safety and essential performance and IEC 60601–1-2, Medical electrical equipment – Part 1–2: General requirements for basic safety and essential performance – Collateral Standard: Electromagnetic disturbances – Requirements and tests.

- Device preparation notes.
- Use and disposal procedures.
- The function(s) of any user interface components (buttons, switches).
- Information regarding what any warning audible or visual indications mean.
- Definitions and descriptions of any symbols that appear on the labels.
- Identification of what additional devices need to be used along with your device (i.e., accessories).
- Whether user training is required (if any).
- Company contact information.
- Information concerning how users and customers can report adverse events to the medical device manufacturer.

abeling is an important part of a device's risk management strategy/plan. Many ser risks can be mitigated through effective labeling because labeling is a key part f user training, especially training regarding a device's contraindications, warngs, precautions, and potential adverse events.

Labeling changes that are made after a device is cleared or approved need to go rough a thorough internal regulatory and quality review with the medical device anufacturer, and then the manufacturer may be required to notify the appropriate gulatory agencies. For example, depending on the change, the regulatory agency ay need to review additional clinical or safety data to support the labeling claims .g., biocompatibility, transportation verification, sterilization) before approving anges.[64]

Medical device labels also contain Unique Device Identifiers (UDI).[65] A UDI consts of device identifiers (unique codes based on the model of the device) and production identifiers (such as the lot number, serial number, date of manufacture, se by date or expiration date) in readable format and automatic information and ata capture format: 2D or 3D barcodes. UDIs will assist regulatory agencies with e unique identification of medical devices for recalls, adverse event reporting, nd will assist manufacturers with distribution and customer records.

Marking a device's identifier directly onto the device is required for 1) implantble devices, 2) devices indicated for more than a single use (these devices are used ith multiple patients), or 3) devices that require reprocessing for repeated use.[66] irect marking is not required if the technology does not exist for direct marking nto the device (the size or base material of the device does not permit direct markg), or if there will be a risk to patient safety or device performance. Class II

4 FDA Guidance, Deciding When to Submit a 510(k) for a Change to an Existing Device, 2017.
5 21 Code of Federal Regulations Part 830, Subpart B: Requirements for a Unique Device entifier.
6 FDA Guidance, Unique Device Identification: Direct Marking of Devices, 2017.

devices (nonimplantable devices that do not sustain life) do not require direct marking. For Class II devices, UDIs can be used.

Labeling information needs to be tightly controlled and managed by the medical device manufacturer. All labels have artwork dates (i.e., a form of version control for labels that include the calendar date that the label was approved)[67] and/or revision information that is stored with the labeling files in the QMS. This labeling information is communicated throughout the supply chain, from suppliers and distributors down to the individual device user. The ownership of labeling specifications and labeling files and changes needs to reside with the medical device manufacturer, even when labeling operations may be outsourced and conducted by a supplier. This way, the right label stock or paper that the label is printed on is procured and used, and regulatory risk is owned and mitigated by the medical device manufacturer.

12.4.3 Medical Device Clinical Trials

Clinical trials may be required to confirm that your medical device performs as intended in your target patient population and is safe and effective. Data from clinical trials should be used strategically to verify and expand device indications for use through additional medical procedures. Data from clinical trials should also be used to expand your target market:

- For example, through clinical trials, you determine that your device is safe and effective with juveniles or pediatrics in addition to adults.
- For example, through clinical trials, you determine that additional adult patient subpopulations can receive your device, such as those with different body mass indices or body weight, those with additional prior medical conditions, those that are considered high-risk patients due to heart or organ failure.
- For example, clinical data can confirm that new device sizes can be added to your portfolio.

Depending on the classification of your device and whether predicates exist, clinical data may be required as part of a 510(k) submission.

Note: Class III devices will require clinical data as part of the pre-market approval (PMA) application.

---

67 The artwork (label) date is the calendar date that the product label was last approved and released in the company's QMS. It is typically written in the format DD/MM/YY.

 clinical data is not required as part of the regulatory application, and if you have low-risk device, think about applying for a nonsignificant risk (NSR) study in the nited States.[68] An NSR study could help promote clinical and market acceptance f the device. NSR studies still require Institutional Review Board (IRB)[69] approval nd follow abbreviated US 21 Code of Federal Regulations (CFR) Part 812 require- ients, that is, GMP, labeling, quality controls, financial disclosure of investigators, iformed consent, site activation, investigator training, and site monitoring are still equired. Depending on the clinical site, especially with academic institutions, iere could be a lengthy regulatory review process and multiple legal contracts to egotiate. Never bet that IRB approval will come quickly, and communicate openly vith your Board of Directors about timelines.

A medical device company should utilize their Scientific or Clinical Advisory oard to define the key summary points of the clinical protocol (refer to Table 12.2).

The Scientific or Clinical Advisory Board can also help identify clinical trial ites where there is guaranteed procedure volume and high patient enrollment and an help vet clinical trial investigators. Advisory Board members can even serve as ome of the investigators, but some institutions may require alternate principal in- estigators to avoid conflict of interest. The Advisory Board can also help to write ie clinical report and have it published in a reputable, peer-reviewed journal, vhich is a huge valuation milestone for medical device startups.

Clinical trials need to be strategic. For instance, the centers or sites that you tilize for your company's clinical trial can become some of your first customers. he physicians at these centers will be very familiar with your product. If the physi- ians have a good experience with your company and they like the results from the ial, they will advocate on your behalf to the center's medical device procurement gents.

A mix of private and public centers may be necessary to meet clinical trial mile- tone targets, because some IRBs and contract arrangements will take longer to ap- rove/close than others, as stated above. Key clinicians that you want to participate i your study may operate in more than one location, so if you have a clinician at a vell-known, prestigious center where IRB approval is delayed, that same clinician

8 FDA Guidance, Significant Risk and Nonsignificant Risk Medical Device Studies – Information heet, 2006.

9 According to US 21 CFR Part 56, an Institutional Review Board (IRB) is "any board, committee, r other group formally designated by an institution to review, to approve the initiation of, and to onduct periodic review of, biomedical research involving human subjects. The primary purpose of uch review is to assure the protection of the rights and welfare of the human subjects. The term as the same meaning as the phrase institutional review committee as used in section 520(g) of the ct." An IRB may be responsible for reviewing the biomedical research that is conducted at one articular medical institution or research hospital (i.e., a local IRB), or may be responsible for re- iewing the biomedical research at many institutions (i.e., a central IRB).

Table 12.2: Example sections within a clinical protocol.

Type of Study	– Is the study an acute study or chronic study? – Is the study a prospective or retrospective study? – Is the study mostly an observational study?
Study Endpoints	– What are the protocol's acceptance (pass/fail) criteria at the primary and secondary endpoints? – If you are testing an in vitro diagnostic (IVD), what is the sensitivity and specificity you are desiring?
Study Design	– How are you designing your controls? If it would be unethical to deny treatment in a control group, are there ways for your test subjects to also serve as controls, depending upon how and when the data is collected? – Are there ways to minimize costs – for instance, using hand-held ultrasound at the patient's bedside instead of requiring a computed tomography scan?
Study Procedures	– What are the clinical procedure(s) that need to be followed, and is this different from standard clinical practice? – What are the outcomes that are being monitored? – What patient vitals need to be monitored?
Inclusion and Exclusion Criteria	– What are the patient inclusion and exclusion criteria, and is there a way to stratify either to report your results in ways that complement your goals?
Study Size	– Will enough patients sign up to participate in the trial? – What is a statistically significant sample size?

may also operate part-time at a smaller, lesser-known clinic where IRB approval may be faster, and perhaps the patient enrollment rate is similar.

Clinicians operating out of prestigious institutions and centers, who are known for ground-breaking research and exceptional care, should be utilized for podium talks at conferences and symposiums to raise the credibility and visibility of the clinical trial. But do n'ot forget that a balance of well-known and lesser-known clinicians and sites are required to ensure that schedule and budget milestones are achievable.

If leadership at your medical device company determines that internal resources do not have the clinical trial management expertise, do not hesitate to engage with a reputable Clinical Research Organization (CRO). In this example, a CRO is a third-party vendor that engages with clinical sites on behalf of the medical device clinical trial sponsor (i.e., typically the medical device manufacturer) to ensure that the trial is planned and executed smoothly. CROs can assist with clinical trial protocol writing, clinical site IRB applications and approvals, clinical protocol training with clinical sites, clinical site enrollment, clinical data monitoring, medical device safety reporting, clinical trial data analysis, and so on. Medical device clinical trial sponsors should select a CRO that:

Has plentiful experience with medical device clinical trials.
Has a good rapport with your team and clinical centers.
Has many contacts in the device sector.
Can introduce your company to other service providers (e.g., medical device distributors, medical device marketers, medical device decontamination, and return labs) to assist with clinical trial preparation.
Is willing to work with your startup to be creative in how to manage resources effectively, such as clinical electronic data capture tools are scalable, and a more complex solution should n'ot be used just because it has fancier database functionality.

clinical trials that are sanctioned by regulatory agencies, for example, Investiga-onal Device Exemption (IDE) studies in the United States (refer to Figure 12.3), edical Device Reporting requirements are in effect (refer to Section12.4.4). There-re, if medical device manufacturers are made aware of a device that is suspected causing serious injury/illness or death, or if a device malfunction was investi-ated and acknowledged to likely cause serious injury/illness or death, then regula-ry agencies must be notified.

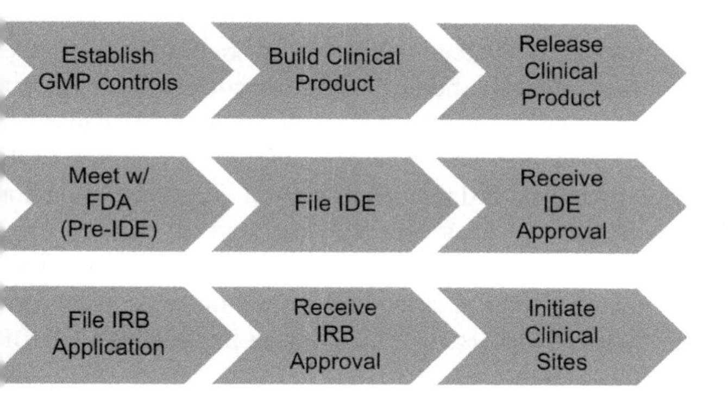

gure 12.3: Key manufacturing, IDE, and IRB milestones in clinical trial preparation.

.4.4 Medical Device Reporting

gulatory agencies require additional data (i.e., data surveillance) following market-g clearance/authorization, to ensure that the clinical benefits that were described the submission or application outweigh the risks to patients and users, and that e probable benefits to the patient outweigh the probable harms. For example, with nplantable devices, regulatory agencies will want to know about the long-term per-rmance of the device, whether risks have a higher probability of occurrence over time, and whether there are new, previously unreported adverse effects or nonadverse effects with the device implant. Regulatory agencies also want to ensure that the risk mitigations, described in the submission or application, are in fact effective, and that failure modes will not go unnoticed. For example, if one of the described primary device risk mitigations is physician training or training on the device IFU, especially if the medical procedure to use or implant the device is complex, then the effectiveness of that training will be monitored. Medical device manufacturers also can use postmarket surveillance data to prove that the device works in additional patient populations and in additional indications or medical procedures.

Medical device manufacturers need to have processes in place to manage complaints, investigate complaints, submit reports to regulatory agencies, and complete high-quality postmarket surveillance. Any user complaints from physicians, patients, caregivers, surgical technicians, healthcare professionals, and so on that are received by medical device manufacturers need to be tracked and reported to the regulatory agencies in the countries where the device is marketed commercially and/or made available for clinical use within a certain period of time.[70] Also, it does n'ot matter how a complaint is received and who in the company receives the complaint – from the junior quality assurance analyst to the head of Regulatory or Clinical Affairs – complaints need to be escalated, assessed, and reported (when deemed appropriate).

During clinical investigations, an adverse medical event[71] needs to be assessed as to whether it is a "reportable" event by the device manufacturer. These medical events need to be investigated whether the adverse medical event was caused by the performance, or lack of performance, of the device, before it is reported to the regulatory agencies. The investigation could include follow-up questions to the hospital or clinic where the medical device was used and the adverse medical event occurred, a return of the device (if possible) to the device manufacturer for a complete root cause analysis, a review of similar complaints in regulatory agency databases (e.g., the FDA's Manufacturer and User Facility Device Experience (MAUDE) database[72]), or a review of clinical literature.

In the United States, medical device reporting to the FDA needs to be completed in 5 work days or 30 calendar days, depending upon the type of adverse effect (e.g., major, serious, unforeseen, or unanticipated) and/or the probability that the risk can impact the widespread patient population because the risk cannot be mitigated

70 See 21 Code of Federal Regulations (CFR) Part 803 Medical Device Reporting.

71 According to ISO 14155, Clinical Investigation of Medical Devices for Human Subjects – Good clinical practice, an adverse event is "Any untoward medical occurrence, unintended disease or injury, or untoward clinical signs (including abnormal laboratory findings) in subjects, users or other persons, whether or not related to the investigational medical device."

72 The FDA's Manufacturer and User Facility Device Experience (MAUDE) database is accessible online at https://www.accessdata.fda.gov/scripts/cdrh/cfdocs/cfmaude/search.cfm.

nd the risk will reoccur in other medical devices being used in other patients, that , which propagates the necessity of a device recall.

2.4.5 Medical Device Marketing

evice marketing claims need to comply with the indications (claims) that have een approved by the regulatory agencies in the markets where applications have een filed. And if a medical device manufacturer wants to advertise and promote its roducts to prospective users and patients prior to receiving regulatory approval to arket those claims (e.g., via product videos that are posted online, website con- nt that discusses product features, brochures that are disbursed at medical tech- ology conferences, press releases, conference presentations), then the marketing aterials need to indicate its regulatory status. Marketing materials can include the llowing disclaimers: "This product is not approved for use in humans," or "This evice is for investigational purposes only," or "This device is not cleared for mar- et approval."

Medical device companies can also utilize social media channels to reach pro- pective users and patients by providing company updates (e.g., key company lead- rship hires, fundraising goals and progress, program milestone successes) and levant findings in medical journals that support their medical technology. For ex- mple, if your device minimizes bleeding, then a good topic to raise on social edia channels would be the incidence of bleeding in certain types of procedures, specially procedures that are related to your device indication.

Company Scientific Advisory Board members and clinical trial Principal Investi- ators can help market clinical devices, because they can inform the medical com- unity, especially providers,[73] about the device prior to regulatory approval. owever, Board members and Principal Investigators cannot promote the sale of e device or convey false or misleading information about the device's effective- ess, that is, they cannot promote a use or an indication that has not been proven nd approved.

Medical device manufacturers should have QMS work instructions in place efer to Chapter 11) for creating, reviewing, and approving marketing content, spe- ifically if the content would be determined by regulatory authorities to contain edical product claims or promote the use of medical products.

---

ote: These work instructions would be referenced in a higher level Medical Device Labeling Stan- ard Operating Procedure within the QMS.

---

3 Most devices can only be prescribed by a physician for use in patients, so most of the marketing ontent is written for providers, who determine which devices to use in their patients.

Medical device marketing content (such as website pages, social media videos) are created and approved as marketing specifications (with an associated artwork date) within the QMS. Medical device marketing content and medical device product-related content (such as medical device press releases, sales force training materials) should be reviewed through an internal advertising and promotion review process. In this process, the company's Chief Medical Officer and Regulatory representative can review and sign off on this content to ensure that the right disclaimers are included in the content, and that claims for which the manufacturer does not have approval are not included in the content. As stated above, advertising and promotion reviews can be performed for medical device manufacturer websites, medical product literature and videos, sales force training literature, and content that is written for physicians and users of your device.

Lastly, individual governments also have a say in how medical device marketing information is shared with prospective users and patients. For instance, in the EU, direct to consumer marketing is allowed in some countries and not in others.

12.4.6 Medical Device Reimbursement in the United States

As a medical device manufacturer, how do you intend to get paid for your device? A common pathway for most medical devices is to get approved for reimbursement from medical device payers or insurance companies. In the United States, reimbursement is based on CPT codes, which are dictated by the CMS at the advice of the AMA (or AMA-related specialty association).

Reimbursement is a process by which providers (i.e., healthcare providers, physicians, hospital systems, etc.) are reimbursed by payers (i.e., insurance companies) for the medical treatments, procedures, and services they prescribe to patients. Providers assign CPT outpatient and office healthcare procedure billing codes (after patients receive their treatments, procedures, and services) and then file insurance claims to payers on behalf of their patients. Payers review patients' insurance plan requirements (i.e., payers determine if patients are eligible to receive compensation benefits for medical services received) and determine how much providers will be reimbursed by the payer and how much patients are required to pay to cover the balance of the claim.

Reimbursement gets really tricky when a subset of patients, who are eligible to receive the device/treatment and fall under one healthcare plan, can receive the device because the provider can file a claim with the patient's healthcare plan so that the provider/healthcare system can be reimbursed for the cost of the device/treatment that the patient received. But there can be other subsets of patients whose healthcare plans indicate they are not eligible for reimbursement.

For the patients who are not eligible for reimbursement, these patients can decide to receive the device/treatment anyway. However, the provider's services will ɔt be reimbursed, the hospital will not be able to file a claim, and the healthcare ɔsts will most likely be paid out-of-pocket by the patient. When a device or treat-ent does not have a CPT code that is generally reimbursed by a majority of payers, e lack of a reimbursable CPT code will limit the device's TAM, because a percent-ge of the eligible population will decide not to use the device since their insurance ill not pay for it.

The reimbursement landscape with insurance companies (i.e., payers) in the nited States can be challenging to navigate. Do you intend to apply to utilize the ealthcare Common Procedural Coding System or existing CPT codes, because your evice will be used in a similar fashion/function as devices that are used in existing edical or surgical procedures? CPT codes are divided into three Categories:

Category 1 – standard codes that are used with commonly performed procedures and are deemed standard clinical practice nationwide.
Category 2 – supplemental codes that are used as performance measurements.
Category 3 – temporary codes that are used for new and emerging technologies.

ifferent levels of clinical evidence are required for Category 1, and Category 1 odes are used for procedures that are approved by the FDA.

Or do you intend for patients to pay out-of-pocket to use your device, or utilize SAs or high deductible healthcare plans? The percentage of employers that offer SAs, patients who take advantage of FSAs, and healthcare consumers who elect r higher deductible plans are on the rise.

It is challenging for medical device manufacturers to receive approval to use PT reimbursement codes for new devices and technologies. For example, it typi-lly takes several years after regulatory approval in the United States for CPT code pproval to be granted to a medical device manufacturer, and only once enough ata has been collected through pivotal trials and clinical adoption to prove the inical benefits and utility of the device. Remember that value needs to be demon-rated for all customers, for example, patients, providers, and payers.

If a new CPT code needs to be created, that is, if an existing CPT code does not xist for your device/technology, then this is even more challenging, because medi-l device manufacturers will need to lobby the AMA and individual healthcare ayers to accept the new code. These negotiations take time, and the first CPT code ontracts are always the hardest for payers to sign.

If a CPT code does not currently exist for a new device or technology you are eveloping (e.g., your company is waiting to receive a temporary CPT Category 3 ode while clinical data is being collected), then perhaps the product, at a lower verage selling price, can be bundled into a Diagnosis Related Group (DRG)-Bun-led Facility Payment for in-patient medical procedures. DRG-bundled payments over in-patient diagnoses, imaging, and therapy costs. DRG-bundled payments are etermined by the assignment of International Classification of Diseases (ICD), for xample, ICD-10 diagnoses and operating procedure codes, and DRG payments cover all costs related to a hospital stay. The CMS governs DRGs, and new DRG applications get directed to the AMA. Once a new DRG is granted, DRGs then need to be negotiated with the healthcare systems. Healthcare systems need to be convinced that your device, even though it will initially cost the hospital to procure, will provide added benefits and decrease the overall costs of healthcare, thus warranting its inclusion in the DRG-bundled payment.

For example, if your company is developing a new safety device that is intended to address soaring healthcare costs, device payment may be included in procedural costs that will be covered by the hospital (e.g., catheter lab procedure cost). Usage of your safety device can reduce the hospital's surgical procedural risks (e.g., during Transcatheter Aortic Valve Implantation procedures), which positively impact patient health outcomes. Therefore, your safety device will reduce costs incurred by the hospital, for example, by decreasing patient mortality rates, patient length of stay costs, patient readmission rates, and postoperative patient complications.

With medical device reimbursement, medical device manufacturers need to influence and convince provider associations, for example, the AMA, American Urological Association, American Association of Orthopaedic Surgeons, and so on, that reimbursement is warranted for new devices and technologies. Provider associations and medical specialty committees will then lobby healthcare payers on behalf of the medical device manufacturers. That is another reason why it is so important for medical device manufacturers to establish a good Scientific or Clinical Advisory Board who can encourage other physicians to complete medical device user surveys, assist with physician training, and advocate for your device with the different provider societies and associations. This also demonstrates the need for medical device manufacturers to acquire sound clinical data to publish in prestigious, peer-reviewed journals and present at medical conferences to inform and influence provider associations.

12.5 Conclusion

In closing, you are never operating alone as a medical device manufacturer – there is a thriving MedTech industry that wants you to succeed. Think outside the confines of your company to address any challenges that you will inevitably face. For example, if market penetration itself is an issue, why go at it alone? Is there an opportunity to partner with a different device manufacturer that is marketing a portfolio of devices that are causing clinical complications, to which your device was designed to address an unmet clinical need? Is there an opportunity for both companies to conduct clinical studies together? And, when given the data to support these claims, is there an opportunity to market these devices together as complementary products?

The creations of medical device inventors and entrepreneurs save lives, im-
ove the quality of human health, and give hope to millions. The road to medical
vice commercialization is a long one, requiring perseverance and patience, but,
the end, is absolutely worth it.

List of Acronyms and Initialisms

2D	Two-Dimensional
3D	Three-Dimensional
AAOS	American Association of Orthopaedic Surgeons
AI	Additional Information
AIDC	Automatic Information and Data Capture
AMA	American Medical Association
ASP	Average Selling/Sales Price
AUA	American Urological Association
AVL	Approved Vendor List
BOM	Bill of Materials
CAD	Computer-Aided Design
CAGR	Compounded Annual Growth Rate
CAM	Computer-Aided Manufacturing
CAPA	Corrective and Preventive Action
CE	Conformité Européene
CEO	Chief Executive Officer
CFO	Chief Financial Officer
CFR	Code of Federal Regulations
CMS	Center for Medicare and Medicaid Services
COO	Chief Operations Officer
COGS	Cost of Goods Sold
COPQ	Cost of Poor Quality
CNC	Computer Numerical Control
CPT	Common Procedural Terminology
CRO	Clinical Research Organization
CT	Computed Tomography
CTO	Chief Technology Officer
DCOV	Define Characterize Optimize Verify
DFA	Design For Assembly
DFM	Design For Manufacturing
DHF	Design History File
DHR	Device History Record
DMADV	Design Measure Analyze Design Verify
DMAIC	Define Measure Analyze Improve Control
DMR	Device Master Record
DoD	Department of Defense
DPMO	Defects Per Million Opportunities
DRG	Diagnosis Related Group
DUNS	Data Universal Numbering System
EDC	Economic Development Corporation
EU	European Union
FD&C	Food Drug and Cosmetic
FDA	Food and Drug Administration
FMEA	Failure Mode Effects Analysis
FSA	Flexible Spending Account
FTO	Freedom To Operate
FY	Fiscal Year

https://doi.org/10.1515/9783110521900-014

A General and Administrative
P Good Manufacturing Practices
Hospital Acquired Infection
PCS Healthcare Common Procedural Coding System
HP High Deductible Healthcare Plan
AA Health Insurance Portability and Accountability Act
International Classification of Diseases
Investigational Device Exemption
Instructions for Use
Intellectual Property
Initial Public Offering
Institutional Review Board
Installation Qualification
International Organization of Standardization
Information Technology
In Vitro Diagnostic
Juris Doctor
L Key Opinion Leaders
Key Performance Indicators
Light Emitting Diode
Limited Liability Company
Limited Liability Partnership
P Minimally Acceptable Product
UDE Manufacturer and User Facility Device Experience
A Master of Business Administration
R Medical Device Regulation
SAP Medical Device Single Audit Program
CE Mutually Exclusive and Collectively Exhaustive
Magnetic Resonance Imaging
P Minimum/Minimally Viable Product
CS North American Industry Classification System
A Non-Disclosure Agreement
National Institutes of Health
E Not Substantially Equivalent
F National Science Foundation
R Non-significant Risk
M Original Equipment Manufacturer
HA Occupational Safety and Health Administration
Operation Qualification
L Profit and Loss
T Patent Cooperation Treaty
SA Plan Do Study Act
G Percutaneous Endoscopic Gastrostomy
Principal Investigator
A Premarket Approval
O Project Management Office
C Proof-of-Concept
Performance Qualification
Quality Assurance

QC	Quality Control
QMS	Quality Management System
R&D	Research and Development
R&R	Repeatability and Reproducibility
RACI	Responsible, Actionable, Consulted, Informed
RADIO	Risks, Assumptions, Dependencies, Issues, Opportunities
RFP	Request for Proposal
ROI	Return on Investment
RPN	Risk Priority Number
SBA	Small Business Administration
SBDC	Small Business Development Center
SBIR	Small Business Innovation Research
SE	Substantially Equivalent
SEC	Securities and Exchange Commission
SIP	Strategic Implementation Plan
SIPOC	Suppliers Inputs Process Outputs Customers
SMART	Specific, Measurable, Actionable, Realistic, Time-bound
SOP	Standard Operating Procedure
SOW	Statement of Work
SPC	Statistical Process Control
STTR	Small Business Technology Transfer
SWOT	Strengths, Weaknesses, Opportunities, Threats
TAM	Total Addressable Market
TAVI	Transcatherer Aortic Valve Implantation
TQM	Total Quality Management
UDI	Unique Device Identifiers
USPTO	United States Patent and Trademark Office
UI	User Interface
UV	Ultraviolet
UX	User Experience
VC	Venture Capital/Capitalist
VOC	Voice of the Customer
V&V	Verification and Validation
WI	Work Instruction
WIP	Work in Progress

urther Reading

usiness Perspectives

] Blank, S. The Four Steps to the Epiphany: Successful Strategies for Products that Win. Quad/Graphics, 5th edition, 2013.

] Cameron, E., & Green, M. Making Sense of Change Management: A Complete Guide to the Models, Tools & Techniques of Organizational Change. Kogan Page Limited, Reprint edition, 2004.

] Christensen, C. The Innovator's Dilemma: When New Technologies Cause Great Firms to Fail. Harvard Business Review Press, Reprint edition, 2016.

] Christensen, C., Anthony, S., & Roth, E. Seeing What's Next: Using the Theories of Innovation to Predict Industry Change. Harvard Business Review Press, 2004.

] Collins, J. Good To Great: Why Some Companies Make the Leap. . .and Others Don't. HarperCollins Publishers Inc., 2001.

] Collins, J., & Porras, J. Built to Last: Successful Habits of Visionary Companies. HarperBusiness, 3rd edition, 1994.

] De Bono, E. Six Thinking Hats: An essential approach to business management from the creator of *Lateral Thinking*. Little Brown and Company, 1985.

] Halt, G., Donch, J., & Stiles, A. Intellectual Property in Consumer Electronics, Software and Technology Startups. Springer, 1st edition, 2014.

] Halt, G., Donch, J., Stiles, A., & Fesnak, R. Intellectual Property and Financing Strategies for Technology Startups. Springer, 1st edition, 2017.

0] Isaacson, W. The Innovators: How a Group of Hackers, Geniuses, and Geeks Created the Digital Revolution. Simon & Schuster, Reprint edition, 2015.

1] Osterwalder, A., & Pigneur, Y. Business Model Generation: A Handbook for Visionaries, Game Changers, and Challengers. John Wiley & Sons, Inc., 2010.

2] Pitching Hacks!: How to pitch startups to investors, Venture Hacks Inc., 2008.

3] Resources for Entrepreneurs, S3 Ventures, (Accessed January 29, 2019 at http://www.s3vc.com/resources-for-entrepreneurs/).

4] SCORE Business Plan Template for a Startup Business, (Accessed January 29, 2019 at https://www.score.org/resource/business-plan-template-startup-business).

5] Wilmerding, A. Term Sheets & Valuations: An Inside Look at the Intricacies of Term Sheets & Valuations. Aspatore, Inc., 3rd edition, 2003.

perations

] A Guide to the Project Management Book of Knowledge (PMBOK® Guide), Project Management Institute, Inc., 5th edition, 2013.

] Belz, A. The McGraw-Hill 36-Hour Course: Product Development. The McGraw-Hill Companies, Inc., 2011.

] Dekker, S. Drift Into Failure: From Hunting Broken Components to Understanding Complex Systems. Ashgate Publishing, 2011.

] Franchetti, J. Lean Six Sigma for Engineers and Managers: With Applied Case Studies. CRC Press, 2015.

] George, M., Rowlands, D., Price, M., & Maxey, J. The Lean Six Sigma Pocket Toolbook: A Quick Reference Guide to Nearly 100 Tools for Improving Process Quality, Speed, and Complexity. The McGraw-Hill Companies, Inc., 2005.

tps://doi.org/10.1515/9783110521900-015

[6] Longman, A., & Mullins, J. The Rational Project Manager: A Thinking Team's Guide to Getting Work Done. John Wiley & Sons, Inc., 2005.
[7] Pyzdek, T., & Keller, P. The Six Sigma Handbook. McGraw-Hill Education, 5th edition, 2018.
[8] Ries, E. The Lean Startup: How Today's Entrepreneurs Use Continuous Innovation to Create Radically Successful Businesses. Crown Business, 2011.

Medical Devices and Technologies

[1] Edited by Gropp, M., & Takes, P. Global Medical Device Regulatory Strategy. Regulatory Affairs Professional Society, 2016.
[2] Edited by Yock, P., Zenios, S., Makower, J., Brinton, T., Kumar, U., & Watkins, F.T. Biodesign: The Process of Innovating Medical Technologies. Cambridge University Press, 2nd edition, 2015.
[3] Halt, G., Donch, J., Stiles, A., Jenkins VanLuvanee, L., Theiss, B., & Blue, D. FDA and Intellectual Property Strategies for Medical Device Technologies. Springer, 1st edition, 2019.

ubject Index

)(k) 341, 353, 365, 366, 366n54, 367, 368, 369, 373n64, 374

celerator 76, 77, 78
quisition 15, 18, 27, 28, 38, 40, 77, 120, 147, 148, 149, 151, 192, 348, 351
gel investors 76, 127, 133, 147, 150, 155, 156, 165
proved vendor list 175, 192, 196, 329, 351
'ards 78, 147, 152, 162, 235

l of Materials 283, 286, 287, 297, 312, 316, 317, 384
otstrapping 147, 150, 153, 154, 156
siness Development Process 2, 3, 4, 5, 6, 7, 8, 11, 16, 65, 66, 83, 130, 131, 147, 148, 176
siness model 4, 18, 19, 22, 34, 35, 36, 39, 40, 41, 77, 83, 90, 110, 117, 149, 153, 156, 166, 184, 203, 349, 351, 352, 354, 356, 359
siness plan 5, 6, 8, 31, 33, 83, 85, 89, 98, 102, 103, 108, 109, 117, 119, 120, 121, 122, 123, 125, 126, 127, 128, 166, 171, 173, 174, 175, 176, 178, 180, 354, 356
siness validation 4, 5, 34, 35, 36, 39

marking 319
ange management 209, 223, 235, 236, 240, 249
ange review board 209
ass I 341, 360n42, 365
ass II 23n6, 229, 341, 360, 365, 366, 373
ass III 23n6, 341, 360, 365, 365n50, 366, 374
mmercialization 10, 99, 100, 110, 114, 149
mmunication channels 175, 199, 200, 207
mpetition 24, 34, 37, 38, 43, 86, 93, 103, 121, 133, 147, 164, 165, 191, 231, 255, 261, 262, 272
mpetitors 15, 20, 23, 24, 28, 29, 31, 34, 35, 37, 38, 38n10, 40, 43, 45, 49, 86, 87, 90, 93, 102, 103, 107, 129, 130, 132, 138, 143, 152, 165, 191, 202, 231, 261, 273
ncept generation 173, 253
ncept selection 100, 249, 253
ncept validation 4, 34, 40, 40n11, 42, 43, 45, 49, 253, 257, 266

consulting firms 15, 20, 27
convertible notes 147, 158, 159
copyright 137, 140, 146
cross-licensing 129, 145, 146
crowdfunding 147, 150, 151, 155
customer relationships 85, 175, 199
customer segments 52, 83, 84, 85, 87, 93, 94, 200

dashboards 209, 214, 226, 249
data acquisition firms 15, 27
design changes 287, 288
design controls 100, 288, 295, 311
design for assembly 253
design for manufacturing 253, 306
Design Freeze 9, 10, 101, 289, 292, 296, 302, 302n32, 304, 306, 307, 313
design history file 288, 299, 310, 314, 315, 384
design inputs 41, 101, 288
design outputs 101, 288, 312
design patent 132, 353
design transfer review 288
device history record 288, 311, 316, 384
device master record 288, 292, 310, 314, 384
donut example 280, 281, 283, 296, 317
due diligence 2, 4, 15, 17, 18, 19, 20, 21, 22, 23, 24, 25, 27, 28, 31, 32, 43, 46, 93, 102, 144, 149, 151, 152, 153, 154, 160, 171, 172, 173, 175, 190, 191, 192, 210, 253, 257, 261, 263, 266, 272, 284, 293, 302, 354, 356
due diligence stage 2, 15, 154, 210

early-stage funding 147, 150, 151
Economic Development Corporations 65, 73, 74, 75
equity 67, 70, 74, 76, 77, 78, 81, 82, 147, 150n15, 151, 154, 155, 156, 157, 158, 161, 162, 164, 164n21, 172, 201
exit 15, 16, 18, 69, 119, 121, 130, 147, 148, 149, 150, 151, 210, 236, 249, 348, 350, 351
expansion plan 6, 84
expansion stage 3, 16, 151, 210
expansion strategy 6, 83, 84, 202

:ps://doi.org/10.1515/9783110521900-016

financial models 83, 125, 126, 127, 128, 180, 298, 354
freedom to operate 129, 144, 146, 152
funding 2, 7, 8, 16, 18, 20, 36, 71, 73, 74, 76, 78, 82, 120, 147, 148, 149, 150, 151, 152, 153, 154, 155, 156, 157, 158, 161, 162, 163, 164, 165, 166, 171, 173, 174, 176, 178, 188, 195, 214, 232, 236, 302n32, 352, 353, 354, 356, 357n39
Funding Stages 3, 151

Gantt chart 114, 178, 180, 227
growth funding 147
growth stage 3, 16, 151, 161, 162, 210

high-level validation 253, 262
human factors 34, 48, 53, 266, 291, 297, 359

incubator 28, 76, 77, 78
independent discovery 129, 139
in-depth validation 253, 263
infringement 6, 98, 129, 144, 146
initial public offering 147, 148, 151, 173
installation qualification 288, 316
intellectual property 6, 18, 20, 21, 24, 25, 31, 38, 45, 83, 84, 88, 98, 103, 107, 125, 129, 130, 131, 133, 135, 137, 138, 139, 140, 143, 144, 156, 164, 172, 231, 253, 257, 263, 266, 277, 350, 353, 356
intellectual property (IP) strategy 6, 84, 98
intellectual property portfolio 6, 130, 143, 172
IP plan 6, 98, 140
IQ/OQ/PQ 288, 293, 310, 316

Key Opinion Leader 101

Lean Six Sigma 319, 333, 339
licensing 6, 18, 25, 45, 98, 129, 130, 132, 133, 136, 143, 145, 146, 152, 172, 177, 178, 232, 350, 353
licensing agreements 145
life cycle 2, 15, 16, 18, 19, 20, 22, 40, 82, 83, 99, 100, 101, 102, 103, 110, 111, 134, 149, 150, 151, 154, 161, 162, 173, 210, 309, 362n45
limited liability company 65, 69, 71, 150

market expansion 19, 34, 98, 202, 204
market growth 34, 37, 38, 88
market reports 15, 25, 26, 27, 28, 29
market research 4, 15, 24, 25, 26, 27, 29, 31, 32, 39, 41, 43, 93, 231, 255, 256, 272, 290, 292, 356
market segments 15, 22, 23, 30, 32, 34, 36, 37, 53, 197, 199, 202, 203, 231
market share 25, 34, 37, 38, 39, 103, 203, 334, 340, 356
marketing and sales 84, 85, 89, 90, 140, 166, 171, 175, 176, 197, 199, 201, 202, 204, 208, 292
marketing plan 6, 83, 84, 85, 89, 90, 91, 92, 95, 96, 97, 98, 121, 178
marketing strategy 6, 29, 31, 59, 64, 83, 84, 85, 86, 88, 89, 102, 185, 200
maturity 15, 16, 79, 151, 159, 173, 184, 210, 322
medical device reporting 341, 378
Minimum Viable Product 10, 22, 44, 153, 249, 279, 284, 287, 288, 289, 292, 293, 296, 297, 301, 304, 306, 356
misappropriation 129, 138, 139

Needs Finding 253, 255, 257, 258, 260, 268, 276
needs gathering 253
needs refinement 253
non-provisional patent application 129, 133
not substantially equivalent 341, 366

operational qualification 288, 316

pain point 21, 253, 257
patent xviii, 18, 129, 130, 131, 132, 133, 134, 135, 138, 140, 143, 144, 145, 146, 148, 152, 161, 180, 253, 255, 257, 263, 266, 275, 278, 353
performance qualification 288, 316
phase gates 209
pitching to investors 7
post-market launch customer surveillance 175, 176
predicate device 22n6, 341, 367, 368
pre-market approval 341, 366, 374
pre-seed funding 150
pre-submission meetings 341, 368
Product Development 10, 43, 44, 187, 278, 279, 280, 284, 287, 288, 289, 290, 291, 292, 293, 294, 296, 297, 301, 302, 304, 306, 307, 309, 310, 311, 314, 318

uct/Technology Development Process 2, 3, 4, 9, 10, 11, 16
ct charter 209, 214, 215, 238, 249
ct management xviii, 8, 39, 117, 182, 209n27, 210, 211, 212, 217, 222, 223, 234, 236, 249, 293
ct scorecards 209
f-of-concept 9, 101, 113, 116, 153, 242, 249, 253, 257, 280, 293, 357
type 2, 9, 16, 17, 20, 22, 23, 44, 49, 80, 101, 111, 113, 116, 127, 147, 153, 154, 156, 172, 182, 194, 230, 242, 249, 253, 257, 276, 278, 279, 279n30, 280, 281, 283, 284, 286, 287, 288, 289, 290, 292, 293, 296, 297, 301, 302, 304, 306, 309, 310, 314, 357n39
sional patent application 133, 134, 266, 275
matrix 253

ity Assurance 312, 321, 322, 324, 385
ity Control 321, 329, 386
ity Management Systems 10, 197, 318, 319, 320, 322, 323, 324, 327, 328, 330, 340, 369, 370, 379

O 209, 214, 217, 218, 219, 220, 245, 386
bursement 40, 88, 118, 127, 341, 348, 350, 353, 356, 364, 380, 381, 382
se engineering 129, 139
ion history 288, 315, 316, 317
management 206, 209, 217, 218, 220, 222, 235, 373
maps 87, 178, 209, 227, 249, 337

s plan 6, 85, 89, 90, 92, 121, 171, 178
s strategy 6, 41, 85, 86, 89, 166, 199, 204
e creep 175, 177, 195, 230, 248
e management 209
UM boards 209, 224, 225, 249
funding 147, 150, 155
ction criteria 101, 253, 261, 262, 263, 265, 266, 268, 269, 271, 272
s financing 147
le-source vendor 175, 193, 195
Sigma 10, 319, 331, 333, 335, 337, 339, 339n35
ll Business Administration 26, 65, 73, 386
Small Business Development Centers 38n10, 65, 72, 74, 126
sole proprietorship, partnership 65
sole-source vendor 175
sourcing 193, 253, 283, 284, 286, 290, 291, 301, 306, 314
stakeholder communication 209, 236, 239
stakeholder mapping 29, 34, 46, 47
stakeholders 4, 24, 34, 35, 36, 40, 41, 43, 44, 45, 46, 47, 48, 49, 197, 200, 210, 213, 214, 215, 217, 220, 221, 222, 223, 224, 225, 226, 227, 233, 236, 237, 238, 239, 240, 241, 249, 250, 255, 257, 267, 269, 291, 349
startup stage 2, 15, 18, 151, 161, 182, 210, 282
statement of work 175, 194, 209
storyboarding 101, 209
Strategic Implementation 8, 175, 175n22, 176, 177, 178, 288, 293, 302n32, 304, 309, 310, 386
strategic implementation plan 8, 107, 109, 122, 163, 176, 293
Strengths, Weaknesses, Opportunities, Threats (SWOT) analysis 4
substantially equivalent 341, 353, 365n49, 366
SWOT analysis 34, 39, 83, 90, 237

target consumer 15, 29, 30, 31
target market 25, 26, 27, 29, 30, 32, 37, 38, 40, 55, 56, 57, 83, 90, 92, 93, 94, 95, 97, 109, 121, 165, 200, 231, 346, 374
technology development and implementation plan 6, 84
technology development plan 83, 122, 171
technology life cycle 99
technology management 8, 210, 234, 236
total quality management 319, 330
trade secret 6, 137, 138, 139, 140, 172
trademark 130, 135, 136
Transfer to Manufacturing 10, 278, 279, 284, 291, 292, 296, 297, 301, 304, 306, 307, 309, 314, 318

unauthorized disclosure 129, 138, 139
unmet need 9, 32, 92, 103, 109, 116, 253, 255, 257, 258, 260, 261, 263, 264, 276, 278, 292
user requirements 4, 34, 40, 41, 43, 46, 101, 279, 280, 292, 293, 296, 333, 367
utility patent 132, 353

value proposition 15, 32, 33, 40, 44, 83, 88, 90, 91, 92, 94, 165, 197, 346, 347, 348, 356
vendor management 175, 176, 188, 191, 196, 355, 356
vendor risk management 175, 190
venture capital 147, 156, 162, 173
Verification and Validation 9, 10, 149, 296, 301, 304, 305, 386
voice of the consumer 34

CPSIA information can be obtained
at www.ICGtesting.com
Printed in the USA
LVHW011333031221
704911LV00001B/2